D1179435

The Biology of *Xenopus*

This is a volume in the series
Symposia of the Zoological Society of London

Number 68

From a Symposium held at the Zoological Society of London
on 10th and 11th September 1992

The Biology of *Xenopus*

Edited by

R.C. TINSLEY

School of Biological Sciences
University of Bristol

and

H.R. KOBEL

Laboratoire de Génétique Animale et Végétale
Université de Genève
Switzerland

Published for THE ZOOLOGICAL SOCIETY OF LONDON

by CLARENDON PRESS · OXFORD

1996

Oxford University Press, Walton Street, Oxford OX2 6DP
Oxford   New York
Athens   Auckland   Bangkok   Bombay
Calcutta   Cape Town   Dar es Salaam   Delhi
Florence   Hong Kong   Istanbul   Karachi
Kuala Lumpur   Madras   Madrid   Melbourne
Mexico City   Nairobi   Paris   Singapore
Taipei   Tokyo   Toronto
and associated companies in
Berlin   Ibadan

Oxford is a trade mark of Oxford University Press

Published in the United States
by Oxford University Press Inc., New York

A catalogue record for this book is available from the British Library

Library of Congress Cataloging in Publication Data
The biology of Xenopus / edited by R.C. Tinsley and H.R. Kobel.
(Symposia of the Zoological Society of London ; no. 68)
Papers originally presented at a symposium held at the Zoological
Society of London on Sept. 10–11, 1992.
Includes bibliographical references and index.
1. Xenopus–Congresses.   I. Kobel, H. R.   II. Tinsley, R. C.   III. Series.
QL1.Z733 no. 68
[QL668.E265]   591 s–dc20   [597.8'4]   95–20387
ISBN 0 19 854974 1

Typeset by Palimpsest Book Production Limited,
Polmont, Stirlingshire

Printed in Great Britain by
Bookcraft (Bath) Ltd
Midsomer Norton, Avon

Preface

Xenopus has been employed intensively in laboratory-based research for over 50 years. Much of the earlier work was concerned with physiology, biochemistry, endocrinology and developmental biology, and was based almost exclusively on a single species, *Xenopus laevis*, from South Africa. The extensive literature which was generated contributed significantly to the foundation of contemporary understanding in these disciplines. However, these important advances emerged against an almost total lack of information on ecology and species diversity. Despite the occurrence of *Xenopus* throughout more than 45° of latitude in sub-Saharan Africa, in habitats ranging from rain forest to semi-desert and from lowland swamps to alpine lakes, the rest of the genus was considered relatively uniform and uninteresting. At the start of the 1970s, six species were recognized of which only two had been described this century.

A new phase of research interest in *Xenopus* developed in the early 1970s. This took two directions: on the one hand, major developments in cell and molecular biology became possible through utilization of *Xenopus* within the test tube (see the Introduction to this volume, by Professor John Gurdon); on the other hand, interests in systematics and genetics arose out of studies of *Xenopus* in the field, primarily in relatively remote areas of Africa. This latter work has transformed knowledge of the genus: there are currently 17 named species and, with further work in progress, this total will soon exceed 20. The pace of these advances has resulted in significant confusion amongst non-specialists, reflecting both the lack of basic information on the biology of *Xenopus* and the considerable difficulties of species identification.

Xenopus is referred to as either the African clawed toad or frog but is actually very distant from the anuran lines with which the familiar names toad and frog are associated. The literature on the lifestyle of *Xenopus* in its natural environment is scattered, much of it buried in old and relatively obscure sources. These contain clues to natural history including abilities for aestivation and overland migration, tolerance of starvation and the bio-chemical and physiological processes accompanying water shortage. However, there are no thorough ecological studies of population biology, interactions within communities, generation times, growth rates, etc. derived from long-term field-based research. Even records of the diet of *Xenopus* and its temperature and water chemistry preferences in natural habitats are sketchy. Clearly, the success of *Xenopus* as a 'laboratory animal' demonstrates its adaptability and tolerance, but it is interesting that research in physiology and biochemistry proceeded without reference to the appropriateness of the laboratory environmental conditions in which *Xenopus* was maintained.

Information on specific aspects of *Xenopus* has been reviewed in the past few years by Kay & Peng (1991)[1] (cell and molecular biology, genetics, early development), Tymowska (1991)[2] (cytogenetics), Hausen & Riebsell (1991)[3] (early development), and Cannatella & De Sa (1993)[4] (phylogeny). The subject matter of this symposium is distinct from these: our approach has been to focus on those areas of research which contribute to an understanding of the life of the *Xenopus* species. We have not considered research primarily based on *using Xenopus* (as a source of cellular components) to answer fundamental questions of molecular and developmental biology. We have focused on the diversity of knowledge relating to *Xenopus*, information which has not previously been reviewed comprehensively and which will represent a basic source of literature reference. We hope that this will achieve a relatively comprehensive appraisal of a 'type example' in vertebrate zoology. To this end, authors have been encouraged, within the limitations of space, to provide comprehensive bibliographies for their respective fields so that this volume should serve as a starting point for future lines of *Xenopus* research. We have intended that specialists in particular fields will see the present state of knowledge across the range of research involving *Xenopus*. By combining reviews of this diversity in a single volume, we hope that future studies will be directed to the potential offered by the special features of *Xenopus*. For example, the theme of allopolyploidy recurs in a range of subject areas (including biochemistry, evolutionary ecology and parasitology): further studies of phenotypically identical species which have 36, 72 or 108 chromosomes within each cell should have applications in many areas of research. Above all, the selection of subject matter reflects our own relatively wide-ranging interests in this fascinating animal.

We initially drew up a list of key aspects of biology which should be included in a comprehensive review, but we failed to locate current research in a few of these areas, most notably reproductive endocrinology. Ultimately, some aspects of endocrinology are included in Kelley's account of sexual differentiation and Just & Kraus-Just's chapter on metamorphosis, but it appears that the intensive research which previously addressed the reproductive biology of *Xenopus* has now moved on to other models and other questions.

It was not our original intention to write chapters on several of the themes ourselves. We were convinced that future workers should have access to information on ecology and ecophysiology, but we found that this area is

[1]Kay, B.K. & Peng, H.B. (Eds) (1991). *Xenopus laevis*: practical uses in cell and molecular biology. *Methods Cell Biol.* 36: 1–718.
[2]Tymowska, J. (1991). Polyploidy and cytogenetic variation in frogs of the genus *Xenopus*. In *Amphibian cytogenetics and evolution*: 259–297. (Eds Green, D.M. & Sessions, S.K.). Academic Press, San Diego.
[3]Hausen, P. & Riebsell, M. (1991). *The early development of* Xenopus laevis: *an atlas of the histology*. Springer Verlag, New York.
[4]Cannatella, D.C. & De Sa, R.O. (1993). *Xenopus laevis* as a model organism. *Syst. Biol.* 42: 476–507.

still completely neglected. Eventually, we decided to review this area with Catherine Loumont, building on our experiences from a combined total of over a dozen fieldtrips to study *Xenopus* (in 18 countries of sub-Saharan Africa). We are conscious that our background in ecology, although based on first-hand field observations, is nevertheless dependent on snapshots obtained during relatively short fieldwork visits. There remains a vital need for long-term studies by workers based in areas of *Xenopus* occurrence. For similar reasons, we decided to combine our experience to produce a synopsis of the characteristics of the extant species of *Xenopus* (not least because we have been involved in describing seven of them). In the opening chapter, we have responded to the repeated requests for a key to species, but reliable species differentiation still remains a challenge to taxonomists who study preserved specimens, because characters of major biological significance require living material (for data on chromosome number, mating calls, etc.)

With this background of plans, we were delighted to bring together a total of 22 papers for the symposium, and these form the basis of this volume. The first five chapters were designed to set the scene in terms of systematics, ecology, distribution, species interactions. The three-fold increase in the number of described *Xenopus* species in the past 20 years has inevitably led to some bewilderment for many anuran taxonomists, not least because many of the species are morphologically very similar or even identical! The chapter on ecology provides a review of the current state of published knowledge and is intended to highlight areas which require in-depth fieldwork. Recent biogeographical research, especially involving pollen analysis, has transformed ideas on the presumed stability of tropical rainforests: it now emerges that the distributions of the *Xenopus* species will have undergone major cyclical changes during the Quaternary, corresponding with climatic fluctuations. The resulting dynamic zones of sympatry between the *Xenopus* species will have had a profound influence on evolution through interspecies hybridization and polyploidization. This recent insight into the biogeographical history of *Xenopus* also helps to explain the sudden increase in the numbers of known species: our fieldwork in the remote highlands of Cameroon and Central Africa has involved areas which acted as refugia during successive climatic oscillations, concentrating the species and promoting differentiation and hybridization.

With so few published ecological studies, we decided to include one recent case study (by Picker, Harrison & Wallace) based on fieldwork: the interaction between *X. l. laevis* and *X. gilli* in areas of sympatry on the Cape Peninsula. This has important evolutionary significance because hybridization between these species has been shown to lead to introgression and, potentially, to further species evolution through allopolyploidy. The genetic basis of this mechanism, determined through laboratory studies, is of considerable interest and prompted us to include a paper by Kobel explaining the outcome of experimental hybridization. In line with our aim to focus on the special features of *Xenopus* biology, the final chapter in this section (Tinsley & McCoid) highlights the capacity of *Xenopus* to become established in environments far removed from its current natural distribution. The success of *Xenopus laevis* as a colonizing species is illustrated by a series of independent introductions in California where attempts to control the dramatic spread of feral populations have failed. *X. laevis* is also established in other regions, including the United Kingdom. This adaptability will not surprise those familiar with *Xenopus*; however, it is intriguing that the recent intercontinental extension to its biogeographical range has actually involved reintroduction to some regions of former occurrence (South America) in which the genus has long been extinct.

The second theme of this volume brings together exciting current research on the behaviour, sensory perception and development of *Xenopus*. Elepfandt's studies demonstrate the remarkable precision of both the auditory and lateral line systems for localizing underwater stimuli and for analysing complex environmental input. Yager has elucidated the unique sound production mechanism which provides a means of communication between individuals. It is particularly interesting that field studies by Elepfandt and Yager are now revealing that interactions within populations include the maintenance of territories in the underwater habitats. Sound production for mating interactions is taken further by Kelley in studies of the endocrine control of sexual differentiation, and her work explores the molecular basis of development.

One of the remarkable features of the Anura is the fundamental difference in body organization and life style between larval and adult stages. Anuran metamorphosis is so familiar that most biologists give little thought to the highly complex reorganization affecting virtually all body functions, from the digestive tract to the immune system. To illustrate these aspects of *Xenopus* biology, Wassersug reviews information on the tadpole, including its functional specializations and behaviour, and Just & Kraus-Just consider the endocrine control of metamorphosis.

The next section focuses on infections and defence. An exceptionally rich fauna of parasitic organisms is associated with *Xenopus*. There are over 25 genera from seven invertebrate groups whose relationships reflect dual influences: on the one hand, the ecological overlap of aquatic *Xenopus* with fish, and, on the other hand, a long phylogenetic isolation from other anurans. There is clear evidence of tight evolutionary linkage between the host and its highly specific parasites. In the context of microbial infection and pathology, most laboratory workers will be familiar with the extraordinary resilience of *Xenopus* to disease. Kreil reviews the remarkable diversity of peptides and biogenic amines which are abundant in the skin of *Xenopus*. Some of these have powerful antimicrobial properties, others have structural and functional homology to mammalian hormones and neurotransmitters, others are opioid peptides. In another rapidly-moving research field, studies of the immune system of *Xenopus* also reveal great complexity. Horton and Du Pasquier and their co-workers consider, respectively, T cell and B cell biology. Non-specialists will be surprised that most elements of the mammalian immune system are recognizable, and both chapters consider the ontogeny of the immune response, particularly in relation to the unique challenge created by metamorphosis.

The final group of chapters addresses phylogenetic considerations, and we decided to preface this with a wider review of the evolutionary relationships of the Anura (by Sanchiz & Roček) in order to put *Xenopus* evolution into the context both of time and of amphibian diversity. The current findings of Baez on the fossil record of *Xenopus* must excite all those who work with extant species, particularly her superb preparations of *Xenopus*-like pipids up to 100 million years old from Patagonia. Morphological adaptations for aquatic life are considered further by Trueb who documents, with meticulous illustrations, how *Xenopus* and relatives have diverged from the generalized anuran morphotype. The recent intensive research on biochemical and molecular characteristics of *Xenopus* has led to a series of assessments of relationships: Graf presents a critical review of the molecular evidence for the phylogeny of *Xenopus*. *Xenopus* speciation is characterized especially by the involvement of allopolyploidy, a mechanism otherwise rare in vertebrates: Kobel analyses the requirements of this mode of evolution. There can be no doubt that the parasites specific to *Xenopus* will have had a close interest in these evolutionary events: in a concluding chapter (by Tinsley) it emerges that diverse groups of parasites, including platyhelminths and arthropods, provide an interpretation of *Xenopus* phylogeny similar to that recently proposed by geneticists and molecular biologists (but of course these parasites have had a long time to form their conclusions!).

Xenopus research reached new heights during the course of the symposium: Richard Wassersug was unable to come to London because he was committed on the same days with a NASA Space Shuttle launch which carried an experiment involving *Xenopus* to study early embryo development in zero gravity—the first free-living vertebrates conceived in space.

With such a (deliberately) diverse spectrum of chapters, we aimed to achieve a critical overview of approaches by asking most authors to referee other manuscripts. In addition, we are grateful for independent reviews of manuscripts which were willingly undertaken by Ronn Altig, Barry Clarke, Margaret Manning, Andrew Milner. Heather Tinsley cheerfully undertook much checking and editing of the texts. We thank John Gurdon for introducing the symposium, and Louis Du Pasquier, Andreas Elepfandt, John Horton, Darcy Kelley and Margaret Manning for chairing the sessions. We are very grateful for financial support of the symposium from The Zoological Society of London, The Royal Society, Merck Sharp & Dohme Research Laboratories, and Singer Instruments (manufacturers of *Xenopus* oocyte micromanipulators), together with the contributors' institutions which provided travel expenses. We thank Lena Clarke for secretarial help in Bristol, and Tim Colborn, also in Bristol, for the beautiful cover drawing of *Xenopus*. We are especially grateful to the Zoological Society of London for hosting the meeting and providing hospitality to the international gathering of contributors, and above all to Unity McDonnell for her patience and understanding, and her careful organization of both the symposium and this volume.

There is major potential for new inter-disciplinary research which examines the exceptional features of *Xenopus*. One aim of this volume is that it will now stimulate both basic studies to fill important gaps in understanding and a cross-fertilization of ideas to develop novel lines of research.

R.C.T.
H.R.K.

Contents

4 Natural hybridization between *Xenopus laevis laevis* and *X. gilli* in the south-western Cape Province, South Africa
M.D. PICKER, J.A. HARRISON & D. WALLACE

5 Reproductive capacity of experimental *Xenopus gilli* × *X. l. laevis* hybrids
H.R KOBEL

6 Feral populations of *Xenopus* outside Africa
R.C. TINSLEY & M.J. McCOID

Behaviour, sensory perception and development

7 Sensory perception and the lateral line system in the clawed frog, *Xenopus*
A. ELEPFANDT

10 Underwater acoustics and hearing in the clawed frog, *Xenopus*

A. ELEPFANDT

11 The biology of *Xenopus* tadpoles

R. WASSERSUG

12 Control of thyroid hormones and their involvement in haemoglobin transition during *Xenopus* and *Rana* metamorphosis
J.J. JUST & J. KRAUS-JUST

Infections and defence

13 Parasites of *Xenopus*
R.C. TINSLEY

14 Skin secretions of *Xenopus laevis*
G. KREIL

15 Immune system of *Xenopus*: T cell biology
J.D. HORTON, T.L. HORTON & P. RITCHIE

Phylogeny and speciation

21   Allopolyploid speciation
H.R. KOBEL

22   Evolutionary inferences from host and parasite co-speciation
R.C. TINSLEY

Index

Contributors

BÁEZ, A.M., Dto. de Geología, Facultad de Ciencias Exactas, Universidad de Buenos Aires, Pabellón II, Ciudad Universitaria, 1428 Buenos Aires, Argentina.

DU PASQUIER, L., Basel Institute for Immunology, CH-4005 Basel 5, Switzerland.

ELEPFANDT, A., Fakultät Biologie, Universität Konstanz, Postfach 5560, D-7750 Konstanz, Germany; *present address* Institut für Biologie, Humboldt-Universität zu Berlin, Invalidenstr. 43, D-10115 Berlin, Germany.

GRAF, J.D., Laboratoire Central d'Examens Biologiques, Hôpital Cantonal Universitaire, 1211 Genève 14, Switzerland.

GURDON, J.B., Wellcome Trust & Cancer Research Campaign, Institute of Cancer & Developmental Biology, Tennis Court Road, Cambridge CB2 1QR, UK.

HARRISON, J.A., Dept. of Zoology, University of Cape Town, Rondebosch 7700, Cape Town, Republic of South Africa.

HORTON, J.D., Dept. of Biological Sciences, University of Durham, South Road, Durham DH1 3LE, UK.

HORTON, T.L., Dept. of Biological Sciences, University of Durham, South Road, Durham DH1 3LE, UK.

JUST, J.J., School of Biological Sciences, University of Kentucky, Lexington, KY 40506–0225, USA.

KELLEY, D.B., Dept. of Biological Sciences, Columbia University, New York, NY 10027, USA.

KOBEL, H.R., Laboratoire de Génétique Animale et Végétale, Université de Genève, 154 Route de Malagnou, CH-1224 Chêne-Bougeries, Geneva, Switzerland.

KREIL, G., Institute of Molecular Biology, Austrian Academy of Sciences, Billrothstrasse 11, A-5020 Salzburg, Austria.

KRAUS-JUST, J., Lexington Clinic, 1221 South Broadway, Lexington, KY 40504, USA.

LOUMONT, C., Muséum d'Histoire Naturelle, Case Postale 434, 1211 Geneva, Switzerland.

MCCOID, M.J., Caesar Kleberg Wildlife Research Institute, Texas A&M University, Kingsville, Texas 78363, USA.

PICKER, M.D., Dept. of Zoology, University of Cape Town, Rondebosch 7700, Cape Town, Republic of South Africa.

RITCHIE, P., Dept. of Biological Sciences, University of Durham, South Road, Durham DH1 3LE, UK.

ROBERT, J., Basel Institute for Immunology, CH-4005 Basel 5, Switzerland.

ROČEK, Z., Palaeontological Department, Academy of Sciences, Rozvojová 135, 165 00 Prague 6, Czech Republic.

SANCHIZ, B., Museo Nacional de Ciencias Naturales, CSIC, J. Gutierrez Abascal 2, Madrid E-28006, Spain.

TINSLEY, R.C., School of Biological Sciences, University of Bristol, Woodland Road, Bristol BS8 1UG, UK.

TRUEB, L., Natural History Museum and Dept. of Systematics & Ecology, The University of Kansas, Lawrence, Kansas 66045–2454, USA.

WALLACE, D., Dept. of Zoology, University of Cape Town, Rondebosch 7700, Cape Town, Republic of South Africa.

WASSERSUG, R.J., Dept. of Anatomy & Neurobiology, Sir Charles Tupper Medical Building, Dalhousie University, Halifax, Nova Scotia, Canada B3H 4H7.

WILSON, M., Basel Institute for Immunology, CH-4058 Basel 5, Switzerland.

YAGER, D.D., Dept. of Psychology, University of Maryland, College Park, MD 20742–4411, USA.

Organizers of symposium

PROFESSOR R.C. TINSLEY, School of Biological Sciences, University of Bristol, Woodland Road, Bristol BS8 1UG, UK.

DR H.R. KOBEL, Laboratoire de Génétique Animale et Végétale, Université de Genève, 154 Route de Malagnou, CH-1224 Chêne-Bougeries, Geneva, Switzerland.

Chairmen of sessions

DR L. DU PASQUIER, Basel Institute for Immunology, CH-4005 Basel 5, Switzerland.

PROFESSOR A. ELEPFANDT, Fakultät Biologie, Universität Konstanz, Postfach 5560, D-7750 Konstanz, Germany; *present address* Institut für Biologie, Humboldt-Universität zu Berlin, Invalidenstr. 43, D-10115 Berlin, Germany.

PROFESSOR SIR JOHN GURDON, FRS, Wellcome Trust & Cancer Research Campaign, Institute of Cancer & Developmental Biology, Tennis Court Road, Cambridge CB2 1QR, UK.

DR J.D. HORTON, Dept. of Biological Sciences, University of Durham, South Road, Durham DH1 3LE, UK.

PROFESSOR D.B. KELLEY, Dept. of Biological Sciences, Columbia University, New York, NY 10027, USA.

PROFESSOR M.J. MANNING, Dept. of Biological Sciences, Plymouth University, Drake Circus, Plymouth PL4 8AA, UK.

Introduction

1 Introductory comments: *Xenopus* as a laboratory animal

J.B. GURDON

Xenopus laevis, the South African clawed frog, has come to be, together with the mouse and the chick, the vertebrate species most widely used for research in the areas of developmental, cell and molecular biology. This somewhat surprising situation arose from the use of X. *laevis* about 50 years ago as an assay for pregnancy in humans; the injection of a few millilitres of urine from a pregnant woman under the female frog's skin, into the dorsal lymph sac, causes it to lay eggs. From around the 1950s, this was replaced by a more convenient form of pregnancy test, but it was apparent that *Xenopus* can be induced to mate and provide fertile embryos by the gonadotrophic hormones (primarily luteinizing hormone) of other vertebrate species. This characteristic is not shared by *Rana* species in which the same result requires the injection of homogenized pituitary glands from several adults, a procedure that is very laborious and extremely wasteful of animals. Even then, fertile eggs can be obtained from *Rana* only during certain months of the year, as is also true of other amphibians such as newts and salamanders (*Triturus*) which were favourite objects of research for great names in the field of experimental embryology including Spemann, Holtfeter, etc. The ability to obtain fertile eggs from *Xenopus* throughout the year by the injection of commercially available hormone was what primarily persuaded M. Fischberg to concentrate his laboratory at Oxford, England, in the mid 1950s, on X. *laevis*. Although *Xenopus* had been used by Witschi, Gallien and some others for sex-determination work, as well as by Nieuwkoop and others for embryological work (Nieuwkoop & Faber 1956, and references therein), Fischberg's laboratory at that time greatly helped to popularize the species as a major organism for research in developmental biology.

There were, however, three additional reasons why *Xenopus laevis* gradually became increasingly popular as a laboratory animal. One is its permanently aquatic life style, enabling it to be kept in water tanks which are much more easily cleaned and changed than the terraria required for most other amphibians. Second is its remarkably robust constitution: it is extraordinarily resistant to disease and infection, a characteristic probably related to its preferred native habitat, which is nutrient-enriched stagnant water. Last, and particularly important, is its relatively short life cycle, such that a fertilized egg can be grown to be a sexually reproducing adult in one year, or less for a male; this may be compared to some four years needed for *Rana pipiens*.

This last point assumes enormous significance when attempts are made to take advantage of genetics. Although a one-year life cycle is much too long for convenient genetic analysis, compared to the three-month cycle of a mouse or the zebra fish, it is acceptable for maintaining and using valuable mutants. It was indeed in Fischberg's laboratory that the first really important *Xenopus* mutation was discovered (Elsdale, Fischberg & Smith 1958), namely the O-*nu* mutant, which has been invaluable as a cell marker in nuclear transplantation and other experiments, and for the analysis of ribosomal genes and their function (Elsdale, Gurdon & Fischberg 1960). A number of other *Xenopus* mutations have been isolated since then, predominantly in Fischberg's laboratory in Geneva.

Since those early days, there has been a steady increase in the popularity of *Xenopus* for research in cell and developmental biology. Apart from the reasons already mentioned, this further popularity mainly results from the large size of *Xenopus* embryos and cells. In the early 1960s, D.D. Brown carried out pioneering work of a biochemical nature, purifying various kinds of RNA and analysing gene expression by molecular methods (e.g. Brown & Littna 1966). Nowadays it is possible to carry out quite sophisticated transcript analyses on single copy genes, e.g. by nuclease protection, on the amount of material contained in two *Xenopus* embryos. The same quantity of material is contained in 8000 mouse embryos. *Xenopus* cells, and especially those of early embryos, are very large, and are very well suited for cell or tissue recombination experiments of a kind which have scarcely been attempted in other vertebrates. Thus it is entirely convenient to carry out a biochemical analysis on a small number of experimentally manipulated *Xenopus* embryos or tissue combinations.

Another major interest in *Xenopus* material arose from the discovery, very surprising at the time, that purified messenger RNA is very efficiently translated when microinjected into oocytes. These are the large (*c.* 1 mm diameter) fully grown egg precursor cells, which accumulate in large numbers in the ovary, but which are still very active in gene transcription in meiotic prophase. Their large size makes it easy to inject 10^8 molecules of RNA into the cytoplasm of one oocyte (Gurdon *et al.* 1971) or, as later discovered, of cloned DNA into the immensely large nucleus or germinal vesicle. In the course of time, it was discovered that all steps of gene expression, including transcription, splicing, transport of mRNA to the cytoplasm, assembly into polysomes, translation into protein, protein cleavage secondary modification and assembly into complex structures, as well as secretion and retrograde protein accumulation into the oocyte nucleus, all take place when genes of a foreign species, vertebrate or invertebrate, are injected into an oocyte nucleus. These processes normally occur more slowly in injected oocytes than in normal cells and this, together with the value of being able to follow the expression of a single gene step by step, has led to the *Xenopus* oocyte being regarded as a 'living test tube' (review by Gurdon & Wickens 1983). A particularly widespread and successful application of oocyte injection has been its use for the identification of genes that encode membrane receptor subunits. For example, a total messenger RNA population from a serotonin-responsive tissue can be injected into an oocyte, which translates the messenger RNAs including those coding for serotonin receptor subunits. The oocyte is then able to assemble the newly synthesized foreign subunits, often using some of its own, to make a functional receptor, which can be recognized electrophysiologically following addition of the ligand, in this case serotonin, to the oocyte (e.g. Barnard, Miledi & Sumikawa 1982). This provides a unique *functional* assay by which to identify mRNAs, hence cDNAs, and hence genes for receptor components, and has resulted in *Xenopus* oocytes being one of the most widely used cell types in molecular biology (see Kay & Peng 1991).

These various reasons why *Xenopus* has become one of the world's most widely used species for biomedical research provided the background to the symposium, the contributions to which constitute this volume. There are numerous meetings around the world, every year, at which experiments on *Xenopus* are described in connection with cell, molecular, or developmental biology or with the genetic basis of human disease. And yet I cannot recall any meeting devoted to the biology of *Xenopus*. All of us who use *Xenopus* as a laboratory animal depend on the reliability with which fertile eggs can be obtained, and on the quality of its eggs and oocytes. In order to maintain *Xenopus* in the laboratory and to obtain the best eggs and oocytes, we need to know the conditions under which they live in the wild. We can also benefit greatly from a further knowledge of the natural genetic variation which exists in the form of many species and subspecies. Last, but not least, there is a real interest in the life-style of these strange aquatic animals; most of us know little about how they catch their food, protect themselves against adverse circumstances, and communicate with each other. Papers on all these topics are included in this volume. Drs Tinsley and Kobel have made many expeditions to Africa, some in association with M. Fischberg. Apart from their own important contributions to the literature, Drs Tinsley and Kobel are among the few in the world who have detailed knowledge of the biology of *Xenopus*. All those of us who use *Xenopus* in the laboratory are indebted to them for having organized this symposium and the publication of its proceedings.

References

Barnard, E.A., Miledi, R. & Sumikawa, K. (1982). Translation of exogenous messenger RNA coding for nicotinic acetylcholine receptors produces functional receptors in *Xenopus* oocytes. *Proc. R. Soc. (B)* **215**: 241–246.
Brown, D.D. & Littna, E. (1966). Synthesis and accumulation of low molecular weight RNA during embryogenesis of *Xenopus laevis*. *J. molec. Biol.* **20**: 95–112.

Elsdale, T.R., Fischberg, M. & Smith, S. (1958). A mutation that reduces nucleolar number in *Xenopus laevis*. *Exp. Cell Res.* **14**: 642–643.

Elsdale, T.R., Gurdon, J.B. & Fischberg, M. (1960). A description of the technique for nuclear transplantation in *Xenopus laevis*. *J. Embryol. exp. Morph.* **8**: 437–444.

Gurdon, J.B., Lane, C.D., Woodland, H.R. & Marbaix, G. (1971). Use of frog eggs and oocytes for the study of messenger RNA and its translation in living cells. *Nature, Lond.* **233**: 177–182.

Gurdon, J.B. & Wickens, M.P. (1983). The use of *Xenopus* oocytes for the expression of cloned genes. *Meth. Enzym.* **101**: 370–386.

Kay, B.K. & Peng, H.B. (1991). *Xenopus laevis*: practical uses in cell and molecular biology. *Methods Cell Biol.* **36**: 299–309.

Nieuwkoop, P.D. & Faber, J. (1956). *Normal table of* Xenopus laevis *(Daudin)*. North Holland Publ. Co., Amsterdam.

Xenopus species and ecology

2   The extant species

H.R. KOBEL, C. LOUMONT and R.C. TINSLEY

Synopsis

As a result of recent detailed analysis of live specimens and also of sampling in more remote parts of Africa, the number of *Xenopus* species has trebled in the last 20 years. Studies of karyotype and DNA-content demonstrate a remarkable series of polyploid species. Based on two different chromosome sets of 10 and 18, they represent ploidy levels of 2:4:8:12. In fact, when compared with other members of the Pipidae, only a single species seems to have remained diploid. Other criteria used to define species are the mate calls, which in almost all cases are species-specific, and experimental hybridization, which reveals genetic incompatibilities (fertilization barriers, viability, sterility, meiotic chromosome pairing).

On the basis of these and additional morphological and biochemical traits, the genus *Xenopus* is composed of two distinct groups, *Silurana (X. tropicalis* 2n=20, *X. epitropicalis* 2n=4X=40) and *Xenopus* (2n=4X=36, 8X=72 and 12X=108). The latter can be divided into (1) the *laevis*-subgroup (tetraploid) comprising *X. laevis* Rassenkreis, *X. gilli* and *X. largeni*; (2) the *muelleri*-subgroup (tetraploid) with *X. muelleri, X. borealis* and *X. clivii*; (3) the *fraseri*-like subgroup with the tetraploid *X. fraseri* and *X. pygmaeus*, the octoploid *X. amieti, X. andrei* and *X. boumbaensis* and the dodecaploid *X. ruwenzoriensis*; (4) the two closely related octoploid *X. vestitus* and *X. wittei* (which exhibit a mixture of characters and fit in none of the three previous subgroups); and (5) the dodecaploid *X. longipes* which also shows mixed affinities including some characters (paired nasals, skin texture) shared with the *Silurana* group. Several other samples have not yet been studied sufficiently to be classified. Despite this genetic diversity, the species of *Xenopus* are relatively uniform in morphology: a number of the recently-described forms are cryptic species with 36, 72 or 108 chromosomes within almost identical phenotypes.

There remain two major taxonomic questions: first, the status of the various *X. laevis* subspecies that are interfertile but whose advertisement calls as well as genetic differences suggest a rather pronounced divergence; and second, whether a separation of *Xenopus* into two genera is justified.

Introduction

Although *Xenopus* occupy almost every kind of waterbody south of the Sahara, from lakes, rivers and swamps to man-made irrigation ditches, waterholes and reservoirs, they are not correspondingly well represented in museum collections of African anurans. This may be due to their aquatic life style, which demands different sampling techniques from those applied to more terrestrial Amphibia. Local people, to whom *Xenopus* are familiar because they catch them for food or to use as aphrodisiacs or fertility medicines, use different kinds of baited traps, a method also efficient in large ponds or lakes where it is otherwise difficult to catch these anurans.

Since many *Xenopus* species are rather indistinct from one another morphologically, identification of specimens is often difficult: this is especially true for preserved material. However, features such as mate calls, karyotype, DNA-content, electrophoresis of proteins and experimental hybridization may reveal differences between taxa that are not reflected by their morphology. As a result of recent analysis of live specimens by means of such non-morphological traits and also because of sampling in more remote parts of Africa, the number of *Xenopus* species has trebled in the last 20 years.

This chapter provides a summary of the characters of the presently recognized species. The major part of this study was based on live specimens that were sampled during our own fieldwork or were generous gifts from colleagues and friends, collected at various localities throughout much of Africa (Cameroon, Central African Republic, Congo, Ethiopia, Gabon, Ghana, Kenya, Liberia, Malawi, Nigeria, Rwanda, Sierra Leone, South Africa, Tanzania, Uganda, Ivory Coast, Zaire, Zambia). For several taxa, a single live sample of only a few specimens was available for study. Many samples or their descendants are maintained alive at the Zoological Department of the University of Geneva. Museum collections consulted are listed by Tinsley, Loumont & Kobel (this volume pp. 35–59).

Criteria for species identification

Morphology

The size (snout-to-vent length, SVL) of adult *Xenopus* varies from about 35 to 130 mm and representatives may be grouped into small (30–45), medium (45–55) and large (55–100) species; only *X. l. laevis* females grow larger than 10 cm. Males are generally 10–30% smaller than females; there seem to be species-specific differences in this sexual dimorphism.

Several external traits distinguish species or species groups such as (1) the prehallux (metatarsal tubercle) which may be reduced or prominent and even armed with a black horny claw, (2) the length of the subocular tentacle, (3) the size of the lower eyelid, (4) the number of lateral line organs, dorsally and ventrally along the body and on the head, especially around the eyes, (5) the number and arrangement of the cloacal lobes, (6) proportions of various body parts (however, these are strongly influenced by allometric variations: see Tinsley, Kobel & Fischberg 1979). There are several distinctive features of the skeleton (Reumer 1985; Cannatella & Trueb 1988), e.g. fused or separate condition of nasals, first and second presacral vertebra, sternum and epicoracoid cartilages, or the oval versus round form of the tympanic annulus, and many other features. Table 2.3 lists some morphological and biochemical characteristics of the species and subspecies recognized.

Skin and skin texture is smooth in all *Xenopus* species; however, *X. (S.) tropicalis, X. (S.) epitropicalis* and *X. longipes*, especially the males, show many pustules on the head and, to a lesser extent, on the back that are not apparent in other species. The dorsal coloration is diverse with a preponderance of olive to brown tints; specimens of *X. borealis* can be steel grey-blue, and *X. vestitus* may have very distinctive bronze-gold coloration. Many species have roundish to angular, small or large melanophore spots irregularly distributed on back and legs. A regular pattern with a parallel arrangement of large oblong spots is characteristic for the majority of *X. gilli* specimens, while in other species the number and distribution of spots is highly variable. In *fraseri*-like *Xenopus*, a dark transverse band marks the limit between head and back, especially in juveniles. Certain species are evenly coloured without dorsal spots. The ventral coloration is generally whitish to grey and numerous small melanophore spots can be present on legs, belly and up to the throat. Wild specimens may have a pronounced yellow to bright orange tint on legs and lower belly due to chromatophores containing carotenoids and pteridines, notably both bufochrome and ranachrome-3 (these are otherwise specific for the respective taxa indicated by their names: Bagnara & Obika 1965). The ventral pattern is extremely variable; even in the offspring of single pairs, the spotting pattern can vary from fleckless to heavily spotted. Thus, ventral coloration is less suitable for distinguishing species though it may be quite typical.

Advertisement calls

Advertisement (mating) calls are significant signals in the reproductive behaviour of Anura. The aquatic *Xenopus* have adapted their vocal apparatus for underwater sound production (Loumont 1981; Yager 1982, 1992 and this volume). Calls of all *Xenopus* species (Vigny 1979a; Loumont 1983, 1986) are based on single pulses of a fundamental frequency between 0.5 and 3 kHz which in most species is also the dominant frequency. Some species, e.g. *X. borealis, X. l. laevis, X. ruwenzoriensis*, have a second frequency component of comparable amplitude, reminiscent of the bimodal signals which have been shown to be important to females for species discrimination in several other anuran species (Littlejohn 1977). On the whole, harmonic frequency spectra are very diverse, reflecting the variety in larynx morphology observed between *Xenopus* species (Loumont 1981). According to species, pulses are emitted singly or in groups (notes) that are

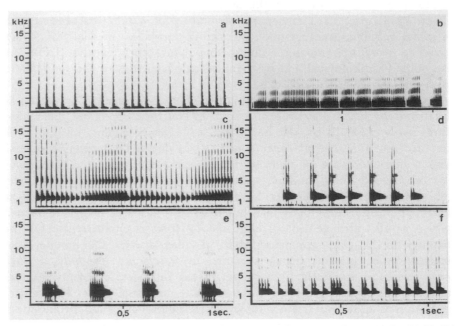

Fig. 2.1. Sonagrams of advertisement calls of *Xenopus* males. (a) *X. (S.) tropicalis*; (b) *X. (S.) epitropicalis*; (c) *X. l. laevis*; (d) *X. l. victorianus*; (e) *X. l. poweri*; (f) *X. l. sudanensis*.

repeated at intervals or linked to longer calls, or pulses may form long trills without distinct notes or trills of notes appearing as variations in sound intensity only. Selected sonagrams of all species (except *X. gilli*, for which a sonagram is given in Picker, Harrison & Wallace, this volume pp. 61–71) are reproduced in Figs 2.1–2.3, and some call characteristics are listed in Table 2.1.

In almost all species, advertisement calls are very distinct in temporal (pulse rate, arrangement of pulses into notes and trills) and in spectral qualities. Cryptic species such as *X. tropicalis–epitropicalis*, *X. fraseri–pygmaeus* and other *fraseri*-like *Xenopus* are most conveniently distinguished by their calls. This is also true for *X. laevis* subspecies which, by this criterion, undoubtedly represent distinct species. However, little is known about the efficiency of advertisement calls as a premating isolation barrier in *Xenopus*. By contrast, calls of several species are very similar (i.e. *X. andrei*, *X. largeni* and *X. vestitus* or *X. amieti* and *X. ruwenzoriensis*). That suggests a closer relationship between species of different ploidy. In the case of *X. amieti–ruwenzoriensis*, the electrophoretic pattern of globins (Bürki & Fischberg 1985) corroborates the impression that the former species is implicated in the evolution of the latter higher polyploid one.

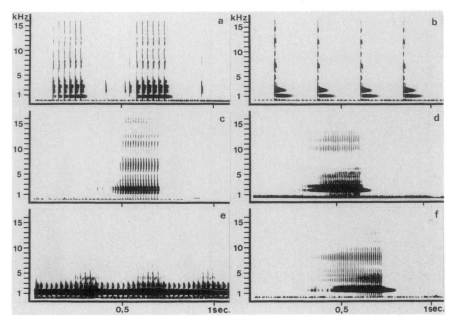

Fig. 2.2. Sonagrams of advertisement calls of *Xenopus* males. (a) *X. muelleri*-East; (b) *X. borealis*; (c) *X. clivii*; (d) *X. largeni*; (e) *X. wittei*; (f) *X. vestitus*.

DNA-content, karyotype, NOR location

The DNA-content as measured on erythrocyte or liver nuclei (Thiébaud & Fischberg 1977; Giorgi & Fischberg 1982; Reumer & Thiébaud 1987) varies between 3.5 and 16.25 pg (Table 2.2). In comparison with the amount of DNA in members of other pipid genera and considering their karyotypes, it was concluded that all but one of the *Xenopus* species are polyploids to various degrees.

The cytogenetics of *Xenopus* has recently been reviewed by its main investigator (Tymowska 1991) and illustrations of karyotypes can be found in that account. *Xenopus* species have multiple sets of two different basic karyotypes, one with 10 and the other with 18 morphologically distinct chromosomes. Since species with 18 pairs of chromosomes have a DNA-content about twice that of the single 20-chromosome species, *X. tropicalis*, or of *Hymenochirus* and *Pipa*, their karyotype should represent a tetraploid constitution. Indeed, its tetraploid nature has been reconstructed by means of replication banding of *X. l. laevis* chromosomes (Schmid & Steinlein 1991); most chromosomes can be arranged into quartets by their similar specific replication pattern. The heterogeneity within the remaining quartets may be due to various chromosome rearrangements.

Secondary constrictions are visible on one or several chromosome pairs,

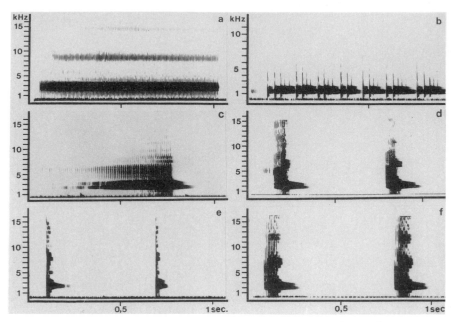

Fig. 2.3. Sonagrams of advertisement calls of *Xenopus* males. (a) *X. fraseri*; (b) *X. pygmaeus*; (c) *X. andrei*; (d) *X. amieti*; (e) *X. boumbaensis*; (f) *X. ruwenzoriensis*.

one of which bears the nucleolus organizer (NOR). Independent of their ploidy level, all species have a single chromosome pair with NORs. Table 2.2 relates NOR location and that of other secondary constrictions to chromosome groups recognizable by conventional staining.

It is apparent that *Xenopus* species fall into a polyploid series with 20, 40, 36, 72 and 108 chromosomes. NOR location and the number and location of other secondary constrictions are typical for species or groups of species. Meiosis at all ploidy levels proceeds with bivalents, and multivalents are rarely observed, in a few spermatocytes of a testis only. Such chromosome behaviour is expected to occur in allopolyploids because chromosomes may have diverged sufficiently during the separate evolution of parental species to prevent homoeologous chromosome pairing. Disomic inheritance ('diploidization') thus may be constitutive in allopolyploids.

Biochemical characters

There are large allelic differences between species, and also between the X. *laevis* subspecies, with respect to electrophoretic mobilities and other properties of a variety of enzymes, albumins and other serum proteins, globins, basic sperm nuclear proteins and other polypeptides (Wall & Blackler 1974; Bisbee *et al.* 1977; Jeffreys *et al.* 1980; Vonwyl & Fischberg 1980; Jaggi, Wyler &

Table 2.1. Sound characteristics of advertisement calls in *Xenopus*.

Species of Xenopus		Fundamental frequency (kHz)	Pulses per second	Pulses per note	Notes per second	Duration of trill
tropicalis		0.5	25	Uniform trill		1–9 s
epitropicalis		0.5	52	8(4–14)	4(2–7)	8–20 s
l. laevis		2	54+38	25(20–37)	1–2	6–28 s
l. poweri		1.8	87	6(4–8)	3–5	
l. victorianus		2.2	60	2(2–4)	7(5–10)	
l. sudanensis		1.8	76 irr.	3(1–7)	11(7–30)	Irregular
gilli		2.8	65	20(16–25)	1–3	
largeni		1.8	102	36	1–2	
muelleri-East	a)	2.2	30	2	8	2 calls:
	b)	1.1	32	8(5–12)	1–3	a or a+b
muelleri-West	a)	2.2	30	2	4	2 calls:
	b)	1.1	32	4(2–11)	1–3	a or a+b
borealis		1.2	3–12	1	1–5	5–35 s
clivii		2	73	12(5–23)	1–3	
fraseri		2.8	156	Uniform trill		0.5–14 s
pygmaeus		1.6	30	4(2–9)	7(6–10)	0.5–9 s
amieti		1.8	167	9(7–11)	1–3	
andrei		2.5	100	45(38–50)	1–2	
boumbaensis		2.5	1–2	1	1–2	
ruwenzoriensis		1.9	150	9(7–10)	1–2	
vestitus		2.0	100	50(26–75)	1–2	
wittei		1.5	36	12(9–24)	1–3	1–8 s

Ryffel 1982; Mann *et al.* 1982; Wolff & Kobel 1982, 1985; Bürki 1985; Bürki & Fischberg 1985; Graf & Fischberg 1986; Graf & Kobel 1991; and other authors). However, since most species are of presumed allopolyploid origin, the complex zymograms of multimeric enzymes are difficult to interpret without extensive family studies. Moreover, not all the duplicated genes are transcribed or have been conserved. A survey for eight characters (five enzymes, albumin, NOR, MHC) in several species of all ploidy levels has shown that tetraploids have conserved 73%, octoploids 53 %, and the dodecaploid *X. ruwenzoriensis* 39% of the expected gene number (Kobel & Du Pasquier 1986). A more accurate figure is known for the tetraploid *X. l. laevis* where 77.3% of the expected genes for 22 enzymes are still expressed (Graf & Kobel 1991). The loss of duplicated genes appears to be species-specific, i.e. each species seems to have conserved its own collection of duplicated genes, as far as the scarce data suggest. Duplicated orthologous genes also tend to acquire a developmental and/or tissue-specific expression which again may be different in different species, e.g. creatine kinase, glucosephosphate isomerase, malate dehydrogenase and other enzymes (Kobel & Du Pasquier 1986; Graf 1989; Robert *et al.* 1990).

The *Xenopus* species thus appear to be genetically rather divergent and therefore well characterized biochemically. However, their complex genetic

Table 2.2. Karyological characteristics in Pipidae: DNA-content, chromosome number and location of secondary constrictions.

Species	DNA (pg/cell)	Chromosome (n/cell)	NOR (N), secondary constrictions (S) on chromosome or chromosome group[a]				
			No. 2	Nos 4–5	Nos 6–7	Nos 10–11	Nos 12,13,18
Hymenochirus (4 species.)	4.2(1)	24					
Pipa (7 species)	4.8(1)	20,22,30					
Xenopus							
Silurana:							
X. (S.) tropicalis	3.6	20					
X. (S) epitropicalis	6.9	40					
Xenopus:							
X. laevis	6.4	36					Np
X. gilli	6.4	36					Np
X. largeni	5.9	36	Npt	Sp			Sp
X. muelleri-East	7.6	36		Npt	Sp		Sp
-West	7.5	36		Npt	Sp		
X. borealis	7.1	36		Npt			
X. clivii	8.5	36		Np			
X. fraseri	6.4	36			Nq(t)		
X. pygmaeus	6.3	36			Nq(t)		
X. amieti	11.4	72		Sp		Sq	
X. andrei	8.8	72		Nq		Sq	
X. boumbaensis	9.3	72		Sp	Np	Sq	
X. ruwenzoriensis	16.3	108		Sp		Npt	
X. vestitus	12.8	72	Sq				Npt
X. wittei	12.6	72					Npt
X. longipes	16	108					Nqt

[a] p = short, q = long chromosome arm; (t), t = (sub-) terminal location

make-up does not facilitate investigations of, for example, the allelic composition of *Xenopus* populations which is, so far, an almost untouched field (Kobel & Du Pasquier 1986). Nevertheless, the few biochemical characters analysed in a sufficiently large number of species (Graf this volume pp. 379–389) readily classify *Xenopus* species into groups and subgroups (Table 2.3).

Interspecific hybridization

Pre-mating isolation mechanisms are difficult to study in the laboratory, especially ethological ones. Very little is known about spatial and temporal isolation during breeding of *Xenopus* species, except that the occurrence of natural hybrids (*X. l. laevis* × *X. gilli*, *X. l. laevis* × *X. muelleri*, *X. l. victorianus* × *X. borealis*) implies an occasional failure of reproductive barriers.

Experimental hybridization, however, can detect gametic and postgametic incompatibilities that unequivocally reveal existing barriers between taxa. Moreover, viable hybrids can be tested for fertility and meiotic chromosome pairing, revealing additional information. Over 100 reciprocal crosses between selected pairs of species have so far been made at the Station Experimentale de Zoologie of the University of Geneva. Usually, different batches of eggs stripped from individual females were simultaneously fertilized with suspensions of spermatozoa from testes of various species including their own.

Gametic incompatibility is generally absent between *Xenopus* species. Exceptions are eggs of *X. tropicalis* that commonly cannot be fertilized by spermatozoa of species having a basic karyotype of 18 chromosomes. However, at elevated temperature, fertilization with *X. l. laevis* spermatozoa was possible and resulted in viable hybrids (Bürki 1985). A unilateral fertilization barrier exists between *X. borealis* and *X. l. laevis* which is less pronounced against spermatozoa of other *X. laevis* subspecies. Spermatozoa of *X. l. laevis* seem to become immobilized at the inner jelly layer of *X. borealis* eggs (Brun & Kobel 1977). Eggs of the *Silurana* and of the *muelleri* group of species appear to be the most difficult to fertilize by alien spermatozoa. Fertilization success, survival of hybrids and sex ratio can be comparable to those of intraspecific crosses, suggesting a broad absence of genetic incompatibilities despite the previously mentioned, rather pronounced, genetic divergence between *Xenopus* species.

Interspecific *Xenopus* hybrids are sterile, the main cause apparently being difficulties in meiotic pairing of homoeologous chromosomes. Univalent meiosis impairs production of spermatozoa and renders eggs aneuploid. Hybrid females, on the other hand, also may produce unreduced polyploid eggs through endoreduplication, a phenomenon dealt with later in this volume (pp. 391–401). A few species combinations, e.g. *X. gilli* × *X. muelleri*, show underdeveloped gonads and ovaries almost devoid of oocytes, indicating other genetic incompatibilities in addition to the problem of chromosome pairing.

Occurrence of homoeologous chromosome pairing in hybrid oocytes has been analysed in more than 25 species combinations (Müller 1977; B. Züst

Table 2.3. Interspecific differences between *Xenopus* species with regard to selected morphological and biochemical characters.

Species	Female size mm(maximum)	Lateral-line organs		Subocular tentacle	Proportion of eye covered by lower eyelid	Prehallux	Cloacal lobes ♀	Albumin (kd)	Globin[a]		
		Eye	Dorsal						s	m	f
tropicalis	43 (55)	8(5–12)	18–23	Medium	< 1/3	Claw	Fused	68	+	–	+
epitropicalis	64 (72)	10(6–14)	18–23	Medium	< 1/3	Claw	Fused	68	+	–	+
l. laevis	110 (130)	17(12–20)	25–34	Short	3/4	Short	3	70 + 74	–	+	+
l. poweri	70 (85)	14(12–16)	19–24	Medium	1/2	Short	3	70 + 74	–	+	+
l. petersi	65 (66)	13(10–16)	20–25	Medium	1/2	Short	3		–	+	+
l. victorianus	62 (78)	14(11–18)	19–25	Short	< 3/4	Short	3	70 + 74	–	+	+
l. sudanensis	62 (64)	13(10–15)	18–24	Short	1/2	Short	3	70 + 74	–	+	+
gilli	55 (60)	13(10–15)	20–24	0	1/2	0	3	70 + 74	–	+	+
largeni	50 (55)	11(11–14)	18–19	0	< 1/3	0	3	70 + 74	–	+	+
muelleri-East	65 (75)	12(9–15)	22–27	Very long	3/4	Prominent	3	70	–	+	+
muelleri-West	53 (90)	11(9–14)	19–25	Long	3/4	Prominent	3	70	–	+	+
borealis	73 (95)	15(13–17)	23–30	Medium	3/4	Prominent	2	70	–	+	+
clivii	70 (82)	14(12–16)	23–28	Medium	3/4	Claw	2	70	–	+	+
fraseri	42 (51)	8(7–10)	18–21	Long	< 3/4	Claw	2	70	+	–	+
pygmaeus	35 (44)	11(9–13)	15–20	Long	1/2	Claw	2				
amieti	53 (57)	11(10–13)	14–23	Medium	1/2	Claw	2	70	+	–	+
andrei	40 (45)	11(9–13)	14–22	Long	1/2	Claw	2	70			
boumbaensis	46 (54)	11(8–13)	17–21	Medium	3/4	Claw	2	70			
ruwenzoriensis	55 (57)	11(9–13)	17–21	Medium	1/2	Claw	2	70	+	+	+
vestitus	47 (55)	10(9–14)	18–28	Medium	1/3	Short	3	70	+	+	+
wittei	46 (61)	10(8–14)	18–25	Medium	1/2	Short	3	70	+	+	+
longipes	34 (36)	10(7–13)	15–24	Medium	1/3	Claw	3		+	+	+

[a]s, m, f: slow-, medium- and fast-migrating globins.

pers. comm.). There is generally an almost complete absence of bivalents (mean number of 1.3 (range: 0–5) bivalents). Exceptions are *X. borealis* × *X. muelleri* hybrids with 4.0 (0–7) and *X. gilli* × *X. l. laevis* hybrids with 10.0 (4–18) bivalents. An interesting case is represented by *X. vestitus* × *X. wittei* hybrids, two species with 2n=4X=72 chromosomes, where about 36 chromosomes form bivalents. A likely interpretation is that these two allopolyploid species share a common 18-chromosome set, whereas the second set originated from a different parental species in either case (Kobel & Müller 1977).

Larval characters

The tadpoles and course of development of all *Xenopus* species are very similar. Nevertheless, developmental stages of the appearance of melanophores on various body parts, their distribution and density, or the general body colour and barbel length, may be distinct and species-specific. Table 2.4 summarizes some of the characteristics of 12 species and subspecies studied by Vigny (1979b).

Taxonomy

The *Xenopus* species can be classified into groups and subgroups, based on the results of the investigations mentioned above and other data. A division into two separate genera has been proposed (Cannatella & Trueb 1988) and this proposal extended to the creation of two corresponding subfamilies of Pipidae. The rank of taxa, however, remains subject to controversy and additional biochemical analysis of this case (De Sà & Hillis 1990) revealed no necessity for elevation of rank in order to render *Xenopus* monophyletic (see also Graf this volume pp.379–389). In the following species descriptions, we opt to conserve the generic name for all *Xenopus* species and to attribute subgeneric status to the two main groups, which may more accurately reflect the respective relationships between various pipid taxa.

In the species descriptions that follow, groups and subgroups of species are headed by a general account of internal and external characters shared by their members. Species descriptions list the type locality and actual geographical distribution, ploidy and chromosome number, the most relevant features of external morphology and coloration pattern; more data are listed in Tables 2.1–2.4. Since advertisement calls are highly specific for most species, special attention should be paid to the sonagrams in Fig. 2.1–2.3. Verbal descriptions of unfamiliar animal sounds are notoriously unsatisfactory; nevertheless, the calls of some *Xenopus* species are reminiscent of familiar noises or, with variable accuracy, may be imitated onomatopoeically. (Pitch of calls: from low pitched—tac—to high pitched—tic, according to the fundamental frequency of the pulses. The timbre of the sounds depends on the harmonics: the more harmonic frequencies there are and the greater their amplitude is, the more metallic and high-pitched the clicks sound. Pulse rate: up to about 30 pulses

Table 2.4. Selected larval characters in 12 species and subspecies of *Xenopus*.

	trop	*l. lae*	*l. pow*	*l. vic*	*gil*	*mue*-E	*bor*	*cliv*	*fras*	*ruw*	*vest*	*witt*
Egg diameter, mm	0.85	1.3	1.0	1.0	1.3	1.0	1.1	1.3	1.0	1.3	1.1	1.3
Disappearance of cement gland, stage	48	49	49	47	49	47	47	47	47	47	46	48
Melanophores:												
Small epidermic, stage	51	No	56	55	No	52	52	52	56	52	52	51
Density, stage 50/60	+	–	++	++	–	+++	+++	+++	+	+	+++	+++
Anal tube, stage	47	53	50	48	49	50	49	49	45	46	46	48
Density, stage 50/60	+++	+	++	+++	+	+	+++	+	+++	+	+++	+
Lower tailfin, stage	51	50	50	48	45	50	49	49	50	48	48	50
Reaching anal tube	53	No	53	52	No	52	51	51	51	50	50	No
Empty area on prox. tail	No	Long	V. long	Short	V. long	Short	Long	Long	Short	No	No	Short
Head ventral, stage	51	No	51	50	No	No	No	No	48	47	46	55
Distribution, stage 50/69	Unif.	–	V	V	–	–	–	–	Unif.	Unif.	Unif.	V
Head dorsal, empty area filled in at stage:												
(a) median to eyes	51	60	51	51	57	50	50	49	50	47	51	47
(b) proxim. pronephros	52	No	53	52	No	No	52	51	50	51	51	49
Barbel length, stage 58	V. long	Short	Short	Medium	Short	Short	Short	Medium	Long	Long	Medium	Short
Yellow body colour	+	No	No	No	No	No	No	(+)	+	+	+	+

V = V-shaped, on the margins of the jaw only.

per second can be heard as individual clicks and as rattling sound; higher pulse rates result in ringing, similar to that of a mechanical telephone bell, and eventually in a continuous tone—raaa . . ., —riii . . . .)

Isolated males, especially when treated hormonally, call under water during the night. Calls are audible outside the aquaria and resemble those recorded by a hydrophone but their generally rattling character sounds somewhat attenuated in the air.

A. *Silurana*-group

This subgenus is distinguished by its basic karyotype with 10 chromosomes and by several morphological characters, e.g. fused presacral vertebrae 1 and 2; oval tympanic annulus; paired nasals; absence of vomers; positions of tendons in thigh; small eyes (diameter less than one-third of interocular distance); cloacal lobes fused ventrally. Albumin 68 kd; no fast-migrating basic sperm nuclear proteins.

1. *Xenopus (Silurana) tropicalis* (Gray, 1864: 316).
Type-locality: Lagos (Nigeria).
Distribution: lowland rain forest of West Africa from the Casamanca (Senegal) to the Cross River (Nigeria); the eastern limit of distribution undetermined.
Diploid 2n=20.
Small to medium-sized: female SVL averages 43 mm (max. 55 mm), males 15% less. Short limbs, fifth toe slightly shorter than tibia; prehallux with black claw; lower eyelid small, covering less than one-third of the eye; number of lateral line organs around eye 7.7 (5–12). Dorsal colour olive to dark brown, no large melanophore spots; ventrally whitish to dark grey, more or less densely spotted with tiny melanophore spots. Dorsal skin, especially on the head of males, with numerous pustules.
Advertisement call a deep rattling trill lasting several seconds: roaroar . . .

2. *Xenopus (Silurana) epitropicalis* Fischberg, Colombelli & Picard, 1982: 53.
Type-locality: Kinshasa (Zaire).
Distribution: lowland rain forest east and south of Mt. Cameroon to eastern border of Zaire and northern Angola; western limit undetermined.
Tetraploid 2n=4X=40.
Female SVL averages 64 mm (max. 72 mm), male 20% less. Morphologically very similar to the former species (in museum collections labelled as *X. tropicalis*), distinguished by a greater number of lateral line organs around the eyes, 9.6 (6–14), and a dorsal colour marbled with tiny yellow, olive and black spots; specimens from Kinshasa uniform grey-brown. Call: long series of low-pitched quacks: whawhawha . . . (about five wha's per second).

B. *Xenopus*-group

Distinguished from the former subgenus by a basic karyotype with 18 chromosomes; eight unfused presacral vertebrae; round tympanic annulus; fused (one exception) nasals; vomers small and fused; large eyes (diameter about half of interocular distance); cloacal lobes not fused, abutting or overlapping (some species with a small dorsal third lobe). Albumin 70 or 70 + 74 kd; complex pattern of fast-migrating basic sperm nuclear proteins.

laevis-subgroup
Tetraploid 2n=2X=36. Medium to large *Xenopus*; strong and relatively long legs; prehallux inconspicuous or small and blunt; subocular tentacle shorter than half eye diameter or absent. Albumins 70 and 74 kd.

3. *Xenopus laevis*.
Several geographical subspecies are recognized that have distinct advertisement calls, differ in size and colour pattern, and are genetically rather divergent. Their rank is uncertain: calls and genetic differentiation suggest separation at the species level but experimental hybrids are fully fertile. Since no investigations have been made in regions (only partly known) of sympatric overlap, the various forms are at present considered as a *laevis*-Rassenkreis comprising at least five subspecies.

3a. *X. l. laevis* (Daudin, 1803: 85).
Type-locality: 'Cape Colony' (South Africa).
Distribution: Namibia except north-west, South Africa, Lesotho, Swaziland, south-eastern Botswana, Zimbabwe, north-western Mozambique. Specimens from the western highlands of Malawi have also been attributed to this subspecies.
Tetraploid 2n=2X=36, NOR on 12p.
The most distinct subspecies. Large, female SVL 110 mm (max. 130 mm), male 25% less. Dorsal colour pattern polymorphic, from finely spotted to marbled or with larger round or irregular spots in yellowish to dark tints; ventrally immaculate white-yellowish to densely spotted. Fifth toe much longer than tibia.
Call: very long trills, pulse rate and number of pulses higher in the first half of notes, second half rattling: riitrrriitrrr . . . (about 1–2 riitrrr per second).

3b. *X. l. petersi* du Bocage, 1895: 187.
Type-locality: Dondo (Angola).
Distribution: north-western Angola, western Congo–Brazzaville.
Tetraploid 2n=2X=36, NOR on 12p.
Female SVL 65 mm (max. 80 mm), male 25% less. Dorsal colour dark grey-brown, dorsal spots few, small and irregularly shaped; ventrally heavily spotted with large irregular spots to nearly immaculate.
Call: unknown.

3c. *X. l. poweri* Hewitt, 1927: 413.
Type-locality: Victoria Falls (Zambia).

Distribution: northern Namibia, Angola except north-west, Okavango (Botswana), Zambia, south-eastern Zaire; specimens from south-west Tanzania, e.g. Uzungwe Mts, may also belong to this subspecies (having resemblance of advertisement calls).
Tetraploid 2n=2X=36, NOR on 12p.
Female SVL 65 mm (max. 79 mm), male 20% less. Dorsal colour dark olive-brown, with 8–15 round dark spots on back and some on legs; ventrally immaculate to heavily spotted with small melanophore spots. Fifth toe only slightly longer than tibia.
Call: short, deep and rattling trills: gra, gra, gra (4–5 gra's per second).

3d. *X. l. victorianus* Ahl, 1924: 270.
Type-locality: Busisi, Lake Victoria (Tanzania).
Distribution: Northern Tanzania, Burundi, Rwanda, eastern Zaire, Uganda and adjacent Sudan, south-western Kenya.
Tetraploid 2n=2X=36, NOR on 12p.
Female SVL 62 mm (max. 78 mm), males 25% less. Dorsal colour yellowish to green-brown, mottled with small dots, some larger irregular spots on back and limbs; ventrally few spots. Fifth toe only slightly longer than tibia.
Call: rapid series of very short trills: drick-drick-drick . . .

3e. *X. l. sudanensis* Perret, 1966: 301.
Type-locality: Ngaoundéré (Cameroon).
Distribution: Jos-Plateau and eastern Nigeria, highlands of west and central Cameroon and of Central African Republic.
Female SVL 57 mm (max. 66 mm), male 20% less. Dorsal colour olive, finely spotted with small and some larger roundish dark spots, small spots on legs dorsally; ventrally few spots on thighs and lower belly.
Call: spattering rattling irregular trills: critr-tr-tr-tritr . . .

3f. *X. l. bunyoniensis* Loveridge, 1932: 114.
Type-locality: Bufundi, Lake Bunyonyi (Uganda).
Loveridge recognized a separate subspecies in the Kigezi district (Uganda) distinguished by relatively long legs. No specimens of this morphology have been found in recent fieldwork by Kobel and Tinsley and all live samples from that region contained *X. l. victorianus* characterized by rather short legs. This form of indeterminate rank is now possibly extinct.

4. *Xenopus gilli* Rose & Hewitt, 1927: 344.
Type-locality: near Cape Town (South Africa).
Distribution: Cape Peninsula and coast to Cape Agulhas. Former records from Citrusdal and Calvinia have not been corroborated by recent sampling.
Tetraploid 2n=2X=36, NOR on 12p.
Female SVL 55 mm (max. 60 mm), male 30% less. Prehallux not apparent; subocular tentacle absent; lower eyelid covering half of the eye. Dorsal colour pattern yellowish-green-brown, often with two or four parallel bands of oblong dark spots behind the eyes, lower back with irregular dark spots,

limbs mottled; ventrally immaculate to heavily spotted, light reticulation, often bright yellow.
Call: short trills: vrii, vrii, vrii (1–3 per second).

5. *Xenopus largeni* Tinsley, 1995: 376.
Type-locality: Sidamo Province (southern Ethiopia).
Distribution: highlands around 2600 m in Bale Mts (southern Ethiopia).
Tetraploid 2n=2X=36, NOR on 2pt.
Medium size, females SVL 55 mm, males *c.* 30% less.
Prehallux not apparent; subocular tentacle absent or a small cone; lower eyelid small, covering less than one-third of the eye. Dorsal colour dark brown, unspotted; ventrally grey-white with irregular small melanophore spots up to throat.
Call: trills of 0.5 s duration: whriiing, whriiing.

muelleri-subgroup
Tetraploid 2n=2X=36. Large sturdy *Xenopus*; female SVL 65–95 mm; prehallux prominent, cone-shaped and pointed, with or without claw; subocular tentacle medium to long, one-third to two-thirds of eye diameter; lower eyelid covering three-quarters of the eye. Albumin 70 kd.

6. *Xenopus muelleri* (Peters, 1844: 37).
Type-locality: Tete (Mozambique).
Distribution: Two geographically separated populations can be distinguished:
X. muelleri-East from regions generally below 800 m altitude in southeastern Kenya (Mombasa, Simba Hills), Tanzania including both shores of Lake Tanganyika, Zanzibar and Mafia Islands, Zambia, Malawi, Okavango (Botswana), Zimbabwe, Mozambique, eastern South Africa north of St. Lucia/Empangeni.
X. muelleri-West from savannas of Upper Volta eastward to southern Sudan.
Tetraploid 2n=2X=36, NOR on 4pt.
Female SVL 65 mm (max. 90 mm), male 20% less. Prehallux cone-shaped and pointed; subocular tentacle two-thirds of eye diameter. Dorsal colour olive to grey-brown with 5–8 large round spots that tend to pale in older specimens; ventrally greyish, immaculate to heavily spotted.
Call: two different calls may be emitted: (a) a repetition of two-pulsed notes sounding like 'a spoon tapping on a pan': tick-tick-tick ... (4–8/s) and (b) notes composed of 5–12 pulses: trra, trra (about 2/s; on sonagrams, each trra is preceded by a tick).
The two geographical forms are of similar appearance but mating calls of the western form are often of type (a) only (and slower), the second call is shorter with fewer pulses, and the parasite fauna differs in the two populations (see Tinsley this volume pp. 403–419). The rank of the two forms remains to be established.

7. *Xenopus borealis* Parker, 1936: 596.
Type-locality: Marsabit (Kenya).

Distribution: Kenya, generally above 1500 m altitude, lower at Marsabit.
Tetraploid 2n=2X=36, NOR on 4pt.
Female SVL 73 mm (max. 95 mm), male 20% less. Prehallux cone-shaped but less prominent than in the former species; subocular tentacle about half of eye diameter. Dorsal colour dark brown to steel-blue, 30–40 irregular black spots, more dense on the lower back and on hindlimbs; ventrally white, more or less spotted.
Call: long series of single loud clicks reminiscent of the sound made by table-tennis balls: tack, tack ... (about 2/s, occasionally accelerating to over 12/s).

8. *Xenopus clivii* Peracca, 1898: 3.
Type-locality: Saganeiti and Adi Caié (Eritrea).
Distribution: Eritrea, Ethiopia, above 1500 m altitude.
Tetraploid 2n=2X=36, NOR on 4p (not terminal).
Female SVL 70 mm (max. 82 mm), male 25% less. Prehallux with black claw; subocular tentacle about one-third of eye diameter. Dorsal colour grey-brown with 15–30 irregular dark spots, limbs with oblong spots, nuptial coloration on male forearm extending to chest (which occurs in no other *Xenopus*).
Call: rolling trills, one per second: qua, qua ...

fraseri-like subgroup
2n=36, 72, 108. Small to medium size. Prehallux with black claw; legs and toes relatively short and thin. Dorsal coloration generally greyish, often and especially in juveniles a typical transverse or elongated median band of darker pigment behind eyes. Albumin 70 kd.
    In museum collections, specimens from the Sanaga-, Ogooué- and Zaire-Basin are labelled as X. *fraseri* according to the original description by Boulenger (1905) of specimens of uncertain origin ('West Africa, probably from Nigeria or Fernando Po'). At present at least six species are recognized. Since the limits of distribution of none of these taxa are known, specimens from localities other than those mentioned below may represent as yet unidentified species.

9. *Xenopus fraseri* Boulenger, 1905: 250.
Type-locality: West Africa. Most investigations have been based on specimens from Sangmelima (Cameroon); since no live samples have been available for analysis, it is uncertain if specimens from Fernando Po belong to this taxon as defined here.
Distribution: inland rain forest of central southern Cameroon and northern Gabon (advertisement calls from Yaoundé, Ebolowa, Sangmelima (Cameroon) and Makokou (Gabon) are similar).
Tetraploid 2n=2X=36, NOR on 6qt.
Small species; female SVL 42 mm (max. 51 mm), male 20–25% less. Subocular tentacle two-thirds of eye diameter; lower eyelid covering half to three-quarters of the eye; lateral line organs around eye 8.7 (7–11); thin limbs. Dorsal colour

grey-brown, frequently with a dark transverse pigment band behind the eyes, lower back and limbs vermiculated; ventrally whitish, thigh reddish.
Call: very long plaintive trills: iii . . . iiing, iii . . . iiing, pulse rate 150/s.

10. *Xenopus pygmaeus* Loumont, 1986: 756.
Type-locality: Bouchia (Central African Republic).
Distribution: rain forest of Central African Republic to north-eastern Zaire.
Tetraploid 2n=2X=36, NOR on 6qt.
Small species, female SVL 35 mm (max. 44 mm), male 5–10% less. Distinguished from the former species by a shorter eyelid covering one-third to half of the eye, a higher number of lateral line organs around the eye, 10.7 (9–13), and a different advertisement call of one-fifth of the pulse rate, 30/s. Dorsal colour grey with a somewhat reddish tint, transverse band often split into two oblong bands, lower back marbled; ventrally grey-white, no spots.
Call: long crackling trills: cracrracrocra . . .

11. *Xenopus amieti* Kobel, Du Pasquier, Fischberg & Gloor, 1980: 920.
Type-locality: Mt. Manengouba (Cameroon).
Distribution: highlands of western Cameroon from Mt. Manengouba to Kumbo.
Octoploid 2n=4X=72, NOR on one pair of chromosome group 4–7.
Medium size, female SVL 53 mm (max. 57 mm), male 25% less. Subocular tentacle shorter than half eye diameter; lower eyelid covering half of the eye. Dorsal colour dark grey-brown, transverse band often present, a few irregular dark spots on back and legs.
Call: short high-pitched metallic trills: cri, cri (about 2/s).

12. *Xenopus andrei* Loumont, 1983: 170.
Type-locality: Longyi (near Kribi, Cameroon).
Distribution: rain forest of coast of Cameroon, northern Gabon and western Central African Republic.
Octoploid 2n=4X=72, NOR on 18qt.
Small size, female SVL 40 mm (max. 45 mm), male 5–10% less. Very similar to *X. fraseri* but lower eyelid only half of eye diameter.
Call: trills of 0.5 s duration: riing, riing.

13. *Xenopus boumbaensis* Loumont, 1983: 169.
Type-locality: Mawa, Boumba Valley (south-eastern Cameroon).
Distribution: rain forest of upper Boumba Valley (Cameroon).
Octoploid 2n=4X=72, NOR on one pair of chromosome group 4–7.
Medium size, female SVL 46 mm (max. 54 mm), male 25–30% less. Lower eyelid covering three-quarters of the eye. Dorsal colour yellowish olive, oblong mediodorsal spot between and behind eyes, numerous small spots on back and limbs.
Call: 1–2 single metallic pulses per second: crick, crick.

14. *Xenopus ruwenzoriensis* Tymowska & Fischberg, 1973: 337.
Type-locality: Bundibugyo, Semliki Valley (Uganda).

Distribution: Semliki Valley (Uganda).
Dodecaploid 2n=6X=108, NOR on 11pt.
Medium size, female SVL 55 mm (max. 57 mm), male 20–25% less. Morphology similar to X. *pygmaeus* but of larger size, subocular tentacle shorter and dorsal colour more brownish with a few large spots on the back, legs spotted ventrally.
Call: short high-pitched metallic trills: cri, cri (about 2/s).

vestitus-wittei subgroup
Octoploid 2n=72. Two closely related sympatric species of similar size and morphology, prehallux small, lower eyelid covering one-third to half of the eye, subocular tentacle one-third eye diameter, number of lateral line organs around the eyes 8–14; similar but not identical electrophoretic pattern of albumins (70 kd) and basic sperm nuclear proteins. However, very distinct in colour pattern and advertisement calls. Bivalent formation between half the chromosomes in hybrid oocytes suggests that both species share a common parental genome.

15. *Xenopus vestitus* Laurent, 1972: 9.
Type-locality: Rutshuru (Zaire).
Distribution: highlands around the Virunga volcanoes, Rwanda, Uganda and adjacent Zaire.
Octoploid 2n=4X=72, NOR on 12pt.
Medium size, female SVL 47 mm (max. 55 mm), male 20% less. Dorsal colour a characteristic marbling of light silver-golden to bronze chromatophores covering a brown background, head lighter and separated from the back by a dark transverse band; ventrally, including throat, heavily spotted except a median light band on the belly. Head wedge-shaped with relatively small eyes (cf. *wittei*).
Call: trills of 0.5 s duration: triing, triing.

16. *Xenopus wittei* Tinsley, Kobel & Fischberg, 1979: 73.
Type-locality: Chelima Forest (south-west Uganda).
Distribution: highlands around the Virunga volcanoes, Rwanda, Uganda and adjacent Zaire.
Octoploid 2n=4X=72, NOR on 12pt.
Medium size, female SVL 46 mm (max. 61 mm), male 20% less. Dorsal colour uniform dark brown to chocolate, no spots; ventrally yellowish-white with few small spots, especially on legs, to heavily spotted, including throat; sharp separation line between dorsal and ventral surfaces. Head rounded with relatively larger eyes (cf. *vestitus*).
Call: long tinkling and ringing trills: trrrirrrirrri . . .

longipes-subgroup
The single species known only from Lake Oku (Cameroon) shows affinities to both *Silurana* and *Xenopus*: paired nasals, straight transverse processes of fourth vertebra, pustules on head and back typical of the former, but eight free

presacral vertebrae, large eyes and karyotype characteristics typical of the latter subgenus. The skull shows peculiarities present in no other *Xenopus*, such as laterally displaced parasagittal crests, lateral location of retractor bulbi muscle scars and an unusual large auditory capsule.

17. *Xenopus longipes* Loumont & Kobel, 1991: 732.
Type-locality and distribution: Lake Oku (Cameroon).
Dodecaploid 2n=6X=108.
Small species, female SVL 34 mm (max. 36 mm), male 15% less. Large eyes; lower eyelid covering one-third of the eye; subocular tentacle one-third of eye diameter; hindlimbs long, fifth toe more than 130% tibia length, prehallux with black claw. Dorsal colour dark caramel to brown, heavily speckled and marbled with small and a few large spots; ventrally also densely speckled with melanophores, belly and throat often almost black on a greyish to bright orange ground coloration.
Call: unknown.

Key to African clawed Anura

In order to identify *Xenopus* specimens, non-taxonomists may require a simple key based on external morphology of the animals. Unfortunately, *Xenopus* species are rather indistinct morphologically; the best criteria (which are indispensable for specimens originating from the margins of their distribution and especially for the *fraseri*-like subgroup) are advertisement calls, DNA content and the exact geographical origin of the specimens. These data together with a few anatomical and morphological traits allow specimens to be ascribed to all presently recognized taxa; in cases of doubt, a more profound analysis of additional biochemical characters may resolve the problems. From the above, it should be evident that identification requires live wild-caught specimens in preference to preserved material (thus, it would be impossible to sort out the *X. fraseri* of museum collections into the six presently known species).

However, since vast regions of Africa are occupied by a single or by a few species only (see Tinsley, Loumont & Kobel this volume pp. 35–59), identification of *Xenopus* samples by morphology alone is in practice relatively simple, provided that their geographical origin is certified: this may allow a majority of species to be eliminated with high probability. Nevertheless, in a few instances knowledge of the DNA content is crucial. Measuring DNA values is most easily carried out on red blood cells with a cell-sorter (an instrument present in many biological and medical diagnostic and research laboratories). Only a very small blood sample is needed, which can be taken from a foot vein without harm to the animal. Since the nuclear volume of cell nuclei depends largely on its DNA content, it is possible to obtain an adequate estimation of DNA values simply by measuring the area occupied by erythrocyte nuclei on a blood smear, using erythrocytes of *X. laevis* as standard (a species which may be found in many research laboratories).

The following polythetic key makes use of a few external characters: skin (texture, colour pattern), lateral line organs around the eye (numbers), lower eyelid (size in proportion to eye diameter), subocular tentacle (length in proportion to eye diameter), hindlimb (length in proportion to snout–vent length), fifth toe (length in proportion to tibia length), prehallux (= metatarsal tubercle: size, shape, presence/absence of a claw). In addition to these external features, DNA values must be obtained in some cases, and the geographical origin of specimens is generally essential for accurate identification.

1. Skin verrucous, dry; hand webbed   2
— Skin smooth, slimy; no webs between fingers   3
2. Lateral line organs not apparent; no movable eyelids   *Hymenochirus*
    (four species, all from the rainforests of the
    Zaire-Basin and adjacent regions)
— Lateral line organs apparent; lower eyelid mobile   *Pseudhymenochirus*
    (one species, rainforests of Guinea-Bissau,
    Guinea, Sierra Leone)
3. Three claws (on toes 1–3), prehallux without claw   4
— Four claws (on toes 1–3), prehallux with a
    black claw   10
4. Prehallux prominent, cone-shaped and pointed   5
— Prehallux small or almost absent or invisible   6
5. Subocular tentacle longer than half eye diameter;
    dorsal skin greenish-brown, few large roundish
    dark spots
    Origin: Upper Volta to southern Sudan   *X. muelleri*-West
    Origin: East Africa from south-eastern Kenya
        to South Africa   *X. muelleri*-East
— Subocular tentacle half or slightly less than eye
    diameter; dorsal skin bluish-grey, many angular
    dark spots. Origin: Kenya   *X. borealis*
6. Subocular tentacle absent or occasionally a small
    cone; prehallux not apparent   7
— Subocular tentacle present; prehallux visible   8
7. Lower eyelid covering one-third or less of the eye
    Origin: southern Ethiopia: Mendebo Mts (Bale)   *X. largeni*
— Lower eyelid covering half of the eye
    Origin: Cape Peninsula and coast to Cape Agulhas   *X. gilli*
8. Hindlimbs (vent to tip of fifth toe) 15–20%
    longer than body (snout to vent) length; fifth toe
    longer than tibia   *X. laevis*
    (several geographical subspecies recognized
    within an extensive distribution from the Cape

to the northern fringes of woodland savanna
(Nigeria to Sudan))

— Hindlimbs as long as or at most 5% longer
than body; length of fifth toe equal to tibia; size
of erythrocyte nuclei much larger than in X. *laevis*   9

9. Dorsal colour on head, back and limbs uniform
olive-brown to chocolate, no melanophore
spots
Origin: Virunga volcanoes and adjacent regions of
Uganda, Rwanda, Zaire at altitudes between 1600
and 2600 m   *X. wittei*

— Dorsal skin on back and limbs irregularly
marbled with light, silvery-golden to bronze
chromatophores on a dark brown background,
head paler than back
Origin: Virunga volcanoes and adjacent regions of
Uganda, Rwanda, Zaire at altitudes between
1200 and 1900 m   *X. vestitus*

10. Cloacal lobes fused ventrally; lower eyelid
covering less than one-third of eye   11

— Cloacal lobes free; lower eyelid covering at least
half of eye   12

11. Size of erythrocyte nuclei much smaller than those
of X. *laevis*
Origin: West from the Cross River (Nigeria)
to Senegal   *X. (S.) tropicalis*

— Size of erythrocyte nuclei equal to that of X. *laevis*
Origin: From Mt. Cameroon to eastern border of
Zaire and southward to northern Angola   *X. (S.) epitropicalis*

12. Fifth toe distinctly longer than tibia   13
— Length of fifth toe about equal to tibia   *fraseri*-like
(The six presently recognized species can be
identified with certainty only by advertisement
calls, DNA-content and biochemical characters:
see species descriptions)

13. Eyelid covering three-quarters of eye; lateral line
organs around eyes 12–16
Origin: Eritrea, Ethiopia   *X. clivii*

— Eyelid covering one-third of eye; lateral line
organs around eyes 7–13; fifth toe at least 130%
of tibia length
Origin: Lake Oku (Cameroon)   *X. longipes*

Acknowledgements

We are greatly indebted to R. Brun, E. Bürki, P. Denny, L. Du Pasquier, J.-D. Graf, H. Hinkel, W.P. Müller, M. Picker, R. Rau, D. Rungger, F. Schütte, Ch.H. Thiébaud, J. Tymowska and D.D. Yager for discussions and for providing live specimens, and to the governments of African states who gave help and permits for collecting *Xenopus* during a series of fieldwork visits. Field and other research was supported by grants from the Fonds National Suisse de la Recherche Scientifique, the Claraz Foundation, The Royal Society and The Natural Environment Research Council, which we gratefully acknowledge.

References

Ahl, E. (1924). Über einige afrikanische Frösche. *Zool. Anz.* 60: 269–273.

Bagnara, J.T. & Obika, M. (1965). Comparative aspects of integumental pteridine distribution among amphibians. *Comp. Biochem. Physiol.* 15: 33–49.

Bisbee, C.A., Baker, M.A., Wilson, A.C., Hadji-Azimi, I. & Fischberg, M. (1977). Albumin phylogeny for clawed frogs (*Xenopus*). *Science* 195: 785–787.

du Bocage, J.V.B. (1895). *Herpétologie d'Angola et du Congo* 1–20. Imprimerie Nationale, Lisbonne.

Boulenger, G.A. (1905). On a collection of batrachians and reptiles made in South Africa by C.H.B. Grant, and presented to the British Museum by Mr. C.D. Rudd. *Proc. zool. Soc. Lond.* 1905 (2): 248–255.

Brun, R. & Kobel, H.R. (1977). Observations on the fertilization block between *Xenopus borealis* and *Xenopus laevis laevis*. *J. exp. Zool.* 201: 135–138.

Bürki, E. (1985). The expression of creatine kinase isozymes in *Xenopus tropicalis*, *Xenopus laevis laevis*, and their viable hybrid. *Biochem. Genet.* 23: 73–88.

Bürki, E. & Fischberg, M. (1985). Evolution of globin expression in the genus *Xenopus* (Anura: Pipidae). *Molec. Biol. Evol.* 2: 270–277.

Cannatella, D.C. & Trueb, L. (1988). Evolution of pipoid frogs: intergeneric relationships of the aquatic frog family Pipidae (Anura). *Zool. J. Linn. Soc.* 94: 1–38.

Daudin, F.M. (1803). *Histoire naturelle des rainettes, des grenouilles et des crapauds.* Levrault, Paris.

De Sà, R.O. & Hillis, D.M. (1990). Phylogenetic relationships of the pipid frogs *Xenopus* and *Silurana*: an integration of ribosomal DNA and morphology. *Molec. Biol. Evol.* 7: 365–376.

Fischberg, M., Colombelli, B. & Picard, J.-J. (1982). Diagnose préliminaire d'une espèce nouvelle de *Xenopus du Zaire*. *Alytes* 1: 53–55.

Giorgi, P.P. & Fischberg, M. (1982). Cellular DNA content in different species of *Xenopus*. *Comp. Biochem. Physiol.* (B) 73: 839–843.

Graf, J.-D. & Fischberg, M. (1986). Albumin evolution in polyploid species of the genus *Xenopus*. *Biochem. Genet.* 24: 821–837.

Graf, J.-D. (1989). Genetic mapping in *Xenopus laevis*: eight linkage groups established. *Genetics, Austin* 123: 389–398.

Graf, J.-D. & Kobel, H.R. (1991). Genetics of *Xenopus laevis*. *Methods Cell Biol.* 36: 19–34.

Gray, J.E. (1864). Notice of a new genus (*Silurana*) of frogs from West Africa. *Ann. Mag. nat. Hist.* (3) **14**: 315–316.

Hewitt, J. (1927). Further descriptions of reptiles and batrachians from South Africa. *Rec. Albany Mus.* **3**: 371–415.

Jaggi, R.B., Wyler, T. & Ryffel, G.U. (1982). Comparative analysis of *Xenopus tropicalis* and *Xenopus laevis* vitellogenin gene sequences. *Nucleic Acids Res.* **10**: 1515–1533.

Jeffreys, A.J., Wilson, V., Wood, D. & Simons, J.P. (1980). Linkage of adult α- and β-globin genes in X. *laevis* and gene duplication by tetraploidization. *Cell* **21**: 555–564.

Kobel, H.R. & Du Pasquier, L. (1986). Genetics of polyploid *Xenopus. Trends Genet.* **2**: 310–315.

Kobel, H.R., Du Pasquier, L., Fischberg, M. & Gloor, H. (1980). *Xenopus amieti* sp. nov. (Anura: Pipidae) from the Cameroons, another case of tetraploidy. *Revue suisse Zool.* **87**: 919–926.

Kobel, H.R. & Müller, W.P. (1977). Zytogenetische Verwandtschaft zwischen zwei tetraploiden *Xenopus* Arten. *Arch. Genet.* **49/50**: 188.

Laurent, R.F. (1972). Amphibians. *Explor. Parc natn. Virunga, deux. Sér.* No. **22**: 1–125.

Littlejohn, M.J. (1977). Long-range acoustic communication in anurans: an integrated and evolutionary approach. In *The reproductive biology of amphibians*: 263–294. (Eds Taylor, D.H. & Guttman, S.I.). Plenum Press, New York & London.

Loumont, C. (1981). L'appareil vocal des males *Xenopus* (Amphibia Anura). *Monitore zool. ital.* (N.S.) (Suppl.) **15** (2): 23–28.

Loumont, C. (1983). Deux espèces nouvelles de *Xenopus* du Cameroun (Amphibia, Pipidae). *Revue suisse Zool.* **90**: 169–177.

Loumont, C. (1986). *Xenopus pygmaeus*, a new diploid pipid frog from rain forest of equatorial Africa. *Revue suisse Zool.* **93**: 755–764.

Loumont, C. & Kobel, H.R. (1991). *Xenopus longipes* sp. nov., a new polyploid pipid from western Cameroon. *Revue suisse Zool.* **98**: 731–738.

Loveridge, A. (1932). New races of a skink (*Siaphos*) and frog (*Xenopus*) from the Uganda Protectorate. *Proc. biol. Soc. Wash.* **45**: 113–116.

Mann, M., Risley, M.S., Eckhardt, R.A. & Kasinsky, H.E. (1982). Characterization of spermatid/sperm basic chromosomal proteins in the genus *Xenopus* (Anura: Pipidae). *J. exp. Zool.* **222**: 173–186.

Müller, W.P. (1977). Diplotene chromosomes of *Xenopus* hybrid oocytes. *Chromosoma* **59**: 273–282.

Parker, H.W. (1936). Reptiles and amphibians collected by the Lake Rudolf Rift Valley Expedition, 1934. *Ann. Mag. nat. Hist.* (10) **18**: 594–609.

Peracca, M.G. (1898). Descrizione di una nuova specie di amfibio del gen. *Xenopus*, Wagl. dell'Eritrea. *Boll. Musei Zool. Anat. comp. R. Univ. Torino* **13** (No. 321): 1–4.

Perret, J.-L. (1966). Les amphibiens du Cameroun. *Zool. Jb. (Syst.)* **93**: 289–464.

Peters, W. (1844). [Nachricht von einigen Fischen u. Amphibien aus Angola u. Mozambique.] *Mber. Kgl. preuss. Akad. Wiss. Berlin* **1844**: 32–37.

Reumer, J.W.F. (1985). Some aspects of the cranial osteology and phylogeny of *Xenopus* (Anura, Pipidae). *Revue suisse Zool.* **92**: 969–980.

Reumer, J.W.F. & Thiébaud, C.-H. (1987). Osteocyte lacunae size in the genus *Xenopus* (Pipidae). *Amphibia-Reptilia* **8**: 315–320.

Robert, J., Wolff, J., Jijakli, H., Graf, J.-D., Karch, F. & Kobel, H.R. (1990). Developmental expression of the creatine kinase isozyme system of *Xenopus*. *Development (Camb.)* **108**: 507–514.

Rose, W. & Hewitt, J. (1927). Description of a new species of *Xenopus* from the Cape Peninsula. *Trans. R. Soc. S. Afr.* **14**: 343–347.

Schmid, M. & Steinlein, C. (1991). Chromosome banding in Amphibia, XVI. High-resolution replication banding patterns in *Xenopus laevis*. *Chromosoma* **101**: 123–132.

Thiébaud, Ch.H. & Fischberg, M. (1977). DNA content in the genus *Xenopus*. *Chromosoma* **59**: 253–257.

Tinsley, R.C (1995). A new species of *Xenopus* (Anura: Pipidae) from the highlands of Ethiopia. *Amphibia-Reptilia* **16**: 375–388.

Tinsley, R.C., Kobel, H.R. & Fischberg, M. (1979). The biology and systematics of a new species of *Xenopus* (Anura: Pipidae) from the highlands of central Africa. *J. Zool., Lond.* **188**: 69–102.

Tymowska, J. (1991). Polyploidy and cytogenetic variation in frogs of the genus *Xenopus*. In *Amphibian cytogenetics and evolution*: 259–297. (Eds Green, D.M. & Sessions, S.K.) Academic Press, San Diego.

Tymowska, J. & Fischberg, M. (1973). Chromosome complements of the genus *Xenopus*. *Chromosoma* **44**: 335–342.

Vigny, C. (1979a). The mating calls of 12 species and sub-species of the genus *Xenopus* (Amphibia: Anura). *J. Zool., Lond.* **188**: 103–122.

Vigny, C. (1979b). Morphologie larvaire de 12 espèces et sous-espèces du genre *Xenopus*. *Revue suisse Zool.* **86**: 877–891.

Vonwyl, E. & Fischberg, M. (1980). Lactate dehydrogenase isozymes in the genus *Xenopus*: species-specific patterns. *J. exp. Zool.* **211**: 281–290.

Wall, D.A. & Blackler, A.W. (1974). Enzyme patterns in two species of *Xenopus* and their hybrids. *Devl Biol.* **36**: 379–390.

Wolff, J. & Kobel, H.R. (1982). Lactate dehydrogenase of *Xenopus laevis laevis* and *Xenopus borealis* depends on a multiple gene system. *J. exp. Zool.* **223**: 203–210.

Wolff, J. & Kobel, H.R. (1985). Creatine kinase isozymes in pipid frogs: their genetic bases, gene expressional differences, and evolutionary implications. *J. exp. Zool.* **234**: 471–480.

Yager, D.D. (1982). A novel mechanism for underwater sound production in *Xenopus borealis*. *Am. Zool.* **22**: 887.

Yager, D.D. (1992). A unique sound production mechanism in the pipid anuran *Xenopus borealis*. *Zool. J. Linn. Soc.* **104**: 351–375.

3 Geographical distribution and ecology

R.C. TINSLEY, C. LOUMONT and H.R. KOBEL

Synopsis

Extant *Xenopus* species principally show a parapatric distribution; many species occupy very large distinct regions of Africa in an exclusive manner, whilst others have limited ranges. Some recently described species have been identified in samples from restricted localities and their geographical ranges are generally ill defined. In contrast to the detailed laboratory studies based on other aspects of *Xenopus* biology, there have been few field investigations and most ecological information is based on inference. For several species, ecological preferences emerge from their geographical distribution: thus, *X. tropicalis, epitropicalis* and certain species of the *fraseri* group are more or less restricted to lowland rain forest, whereas *X. muelleri* is adapted to hot, dry lowland outside the zones of rain forest and wooded savanna. This ecological restriction is particularly well illustrated for the eastern population of *muelleri* whose range lies below 800 m and may be limited by competing species at higher altitudes, and whose southern extension ends at the 18 °C isotherm. While the mutual *laevis–muelleri* exclusion in south-eastern Africa could result from the temperature requirements of *X. muelleri*, *X. laevis* does not seem to be limited by a reciprocal temperature preference since it occupies niches in Namibia and Namaqualand at least as hot as *X. muelleri* inhabits in the east. The northernmost limit of *laevis* is at Jebel Marra, within the *muelleri* zone of the southern Sahara, where *laevis* may have been isolated after a previous pluvial period.

Field samples regularly record three or four species sharing the same confined habitats and it is difficult to assess their ecological preferences. Many *Xenopus* species tolerate a very broad variety of ecological conditions. The present distributions probably reflect the constraints imposed by the cyclical climatic changes during the Pleistocene resulting in expansion and contraction of rain forests. The rich species diversity of highland regions, especially in Rwanda, Uganda and eastern Zaire and Cameroon, accords with these regions acting as refugia in successive cool, dry periods when the mixing of species may have favoured allopolyploidization. The present lowland forest species probably expanded their ranges with the expansion of forest from these centres, and their current distributions, involving local sympatry, may represent labile zones of competition rather than the limits of their respective ecological ranges.

Geographical distribution

Until the early 1970s, when the genus *Xenopus* comprised only six species, the distribution patterns appeared to reflect broad ecological divisions between the taxa (Tinsley 1973). Two of the species recognized at that time, *X. laevis* and *X. muelleri*, occur in savannas; two of the species, *X. fraseri* and *X. tropicalis*, are confined to lowland tropical forest; and two have restricted, perhaps relict, distributions with *X. gilli* confined to the Cape and *X. clivii* to Ethiopia.

In the past 20 years, part of the increase in the numbers of species has come from splitting of pre-existing taxa and division of their geographical ranges. The northernmost subspecies of *X. laevis* has been raised to specific rank, *X. borealis*, restricted to higher elevations in central and northern Kenya (Tymowska & Fischberg 1973). Two *tropicalis*-like forms are now recognised: the diploid (2n=20) *X. tropicalis* to the west of Cameroon, and the tetraploid (2n=40) *X. epitropicalis* to the east and south of Cameroon (Fischberg, Colombelli & Picard 1982; Loumont 1984). Two *fraseri*-like (2n=36) species are distinguished: *X. fraseri*, centred on Cameroon and Gabon, and *X. pygmaeus*, occurring in the centre and east of the Zaire Basin (Loumont 1986). The savanna *X. muelleri* may comprise two allopatric sibling species (see below). However, the major part of the recent increase in species numbers has been the result of more extensive investigation of the *Xenopus* in previously unstudied areas of Africa. This work has led to the recognition of a further eight species specific to two ecological zones. First, the diversity of *Xenopus* in lowland tropical forest has been found to be greater than previously appreciated, with three new species defined in western rainforests (Cameroon, Gabon, Zaire etc). Second, another five new species are associated with montane areas: the mountains of the western Rift Valley, the highlands of Cameroon and the highlands of Ethiopia.

The distribution patterns remain relatively simple in southern Africa and west Africa where the *Xenopus* species occupy a series of broad latitudinal belts. *X. laevis* is distributed over the greatest area with a south–north succession of subspecies, *laevis, poweri, petersi, victorianus* and *sudanensis*, generally corresponding with the relatively cooler highlands between the Cape and the plateaux of Cameroon and Nigeria (Fig. 3.1). This range encircles the Zaire Basin from which there are only isolated records on the fringes of the forest, and *laevis* is also excluded from much of the hotter lowlands of eastern Africa. The systematic separation of these subspecies is poorly defined (see Kobel, Loumont & Tinsley, this volume pp. 9–33) and so too are their ecological relationships. Initially, when these representatives were known from relatively few, scattered localities, it seemed possible that the distinct forms might represent points along a cline of continuous variation. However, Poynton & Broadley (1991) have recognized different subspecies occurring in the same geographical area in Zimbabwe. With this exception, the borders of the subspecies ranges are poorly documented.

In West Africa, distributions are influenced by the latitudinal zonation of

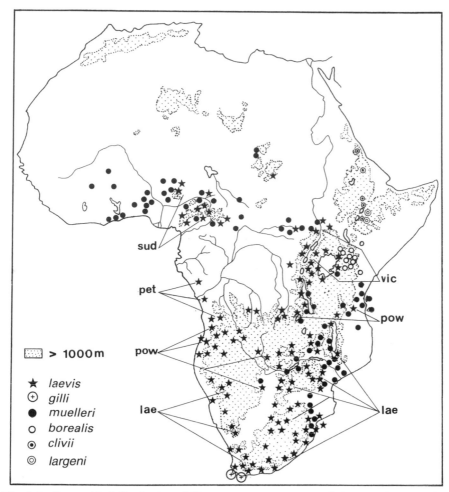

Fig. 3.1. Geographical distribution of *Xenopus* species in savanna habitats. Subspecies of *X. laevis*: lae, *laevis*; pet, *petersi*; pow, *poweri*; sud; *sudanensis*; vic, *victorianus*.

vegetation types. The coastal forest belt is occupied by *X. tropicalis* and the northerly savanna by *X. muelleri*, although this simple pattern is disrupted by forest clearance. *X. muelleri* has the widest geographical range of all the *Xenopus* species, occurring in a broad arc of relatively high-temperature habitats through the grasslands bordering the Sahel and extending south along the coastal belt of East Africa. There is now evidence from comparative biochemistry, molecular biology and parasitology to support the separation of sibling species which are referred to elsewhere in this volume as *muelleri*-East and *muelleri*-West. In these areas, and for much of the range of *X. borealis* in highland Kenya and *X. clivii* in Ethiopia, the *Xenopus* species are largely allopatric; their distributions form a mosaic of interlocking plates, sometimes with overlap along the margins. However, elsewhere there are complex patterns of distribution with extensive zones of sympatry which create fundamental problems for ecological interpretation. At least five *Xenopus* species occur in the lowland rainforests of the Zaire Basin: *epitropicalis, fraseri, andrei, boumbaensis* and *pygmaeus*. Their ranges are poorly defined and their ecological requirements largely unknown. Studies in the Central African Republic show that up to four *Xenopus* species (*epitropicalis, andrei, muelleri* and *pygmaeus*) may be sympatric near the margin of tropical forest (Loumont 1986). The greatest species-richness and complexity of distribution patterns occur in the mountain regions on opposite sides of the Zaire Basin (Figs 3.2 and 3.3).

The Central African highlands represent an important watershed, separating the forests of the Zaire Basin from the savannas of East Africa. In the west, on drainage which passes into the Zaire River and the Atlantic, the two forest species *X. pygmaeus* and *X. tropicalis* occur in lowland habitats (below about 700 m). In the savannas bordering Lake Victoria, on drainage which passes to the Nile, *X. l. victorianus* occupies a considerable area of wooded grassland, principally between about 1000 and 1800 m. In the north, *X. muelleri* extends into the margins of forest from the Sudan zone grasslands. Along these margins there is an overlap in the distribution of forest and savanna *Xenopus* and four species occur in the Garamba region, Zaire (Inger 1968): *X. l. victorianus, X. muelleri, X. epitropicalis* and *X. pygmaeus* (Fig. 3.2).

The major biogeographical interest in this area concerns the fauna of the mountains which border the Western Rift. Three *Xenopus* species are endemic to the region: *X. vestitus, X. wittei* and *X. ruwenzoriensis*. The latter is known only from its type locality in the Semliki Valley, Uganda (Fischberg & Kobel 1978). *X. vestitus* and *X. wittei* are widespread in swamps and lakes in the highlands of Uganda, Rwanda and east Zaire at altitudes generally between 1200 and 2000 m. They occur sympatrically through much of this range, although *X. vestitus* also occurs at lower altitudes in Rwanda, and *X. wittei* has a more extensive distribution at higher elevations, up to 2600 m, on the Kabasha escarpment (Tinsley 1975; Tinsley, Kobel & Fischberg 1979). The region is ecologically complex and rivers draining west to the tropical forests and those draining east and north to the savannas are frequently separated by only a few kilometres.

The most important area for *Xenopus* species diversity centres on Cameroon. Ten species are represented here (Fig. 3.3). Five have wider distributions elsewhere, including *X. pygmaeus* and *X. muelleri*. According to Loumont (1983), the boundary between *X. tropicalis* and *X. epitropicalis* is the Sanaga River in Cameroon, but more recent fieldwork (unpublished) suggests a boundary further to the north-west, probably in the western highlands of Cameroon. *X. laevis* is represented in this region by a distinct subspecies *X. l. sudanensis* (see Perret 1966). Four species are endemic to Cameroon and neighbouring areas: *X. longipes* and *X. amieti* occur in the volcanic highlands of Cameroon; the latter is relatively widespread at elevations of

1100 to 2400 m whilst the former is known only in the crater lake, Lake Oku, at 2200 m (Kobel, Du Pasquier, Fischberg & Gloor 1980; Loumont & Kobel 1991). The other endemic *Xenopus* occur in lowland rainforest and these distributions are not well defined: *X. boumbaensis* is known only from its type locality in southern Cameroon (Loumont 1983), *X. andrei* was first recorded on the Atlantic coast of Cameroon, but is now known also to occur in Gabon and Equatorial Africa (Loumont 1986). Three of these species, *andrei, amieti* and *boumbaensis*, are very similar morphologically to *X. fraseri* and are likely to have been identified as *fraseri* in earlier collections. The recent descriptions now create uncertainty over the distribution of *X. fraseri sensu stricto*. The type locality is unspecified (probably Nigeria or Fernando Po, see Frost 1985) but the recent studies which have defined the characteristics of this species (karyotype, mating call, biochemical features) have all been based on specimens from southern Cameroon (see Loumont 1983). Other collections with similar phenotype cannot be determined precisely and are indicated in Fig. 3.3 as '*fraseri*-like'.

The third region which has contributed to the recent increase in the number of *Xenopus* species is Ethiopia. For nearly 100 years, only a single species

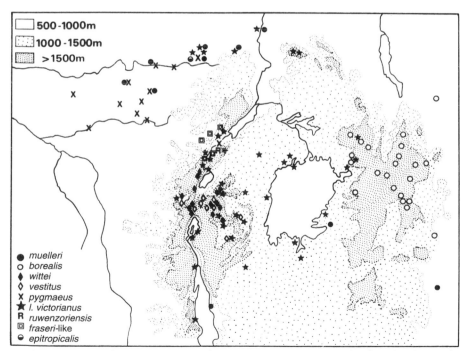

Fig. 3.2. Records of the distribution of *Xenopus* species in central Africa bordering the Western Rift: *X. pygmaeus* and *X. epitropicalis* are specific to lowland tropical forest; *X. laevis victorianus*, *X. borealis* and *X. muelleri* occur principally in savannas; three species, *X. ruwenzoriensis*, *X. vestitus* and *X. wittei*, are endemic to the area bordering the Western Rift.

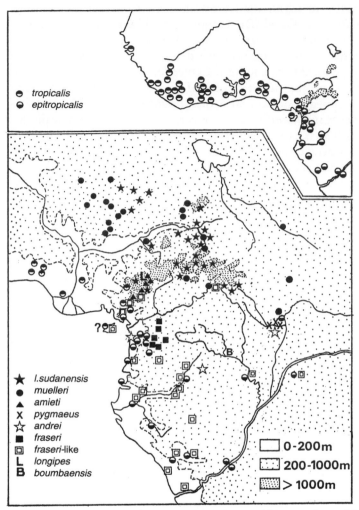

Fig. 3.3. Records of the distribution of *Xenopus* species in west Africa: members of the *tropicalis* and *fraseri* species groups are widely distributed in lowland rain forest, and X. *laevis sudanensis* and X. *muelleri* extend into the fringes of this region in wooded savanna and grassland savanna respectively; however, the most distinctive feature is the rich diversity of species associated with the Cameroon highlands, including at least three endemic species (X. *amieti* and X. *longipes* in the mountains and X. *boumbaensis* in lowland forest).

of *Xenopus*, X. *clivii*, has been known to occur in Ethiopia. Now, a second species, X. *largeni*, has been described from the Bale Mountains, a region noted for the significant numbers of endemic taxa in its flora and fauna (Tinsley 1995) (Fig. 3.1).

The apparent simplicity of the *Xenopus* species distributions recorded in the early 1970s is now replaced by a highly complex pattern with certain regions characterized by great species richness and extensive sympatry. Consideration of the ecological significance of these patterns requires assessment of the environmental biology and habitat selection of the *Xenopus* species.

Ecology

Habitat conditions

Xenopus occurs in virtually all kinds of waterbody in sub-Saharan Africa, including rivers, lakes, dams, swamps, flooded pits, ditches and wells. Within the geographical range of a given species there is little evidence of preference for specific biotypes: individuals appear to occupy whatever aquatic habitats are available. There are some apparent contradictions: the aquatic toads do not often seem to populate large rivers and lakes, and there are indications that they do not tolerate well-established fish communities. On the other hand, several species of *Xenopus* have been reported to occur in large numbers in artificial ponds constructed for fish culture. Lakes which naturally lack fish (volcanic lakes in Uganda and Cameroon, for instance) may support teeming populations of *Xenopus* which occupy the 'fish niche'. For almost all the general characteristics of habitats exploited by *Xenopus*, there are well-documented exceptions. Although *Xenopus* is most common in stagnant or still waters of ponds or sluggish streams, it may also inhabit fast-flowing water. In general, the toads occur in cloudy water where they are hidden from aerial predators once below the surface. However, *Xenopus* can also be found in clear water around the margins of lakes and small streams. Throughout sub-Saharan Africa, *Xenopus* is favoured by human activities: the needs for domestic and agricultural water supplies coincide with the requirements of *Xenopus*, and the toads readily colonize newly constructed dams and wells.

There is wide variation in the water chemistry of habitats occupied by *Xenopus* and insufficient information to assess the influence of this on habitat selection. Within a relatively restricted area, a single species may be found in water with major differences in chemical composition. Thus, in the Kigezi District of Uganda, *X. vestitus* occurs in swamps and lakes rich in minerals of recent volcanic origin: the pH at different localities varies between 5.6 and 8.7 (Visser 1962; Tinsley 1973). However, there are no field data to indicate potential effects on species abundance or breeding success. In the Cape, South Africa, *X. laevis* breeds in both acidic and alkaline water: in the latter case, spawning occurs in artificially fertilized ponds with a pH of 9.0 (Nieuwkoop & Faber 1956). The habitat requirements of one species have been examined in detail and, in this case, water composition is a key determinant of distribution. *X. gilli* occurs in acid 'blackwaters' characteristic of the fynbos biome on the Cape: typically the water has a pH below 5.0 (minimum 3.4) with low nutrients but high dissolved solids, and a peaty brown colour from plant phenols and tannins (Picker 1985). There is field and laboratory experimental evidence that water chemistry accounts for the habitat segregation of X. *l. laevis* and X. *gilli* on the Cape Peninsula. In particular, the survival of tadpoles of X. *laevis* is depressed below pH 5–6, whereas X. *gilli* tadpoles tolerate pH as low as 3.6 (see Picker, Harrison & Wallace, this volume pp. 61–71).

Experimental data for X. *laevis* demonstrate a tolerance of salinity which is (with a few other examples) exceptional amongst amphibians: this species will tolerate 40% seawater for several days (see below).

There is little information on the temperature preferences of the *Xenopus* species. Moron (1947) reported only minor seasonal variation in water temperatures in lowland forest habitats of X. *tropicalis* (in Equatorial Guinea), around 25 °C, and found that this species will tolerate 32 °C for about 12 h. Experience of laboratory maintenance suggests that the lowland rainforest species do not survive well at temperatures below 22 °C. X. *muelleri* is likely to have the highest temperature tolerance but we are not aware of any published records. Exposure to environmental temperatures can be modulated by the behaviour of *Xenopus*. McCoid & Fritts (1980a) observed that in ponds with water temperatures of 30 °C, X. *laevis* excavated pits 30–40 cm deep in the soft bottom mud where water temperature remained around 20 °C. For long-term survival and reproduction in a given locality, the effect of temperature on the tadpoles is likely to be the major limiting factor: Balinsky (1969) cited the upper and lower lethal temperatures for X. *laevis* embryos as 35 °C and 10 °C respectively. The wide geographical and ecological range of X. *laevis* suggests the ability to tolerate a considerable range of environmental temperatures. The distribution of the *laevis rassenkreis* from the Cape, South Africa, to Jebel Marra, Sudan, extends over 45° of latitude and a corresponding diversity of climatic conditions. Within southern Africa, X. *l. laevis* extends from sea level to nearly 3000 m (in the Drakensberg Mountains). This adaptability is tested to its limits in the various feral populations which have established in north temperate areas: X. *laevis* survives in ponds which are ice-covered for several months each year (in Virginia), and in desert ponds exposed to extreme summer heat (in Arizona) (see Tinsley & McCoid, this volume pp. 81–94).

However, the role of temperature tolerance in habitat exploitation may be modified by other factors including potential competitive interactions. The more or less abrupt transition from X. *laevis* to X. *muelleri* along the east coast of southern Africa has been well documented by Poynton (1964). The changeover corresponds with conditions associated with the 18 °C mean July isotherm. However, in southern Namibia (where X. *muelleri* is absent), X. *laevis* occurs in habitats at least as hot as those from which it is apparently excluded in the east (Fig. 3.1).

While the generalized habitat preferences of *Xenopus* might be expected to obscure ecological differences between the species, the segregation of forest and savanna species appears to be more or less profound. The *tropicalis* and *fraseri* groups are strictly specific to tropical forest and there are virtually no records outside the forest boundary. Exceptionally, Inger (1968) recorded *X. tropicalis* (now *epitropicalis*) from the Garamba Park, Zaire, just to the north of the rainforest, and relatively small samples were collected in grass savanna, gallery forest and tree savanna; *X. fraseri* (now *pygmaeus*) was found in a heavily wooded area. Inger commented that these localities are linked with the rainforest by corridors formed by gallery forest, and that the original forest cover has been destroyed by man. That this area represents the extreme edge of the range of the forest species is suggested by the relative numbers of the four *Xenopus* taxa collected: 16 *X. laevis*, 14 *X. epitropicalis* and two *X. pygmaeus*, alongside 422 *X. muelleri* (the species typical of the Sudan–Guinea grasslands).

Equivalent marginal habitats were also recorded in Nigeria by Schiøtz (1963) who found *X. tropicalis* a few kilometres north of the forest savanna border—but in a patch of relict forest. Thus, the *tropicalis* and *fraseri* groups are 'forest species' in the strict sense of Poynton & Broadley (1991), 'specifically tied to the presence of a closed canopy of trees', although it is not clear which niche parameters of these species are dependent on forest cover. The preferences are also found at a local level. Thus, Laurent (1972) recorded that in the same region of Kivu, *X. fraseri* (probably *X. pygmaeus* or another undescribed species) inhabits swamps which are shaded by forest whereas *X. laevis* occurs in swamps exposed to the sun.

The converse of this, the limitation of savanna species to grasslands, may not be so strict. Loumont (1986) recorded *X. muelleri* in the same rainforest swamps as *X. pygmaeus*, *X. andrei* and *X. epitropicalis*. Inger (1968) found *X. muelleri* in every type of aquatic habitat in the Garamba Park, Zaire, although the relative numbers collected (175 in grass savanna, 94 in tree savanna, 22 in gallery forest) confirm a preference for open environments. Some records of *muelleri* and *laevis* within tropical forest might be anticipated given the unstable ecology of areas modified by agriculture. Deforestation may leave behind temporary islands of forest fauna and shifting cultivation creates a mosaic of open and closed canopy: there may be a corresponding flux in the distribution of the *Xenopus* species. Schiøtz (1963) recorded *X. muelleri* in forest 72 km from the savanna border, but the habitat was situated in a cleared area. With some predictable exceptions, therefore, the species of *Xenopus* seem convincingly divided into forest and savanna assemblages: records of *tropicalis* and *fraseri* groups outside forest are very rare; there may be greater extension of *laevis* and *muelleri* groups into forest but this may be linked with the disturbed ecology of forest–savanna boundaries. Certainly, this pattern does not conform with the generalization discussed by Poynton & Broadley (1991) that 'forest species can survive in grasslands, but grassland species cannot survive in forest' (see also Happold & Happold 1989). So, the available evidence suggests that the tropical forest species of *Xenopus* are more or less strictly confined to this biome.

Diet

The food of *Xenopus* appears to include everything available in the aquatic environment. The studies of Inger & Marx (1961), Kazadi, de Bruyn & Hulselmans (1986) and Schoonbee, Prinsloo & Nxiweni (1992) demonstrate the extremely wide taxonomic range of prey items in the stomach contents of *X. laevis*. Not surprisingly, aquatic invertebrates predominate and this distinguishes *Xenopus* from all other anuran species surveyed by Inger & Marx (1961) in the Upemba Park, Zaire: even semi-aquatic species of *Rana* were found to feed almost exclusively on terrestrial prey. This reflects a fundamental ecological separation between pipids and other anurans.

All studies suggest that *Xenopus* are more or less non-selective predators, taking whatever organisms can be ingested. Small crustaceans predominate in the diet of young, recently-metamorphosed *Xenopus* and, predictably, larger individuals have access to a wider range of organisms, with insects as the most common prey. Some vertebrates occur in the diet: Inger & Marx (1961) recorded fish, amphibians and birds in their sample of 180 *X. laevis*, with 35 having eaten amphibians, but they did not specify whether the latter represented interspecific predation or cannibalism. The offspring of *Xenopus* may make a significant contribution to the nutrition of the parental population (Hey 1949; McCoid & Fritts 1980b).

Although tadpoles and post-metamorphic stages of *Xenopus* occur together in the same environment (in contrast to the situation in most other anurans), they exploit fundamentally different resources. The highly specialized filter-feeding mechanism enables tadpoles to extract the finest material in suspension (see Wassersug, this volume pp. 195–211), and rapid growth occurs during plankton blooms. When post-metamorphic *Xenopus* feed on tadpoles, the energy derived from primary production by algae and micro-organisms is very effectively channelled to the adult toads: cannibalism enables the adult population to exploit a nutrient resource which could not be utilized directly.

Xenopus represents a much more formidable predator than most anurans, which rely on the tongue for selective capture of rather small prey items. In *Xenopus*, prey capture employs a combination of toothed jaws which improve the grip on the prey, forelimbs which are used to fork the prey into the mouth, and the strong hindlimbs which can be used to rake the prey with the sharp claws. This shredding action enables *Xenopus* to tackle larger food items than could otherwise be ingested whole; indeed, groups of *Xenopus* may attack the same prey and can tear the body into fragments which can then be ingested. This method of feeding is particularly useful for scavenging.

Scavenging for dead animals is a major adaptation unavailable to most anurans, which rely on the visual stimulus of moving prey. For *Xenopus*, food recognition may involve scent and vibrations as well as vision (see Elepfandt, this volume pp. 97–120). In particular, homing in on olfactory stimuli (which are carried downstream in streams and rivers) enables *Xenopus* to respond to potential food sources over far greater distances than other anurans whose prey location is characteristically very short-range.

The data of Inger & Marx (1961) provide a good example of the opportunistic nature of *Xenopus* feeding. The stomachs of eight *Xenopus*, from one sample, each contained a small, newly hatched bird and this was presumed to have been the outcome of a nest falling into the water.

Two aspects of prey capture by *Xenopus* are controversial and, despite a series of references in the literature, remain unresolved: the extent to which *Xenopus* may prey upon fish, and the ability of *Xenopus* to feed on land.

Some field studies of stomach contents record the presence of fish (Inger & Marx 1961), but McCoid & Fritts (1980b) have doubted whether fish would make a significant contribution to the natural diet. In feral populations in California, *X. laevis* occurs in streams where there are abundant populations of mosquito fish, *Gambusia affinis*, but these were found in the stomachs of only two out of 81 toads. McCoid & Fritts argued that the methods of capturing *Xenopus* may be responsible for erroneous records: when captured in a seine net *X. laevis* continue to feed on organisms restrained with them, and the fish recorded in some previous field studies may have been ingested under these unnatural circumstances. They suggest that, in open water, *Xenopus* could not be fast enough to catch fish. On the other hand, McCoid & Fritts (1980a) recorded how the confinement of *Xenopus* and fish may occur naturally when drought reduces water levels: without any possibility of escape, small fish may readily be consumed.

It remains to be established whether *Xenopus* are important predators of fish (particularly fish fry) in Africa. *Xenopus* seem to be most common in waterbodies where fish are absent. Much of the field data on diet, including the detailed studies of Kazadi *et al.* (1986) and of L. De Bruyn, M. Kazadi & J. Hulselmans (pers. comm.), which recorded an exclusively invertebrate prey range, are not informative on the question of predation on fish. In most cases, the availability of fish in the habitats was not established or, as in the case of L. De Bruyn *et al.* (pers. comm.), the *Xenopus* sample originated in a temporary pond lacking fish. There are, apparently, no reports in the literature of the gut contents of *Xenopus* which inhabit fish hatcheries: such data would establish whether, given maximum opportunity, *Xenopus* may consume fish fry or whether they simply compete with the fish by consuming the same food. The question has important implications and requires careful field observations. However, the suggestion that *Xenopus* is a clumsy predator, unable to capture fast-moving prey (Avila & Frye 1978) is at variance with (personal) field observations which indicate that *Xenopus* is well adapted to ambush-and-snatch techniques of prey capture.

A series of field studies in Africa has recorded a high frequency of terrestrial prey organisms in the stomach contents of *Xenopus* (Noble 1924; Inger & Marx 1961; Tinsley 1973; Kazadi *et al.* 1986; L. De Bruyn *et al.* pers. comm.), and this has raised the question of whether *Xenopus* might feed on land. These authors have considered the possibility that this component of the diet might originate solely from animals which have fallen or been swept from overhanging vegetation, and this must generally explain the occurrence of terrestrial insects such as termites in the diet of tropical fish (see Tinsley 1973). However, this possibility has been thought inadequate to explain the very high proportion of terrestrial prey in the diet (comprising more than 50% of the prey items in a sample of X. fraseri examined by L. De Bruyn et al. (pers. comm.)). There are, to our knowledge, no observations that Xenopus can feed on land during brief excursions out of water. The adaptations of Xenopus for aquatic life seem particularly unsuitable for prey capture on land (the upwardly-directed eyes, mode of locomotion, method of ingestion etc). On the other hand, Xenopus are very adept at detecting prey on or just above the water surface. In laboratory aquaria, submerged Xenopus can readily detect insects flying a few centimetres above the water surface and they will leap partially out of water to capture such prey (see also Elepfandt, this volume pp. 97–120). In confined ponds it is possible that a population of Xenopus will exhaust all potential aquatic prey living in the same habitat. The high proportion of terrestrial prey in stomach contents may then simply reflect the relative lack of invertebrates in the immediate aquatic environment and the toads may become increasingly dependent on organisms which fall into the water.

If Xenopus could feed during short-term excursions onto land, this would represent an important ecological adaptation and a niche overlap with other anurans: there is a major need for unambiguous information on this basic point.

Predators

One would predict that a range of predators similar to those that prey on fish might prey on Xenopus. However, there is remarkably little direct information. Loveridge (1953) described finding eight recently ingested X. muelleri in the gullet and stomach of a hammer-headed stork (Scopus umbretta bannermanni). Loveridge (1942) recovered the feet of one X. laevis bunyoniensis in the stomach of an otter (Lutra m. tenuis). Reliable indirect evidence of predators is provided by parasite life cycles where Xenopus is an intermediate host and the parasite reaches sexual maturity only when transferred by predation to an appropriate final host. Diplostomum (Tylodelphys) xenopodis is one of several metacercarial (fluke) infections of X. laevis (see Tinsley & Sweeting 1974) and has recently been shown to mature in a reed cormorant, Phalacrocorax africanus, and a darter, Anhinga melanogaster (P.H. King & J.G. Van As unpublished). It is likely that predation by these birds must occur relatively frequently in order for this link to have become incorporated into the life cycle (see Tinsley, this volume pp. 233–261). There are several references to the potential of some fish to prey on Xenopus but these relate to introduced species such as largemouth bass (Huro salmoides) (see Hey 1949; Prinsloo, Schoonbee & Nxiweni 1981), and despite suggestions that *Xenopus* cannot co-exist naturally with predatory fish, direct information on predation is lacking.

The skin secretions of *Xenopus* include powerful toxins which would potentially deter predators (see Hey 1949; Zielinski & Barthalmus 1989; and Kreil, this volume pp. 263–277) but there is no information on the effectiveness of this protection, nor which natural predators are affected.

In certain areas of Africa, man is an important predator. The techniques are well established for capture of *Xenopus* from waterholes and streams by means of open baskets as in Cameroon (Kobel, Du Pasquier, Fischberg & Gloor 1980) and Sierra Leone (R.C. Tinsley unpublished) or by draining the water trapped by mud dykes and collecting the imprisoned *Xenopus*, as in the Central African Republic (Loumont 1986). In the volcanic lakes of Uganda, Rwanda and Eastern Zaire large-scale 'fisheries' were formerly based on the *Xenopus* populations, with many thousands of *Xenopus* caught each night in wickerwork traps, baited and set in the reedlined margins of the lakes (Worthington & Worthington 1933; de Witte 1941; Loveridge 1942; Tinsley 1973). Worthington & Worthington (1933) referred to the simple food chains of these lakes which contained no indigenous fish. *Xenopus* formed the principal link between the aquatic invertebrate fauna and the main predators—birds, otters and man. Subsequent ecological changes, including the introduction of fish, have led to a collapse in this productivity, and the 'industry' based both on *Xenopus* and on fish is now almost negligible (Tinsley *et al.* 1979; and recent unpublished data). (Indeed, the *Xenopus* previously very abundant here and regarded as a distinct subspecies, *X. l. bunyoniensis*, by Loveridge (1932) may now be extinct (Tinsley 1981).)

Local customs are important in determining the extent of predation by man; in many areas of Africa this potentially important source of dietary protein is ignored.

Reproductive biology

In favourable conditions in the laboratory, *X. laevis* can reach sexual maturity in eight months from the egg stage, i.e. about six months post-metamorphosis. Environmental conditions in southern California appear to be near optimal with daily water temperatures of at least 20 °C for most of the year, and McCoid & Fritts (1989) recorded ovulation in *X. laevis* females in feral populations about six months post-metamorphosis. Throughout much of southern Africa, with a pronounced cool season, development is likely to be slower, and there are few data for other species elsewhere in Africa.

Although the presence of ripe ova in females and well-developed nuptial pads in males may be recorded almost throughout the year (Schmidt & Inger 1959; Inger 1968), spawning is generally more restricted. In most records, the breeding season coincides with the start of the rainy season: for instance, April–June in the Garamba Park, north-east Zaire (Inger 1968). In the Nairobi region of Kenya, with two rainy seasons, breeding occurs in two periods per year (Wood 1965). In the Transvaal highveld, Balinsky (1969) reported a spring and summer breeding season of six and a half months (early September to mid-March) but it is not clear whether these extreme limits were recorded in the same year. During the breeding season, individual females may spawn more than once: Hey (1949) recorded up to three spawnings per female for *X. laevis* at the Cape.

Oogenesis is a continuous process under favourable environmental conditions and, given suitable temperatures, ovary production is determined principally by food supply: during food shortage, overcrowding and/or insufficient water the ovaries of *X. laevis* regress (Alexander & Bellerby 1938; Bellerby & Hogben 1938). These observations suggest that seasonal cycles of prey abundance (which are determined by rainfall) probably regulate the seasonal breeding cycle of *Xenopus*, and this also explains why some individuals may have well-developed ovaries at almost any time of the year (following local variations in prey availability).

Since spawning occurs intermittently during a breeding season of at least a few months, the stimuli which elicit spawning are likely to be determined by environmental conditions during the preceding day or days. For *X. laevis*, it is considered that temperature must reach about 20 °C, and a series of laboratory and field studies have examined the influence of temperature change and rainfall on spawning (Berk 1938; Hey 1949; Kalk 1960; and others). Balinsky (1969) eliminated the effects of temperature change (both an increase and a decrease had been implicated in previous studies) and concluded that spawning was triggered by a period of rainfall. The ecological effects of rain are indicated by the events reported at *Xenopus* hatcheries (Hey 1949; Du Plessis 1966): when ponds which have remained dry for some time are flooded and the water is fertilized (with fowl manure in the study of Du Plessis), *Xenopus* migrate into the pond and breed *en masse* within 2–3 days. Under natural conditions, the effect of heavy rain would be to wash fresh sediments into ponds, enriching the nutrient status of the water. The selective advantage of this response is that phytoplankton blooms are triggered by the same events, providing a food supply for the hatched tadpoles. Savage (1971) described experimental data to link the spawning of *Xenopus* with the production of specific metabolites by algae and suggested that these provide a natural hormonal stimulus for mating.

There is little other specific field information on reproductive biology but there may be important species differences. Thus, in the winter rainfall area of the Cape, *X. laevis* breeds in spring and early summer (during declining rainfall but rising temperature). However, the sympatric *X. gilli* appears to spawn in mid-winter. Rau (1978) recorded stage 45–46 tadpoles of *X. gilli* (about one week old) in July at a daytime water temperature of 12 °C, and reported that, with much of development at low temperatures, the period to metamorphosis may exceed 17 weeks (about twice that required by *X. laevis*).

Longevity

It is well known from laboratory maintenance that *Xenopus* is long-lived in captivity. Published accounts record up to 15 years for *X. laevis* (see Flower 1936). The catalogue of the Natural History Museum, London, records one preserved *X. gilli* which had lived in the Museum for 27 years. Longevity is certain to be reduced in the wild. Nevertheless, in feral populations in Britain, individually dye-marked *X. laevis* which were adult when first captured were recaptured periodically for up to nine years (see Tinsley & McCoid, this volume pp. 81–94). The reproductive life of *Xenopus* is also considerable: female *X. gilli* aged over 15 years have bred successfully in the laboratory (R.C. Tinsley unpublished).

Survival during unfavourable conditions

Drought

In contrast to the familiar observation that *Xenopus* which escape during the night from laboratory aquaria are dehydrated and dead by the next morning, *X. laevis* is known to aestivate without water for many months in nature. Most evidence is indirect, based on the sudden appearance of *Xenopus* in newly formed ponds at the start of the rains. However, there are also direct observations on *X. laevis* (see Rose 1950; Balinsky, Choritz, Coe & van der Schans 1967), *X. gilli* (see Rose 1962) and *X. muelleri* (see Loveridge 1953) which were found aestivating beneath the cracked mud at the margins of dried-out ponds. Several authors have transferred *Xenopus* to a regime simulating aestivation in the laboratory and recorded prolonged survival: Hewitt & Power (1913) reported that *X. laevis* 'remained in good condition for eight months' in a container of mud without food or water. McCoid & Fritts (1980a) found that the toads burrowed into drying mud and occupied a vertically-orientated chamber communicating with the surface. During a three-month experiment, aestivating toads were not continuously dormant: occasionally they were observed with the tip of the snout projecting from the vertical burrow.

The mechanism of survival includes a switch from the excretion of toxic ammonia to production of urea which accumulates in the blood, liver and muscle during dehydration (Balinsky, Cragg & Baldwin 1961). *Xenopus* also tolerates substantial circulating plasma and liver ammonia levels (Unsworth & Crook 1967). On re-immersion in water, very large quantities of urea are excreted. The metabolic change resembles that employed by aestivating lungfish, and both *Protopterus* and *Xenopus* combine ammonia excretion in aquatic environments with a mechanism of ureogenesis which enables them to survive during drought. The same physiological adaptations enable *X. laevis* to tolerate relatively high salinity (250–280 mM/l NaCl) (Munsey 1972; Romspert 1976; Dejours, Armand & Beekenkamp 1992). Whilst this would enable *Xenopus* to survive (and disperse) in the brackish water of estuaries, it may also be an important factor in ponds subjected to intense evaporation and increasing salt concentrations in the arid environments where *Xenopus* populations occur (the Namib desert, the Sahel etc).

Starvation

Xenopus shows a remarkable ability to tolerate starvation. In addition to the records of eight months' survival during aestivation without water (above), laboratory experiments have monitored physiological effects of starvation for up to 12 months on *X. laevis* maintained in water (Merkle & Hanke 1988). During the first 4–6 months, energy needs are met from stored carbohydrates and lipids and there is no change in plasma glucose. After this, *Xenopus* switches to protein catabolism, especially breakdown of body muscle, and body weight loss is accompanied by a continuous decline in plasma glucose. After 12 months, females had lost 45% and males 35% of their initial body weight (Merkle & Hanke 1988); nevertheless, no deaths occurred. Of course, many other anurans routinely starve during hibernation or aestivation, although at low temperatures energy consumption will be much reduced. Even desert toads such as *Scaphiopus* spp. which may remain dormant for up to 11 months each year actually consume stored reserves at very low rates because of metabolic arrest (Seymour 1973). For *Xenopus*, on the other hand, populations which are stranded in small pools exposed to summer heat remain active, and may have to endure several months without food at temperatures inducing high metabolic rate. Under these conditions, starving *Xenopus* show significant energy conservation: mean oxygen consumption is reduced by about 30% within two weeks of food deprivation (Merkle & Hanke 1988). Nevertheless, it is evident from museum collections and other field samples that *Xenopus* species may be severely emaciated by the end of the dry season in semi-arid regions. Under the most unfavourable conditions, the gonads represent an additional energy reserve (Bellerby & Hogben 1938) although their utilization will obviously delay subsequent reproduction. Nevertheless, in parallel with its extreme ability to survive starvation, *Xenopus* is also capable of rapidly regaining lost body weight once favourable environmental conditions return.

Overland migration

Although *Xenopus* is normally aquatic, and many of its morphological adaptations may be interpreted in relation to life in water (see Trueb, this volume pp. 349–377), there are records of relatively long-distance migrations on land. Exceptionally, Loveridge (1953) observed a mass migration under apparently dry conditions with *X. muelleri* leaving a temporary pond. Hewitt & Power (1913) reported a migration of 'some thousands of individuals . . . away from a dam which was just commencing to dry up'. However, most reports relate to the emigration of previously confined populations at the onset of torrential rain. These include dramatic accounts of 'great numbers . . . invading houses and blocking up irrigation pipes'. It is clear that this behaviour represents a considerable advantage for the colonization of new habitats and enables *Xenopus* to reach ponds which have no communication with waterways. It is less clear how often *Xenopus* may emerge from water as part of its normal activity. There are sufficient references to suggest that short nocturnal excursions onto land occur regularly (see, for instance, Inger 1968; Kazadi *et al.* 1986; Tinsley & McCoid, this volume pp. 81–94). Here, as with other basic aspects of ecology, more field observations are required.

Correlation of ecology and distribution

Although the ecology of *Xenopus* lacks detailed and critical investigation, there is sufficient information to conclude that the species of *Xenopus* are, for the most part, generalists without detectable preferences for habitat type, diet etc. The most obvious exception to this pattern is the apparent mutual exclusion of species specific to tropical forest and those of savanna noted above (although there may be regular overlap on the margins of these biomes). This apart, the absence of major niche differentiation brings into question the factors responsible for the geographical distribution of the *Xenopus* species. The key to understanding much of the current pattern of distribution lies in the environmental history of the tropics, particularly the history of lowland rainforest in equatorial Africa.

The view prevailed until relatively recently that the tropical forests of Africa (and elsewhere) represent an ancient stable ecosystem and that the rich species diversity and complex structure is the result of a very long period of evolution *in situ*. It was considered that the Pleistocene cold stages, the ice ages of higher latitudes, were accompanied in these tropical areas by periods of increased rainfall, the so-called pluvials. Although temperatures were significantly reduced, rainforests were thought to have persisted in the tropical areas throughout the past two million years, largely because of the expansion of montane forest to replace the lowland biome (see, for instance, Moreau 1966). However, it is now well established that, during the last glacial period, the major part of the forest belt disappeared and was occupied instead by grassland similar to that in the savannas today (Hamilton 1982). The explanation lies in two climatic changes: alongside a decrease in temperature (on average around 5 °C cooler than today), there was a marked decrease in precipitation. There is evidence of this reduced rainfall in the depressed water levels of the major African lakes. In the period preceding 12 500 B.P., the water level of Lake Kivu was around 300 m lower than it is today and there was no outflow into Lake Tanganyika. The latter was, in turn, 600 m lower than at present and had no outflow into the Zaire basin. For the best-studied river system, the Nile, it is considered that there was no major contribution of water from the East African lakes between > 40 000 and 12 500 B.P. Without the water supply from highland areas, the rivers of the present equatorial forests greatly declined and, with the general decrease in precipitation, lowland rainforest disappeared from much of the area which it presently occupies. During this period, the Sahara desert increased in area and the line of fixed dunes extended some 400–600 km south of the modern limit of active dunes. The period of most extreme aridity extends from around 25 000 years ago, through the northern hemisphere glaciation maximum (18 000 years ago), until around 12 000 years ago (Hamilton 1982). The same events are well documented and have a similar time-scale in the Amazon rainforests where the lowland tropical forest biome was also largely displaced by cool, dry savanna grasslands (Street 1981).

There is now extensive evidence that, during these arid phases, rainforest persisted in a series of restricted areas—refugia—determined by local conditions of altitude, precipitation and soil type. In Africa, forest refugia persisted throughout the period from 25 000 to 12 000 B.P. in montane areas (including Cameroon, the Central African highlands and Ethiopia) and in certain areas of lowland (coastal margins of Cameroon and Gabon, the foothills of the Rift Valley mountains, and a series of pockets elsewhere) (Fig. 3.4a).

The end of the last glaciation in higher-latitude regions was marked by a rapid rise in global temperature and by an increase in precipitation in the tropics, beginning at about 12 000 B.P. Within a couple of thousand years there was a massive water surplus in the Sahara: Lake Chad reached a maximum level c. 10 000 B.P., forming the so-called 'Mega-Chad' with a surface area five times that of modern Lake Victoria and overflowing into the Benue and Niger Rivers (Hamilton 1982). With this climatic amelioration, lowland tropical forest expanded rapidly to attain a much greater area than today by about 7000 B.P. (the climatic optimum) (Fig. 3.4b) since when there has been a smaller-scale decrease in temperature, precipitation and forest area (Fig. 3.4c). There is some detailed evidence that these environmental changes were not smooth and unidirectional: the increase in forest with warmer, wetter conditions was interrupted locally by a series of climatic fluctuations; there were sudden falls in lake levels accompanied by mass extinctions of fish which, in Ethiopia for instance, occurred at 10 000, 7000 and 4000 B.P.

The evidence of these dramatic environmental changes is recorded in lake sediments and in pollen profiles. However, these are necessarily regional studies and it is difficult to piece together overall patterns which would relate specifically to the habitats of *Xenopus*. Nevertheless, there are some relatively long-term records for specific areas. One is a remarkable sequence for a site at 2100 m in Burundi which provides continuous documentation of vegetation change from 40 000 B.P. to the present (Bonnefille & Riollet 1988). At the start of this record, the environment was cooler but wetter than in the late glacial. Between 30 000 and about 15 000 B.P., grassland greatly expanded during cool, dry conditions. Then, around 13 000 B.P., there is a marked expansion of forest under warm, wet conditions, and the afromontane grasslands retreated again to much higher elevations. The forest

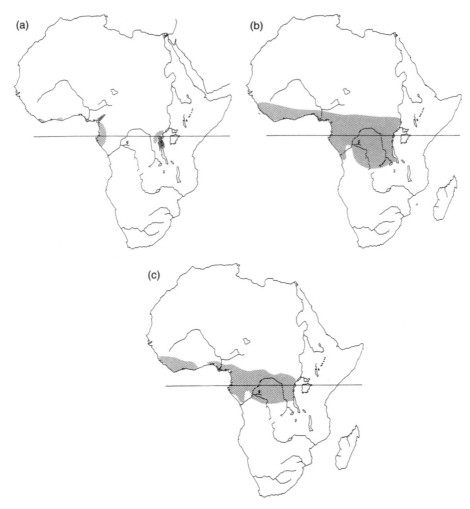

Fig. 3.4. Changes in the distribution of tropical rainforest in west and central Africa, and—by inference—changes in the geographical range of the forest species of *Xenopus*, the *fraseri* and *tropicalis* species groups. (a) At 18 000 years B.P. (the glacial maximum of the Northern Hemisphere), reflecting conditions of greatest aridity, with forest restricted to a series of lowland refugia (hatched areas) and to montane refugia in Cameroon and eastern Zaire (cross-hatched areas) (other forest refugia occurred along the east coast and in the mountains of East Africa and Ethiopia). (b) At 7000 years B.P. (the climatic optimum) with maximum extent of forest (and presumably forest *Xenopus* species) during warm, wet conditions. (c) At the present, showing reduced forest distribution under slightly cooler, more arid conditions and human influence. By inference, the savanna species *X. muelleri* and particularly *X. laevis* will have expanded into the grasslands which replaced forest during the cooler, drier periods of these climatic oscillations; the regions of greatest species richness and endemicity of *Xenopus* coincide with the montane forest refugia of Cameroon and the central African highlands. (Data on environmental changes from Hamilton 1982; Street 1981; and others: see text.)

reached its maximum at around 5000 B.P. and subsequently conditions have been somewhat drier and cooler.

Thus, a vast area of forest will have disappeared from the Zaire Basin during the cool, arid period, with perhaps a belt of lowland forest taking advantage of increased rainfall near the coast and in the foothills of mountains (Fig. 3.4a). This biogeographical history is, of course, directly relevant to interpretation of the distribution and ecology of *Xenopus* since lowland forest can be equated with the occurrence of the *X. tropicalis* and *X. fraseri* groups while grassland savanna indicates *laevis* and *muelleri*. Remarkably, we must envisage that virtually the whole of the Zaire Basin—until recently viewed as an ancient, stable ecosystem —was occupied only 10 000 years ago by the savanna species of *Xenopus*. The *tropicalis* and *fraseri* groups, now ubiquitous throughout the lowland rainforests, have spread out to re-colonize this vast area in much less than 10 000 years (but, of course, this is no more remarkable than that the forests themselves—and all their associated tropical flora and fauna—spread in the same way in the same brief period). Presumably, there will have been repeated contact zones between forest and savanna *Xenopus*, with interspecies interactions leading to local replacement during the course of the species changeover. A glimpse of these circumstances may be gained from Inger's (1968) observations in the Garamba Park (the extreme north-east of Zaire) where the same four *Xenopus* groups—*laevis, muelleri, epitropicalis* and *pygmaeus*—co-occur in fragmented ecotypes near the current border of forest and savanna.

The record of vegetation change—and, by extrapolation, of *Xenopus* distribution—is complicated by a series of minor fluctuations. For instance, at 21 000 B.P. in Burundi, there was a marked increased in wetness and an expansion of forest lasting about 3000 years (Bonnefille & Riollet 1988). It is assumed that the source of these forest trees lay in the local refugia—pockets of forest scattered throughout the area which could expand when suitable conditions arose. It would be a matter of chance whether such local pockets also had persisting populations of *Xenopus* which would expand briefly and then contract again. Certainly the speed of re-colonization of the Zaire Basin would have been accelerated by the occurrence of such small refugia. In general, the increase in tropical forest extended eastwards from the main refugia in Cameroon and Gabon and westwards from east Zaire, and so the advance of the forest *Xenopus* may have followed the same routes. However, as noted by Hamilton (1982), forest expansion should not be envisaged as a wave of forest moving across the region, but rather as the infiltration of forest species and the gradual enlargement and coalescence of forest patches. Gallery forest will have formed important lines of communication for elements of the forest fauna such as the *X. tropicalis* and *fraseri* groups.

All these snapshots of evidence point to periodic and very dramatic flux in the environments inhabited by *Xenopus*.

The longer-term record is increasingly imperfect, but an overall indicator of climate change has come recently from deep ocean cores. These indicate as many as 20 major cold phases separated by warmer, wetter periods in the last 1.75 million years. So, we must envisage that in this time-scale there will have been repeated expansion and contraction in the geographical ranges of aquatic animals such as *Xenopus*, and these longer-term changes will have had an important effect on speciation. This evidence provides explanation for the fact that the forest refugia, particularly in the montane regions, are centres of great species richness for a wide range of plants and animals. Genetic differentiation may have accompanied the isolation of disjunct populations in these areas, so that they now contain a high proportion of endemic species. The large number of species concentrated here during periods of maximum restriction will then have migrated outwards creating gradients of increasing faunal poverty during the 'inter-glacials' (see Diamond & Hamilton 1980).

For *Xenopus* in particular, the periodic fluctuations in distribution and the mixing of species in zones of sympatry will probably have had another fundamental influence on the evolution of recent species: the concentration will have provided circumstances for intensive interspecific interactions including hybridization. Subsequent chapters in this volume (by Kobel, pp. 391–401; Graf, pp. 379–389) discuss the evolutionary implications of the very significant feature of *Xenopus*, that all except one of the known species are almost certainly interspecies hybrids with a polyploid constitution. The events promoting hybridization are illustrated currently on the Cape, South Africa (see Picker *et al.*, this volume pp. 61–71), and the potential for the generation of new polyploid forms following hybridization (between X. *gilli* and X. *laevis*) has been demonstrated by Kobel, Du Pasquier & Tinsley (1981). The interspecies interactions which would have accompanied the last 'glacial period' represent a very brief period in evolutionary time although, of course, allopolyploidization is a mechanism for instant evolution of new species. The evidence from a number of comparative techniques (see Graf, this volume pp. 379–389) suggests that the polyploid species so far studied are considerably older. The wider significance of the biogeographical events now documented for the past 30 000 years is that the accompanying major ecological effects will have occurred repeatedly—at least 20 times in the last two million years. Periodically, during the Quaternary, the contraction and expansion of geographical ranges and the mixing of taxa could have generated a succession of new *Xenopus* species.

This biogeographical evidence also explains why taxonomic knowledge of *Xenopus* has increased so dramatically in the past 20 years. This is a direct result of field studies in remote areas, particularly the highlands of Cameroon and the Rift Valley, which represent the refugia that formed centres for concentration of species during successive climatic oscillations, and centres for differentiation and hybridization (Figs 3.2, 3.3, 3.4a).

This explanation of the factors responsible for the current distributions of the *Xenopus* species also indicates that the distributions are unlikely to be stable. We should not expect that the species necessarily occupy zones representing their exact niche specialization; it is quite likely that *Xenopus* species will occur in competition with others which may limit their potential range. It reflects the lack of good ecological information for *Xenopus* that the dynamic interactions between competing species in current areas of sympatry are as yet undocumented. This represents a rich field for future integrated studies of biogeography, taxonomy, ecology and evolutionary genetics.

Acknowledgements

Fieldwork was supported by grants from The Royal Society and the Natural Environment Research Council (including GR3/6661, GR9/632), the Fonds National Suisse de la Recherche Scientifique, and the Claraz Foundation. We also thank the governments of the many African states who gave help and permission for our fieldwork on *Xenopus*. We are grateful to Dr Heather Tinsley for biogeographical advice. Many of the data reported in this account are based on the invaluable resources of museums: we thank particularly the Natural History Museum, London; the Musée Royal de l'Afrique Centrale, Tervuren; the Field Museum of Natural History, Chicago; the American Museum of Natural History, New York; the Institut Royal des Sciences Naturelles de Belgique, Brussels; the Zoologisches Forschungsinstitut und Museum Alexander Koenig, Bonn; the Muséum d'Histoire Naturelle, Geneva; the Museum of Comparative Zoology, Harvard, and others acknowledged in Loumont (1984).

References

Alexander, S.S. & Bellerby, C.W. (1938). Experimental studies on the sexual cycle of the South African clawed toad (*Xenopus laevis*). I. *J. exp. Biol.* 15: 74–81.

Avila, V.L. & Frye, P.G. (1978). Feeding behavior of the African clawed frog (*Xenopus laevis* Daudin): (Amphibia, Anura, Pipidae): effect of prey type. *J. Herpet.* 12: 391–396.

Balinsky, B.I. (1969). The reproductive ecology of amphibians of the Transvaal highveld. *Zool. Afr.* 4: 37–93.

Balinsky, J.B., Choritz, E.L., Coe, C.G.L. & van der Schans, G.S. (1967). Amino acid metabolism and urea synthesis in naturally aestivating *Xenopus laevis*. *Comp. Biochem. Physiol.* 22: 59–68.

Balinsky, J.B., Cragg, M.M. & Baldwin, E. (1961). The adaptation of amphibian waste nitrogen excretion to dehydration. *Comp. Biochem. Physiol.* 3: 236–244.

Bellerby, C.W. & Hogben, L. (1938). Experimental studies on the sexual cycle of the South African clawed toad (*Xenopus laevis*). III. *J. exp. Biol.* 15: 91–100.

Berk, L. (1938). Studies in the reproduction of *Xenopus laevis*. 1. The relation of external environmental factors to the sexual cycle. *S. Afr. J. med. Sci.* 3: 72–77.

Bonnefille, R. & Riollet, G. (1988). The Kashiru pollen sequence (Burundi) palaeoclimatic implications for the last 40,000 yr B.P. in tropical Africa. *Quat. Res., N.Y.* 30: 19–35.

Dejours, P., Armand, J. & Beekenkamp, H. (1992). Étude chez l'amphibien anoure, *Xenopus laevis*, de quelques caractères du sang et de l'excrétion azotée en fonction du changement de l'osmolarité de l'eau ambiante. *C.r. Seánc. Acad. Sci. Paris Ser. III Sci. Vie* 315: 145–150.

Diamond, A.W. & Hamilton, A.C. (1980). The distribution of forest passerine birds and Quaternary climatic change in tropical Africa. *J. Zool., Lond.* 191: 379–402.

Du Plessis, S.S. (1966). Stimulation of spawning in *Xenopus laevis* by fowl manure. *Nature, Lond.* 211: 1092.

Fischberg, M., Colombelli, B. & Picard, J.-J. (1982). Diagnose préliminaire d'une espèce nouvelle de *Xenopus* du Zaire. *Alytes* 1: 53–55.

Fischberg, M. & Kobel, H.R. (1978). Two new polyploid *Xenopus* species from western Uganda. *Experientia* 34: 1012–1014.

Flower, S.S. (1936). Further notes on the duration of life in animals—2. Amphibians. *Proc. zool. Soc. Lond.* 1936: 369–394.

Frost, D.R. (Ed.) (1985). *Amphibian species of the world: a taxonomic and geographical reference*. Association of Systematics Collections and Allen Press, Lawrence, Kansas.

Hamilton, A. (1982). *Environmental history of East Africa: a study of the Quaternary*. Academic Press, New York & London.

Happold, D.C.D. & Happold, M. (1989). Biogeography of montane small mammals in Malawi, Central Africa. *J. Biogeogr.* 16: 353–367.

Hewitt, J. & Power, J.H. (1913). A list of S. African Lacertilia, Ophidia and Batrachia in the McGregor Museum, Kimberley, with field notes on various species. *Trans. R.Soc. S. Afr.* 3: 147–176.

Hey, D. (1949). A report on the culture of the South African clawed frog *Xenopus laevis* (Daudin) at the Jonkershoek inland fish hatchery. *Trans. R. Soc. S. Afr.* 32: 45–54.

Inger, R.F. (1968). Amphibia. *Explor. Parc natn. Garamba Miss. H. de Saeger* No. 52: 1–190.

Inger, R. & Marx, H. (1961). The food of amphibians. *Explor. Parc natn. Upemba Miss. G.F. de Witte* No. 64: 1–86.

Kalk, M. (1960). Climate and breeding in *Xenopus laevis*. *S. Afr. J. Sci.* 56: 271–276.

Kazadi, M., De Bruyn, L. & Hulselmans, J. (1986). Notes écologiques sur les contenus stomacaux d'une collection de *Xenopus laevis* (Daudin, 1803) (Amphibia: Anura) du Ruanda. *Annls Soc. r. zool. Belg.* 116: 227–234.

Kobel, H.R., Du Pasquier, L., Fischberg, M. & Gloor, H. (1980). *Xenopus amieti* sp. nov. (Anura: Pipidae) from the Cameroons, another case of tetraploidy. *Revue suisse Zool.* 87: 919–926.

Kobel, H.R., Du Pasquier, L. & Tinsley, R.C. (1981). Natural hybridization and gene introgression between *Xenopus gilli* and *Xenopus laevis laevis* (Anura: Pipidae). *J. Zool., Lond.* 194: 317–322.

Laurent, R.F. (1972). Amphibiens. *Explor. Parc natn. Virunga deux. Sér.* No. 22: 1–125.

Loumont, C. (1983). Deux espèces nouvelles de *Xenopus* du Cameroun (Amphibia, Pipidae). *Revue suisse Zool.* 90: 169–177.

Loumont, C. (1984). Current distribution of the genus *Xenopus* in Africa and future prospects. *Revue suisse Zool.* 91: 725–746.

Loumont, C. (1986). *Xenopus pygmaeus*, a new diploid pipid frog from rain forest of equatorial Africa. *Revue suisse Zool.* 93: 755–764.

Loumont, C. & Kobel, H.R. (1991). *Xenopus longipes* sp. nov., a new polyploid pipid from western Cameroon. *Revue suisse Zool.* 98: 731–738.

Loveridge, A. (1932). New races of a skink (*Siaphos*) and frog (*Xenopus*) from the Uganda Protectorate. *Proc. biol. Soc. Wash.* 45: 113–116.

Loveridge, A. (1942). Scientific results of a fourth expedition to forested areas in East and Central Africa. V. Amphibians. *Bull. Mus. comp. Zool. Harv.* 91: 377–436.

Loveridge, A. (1953). Zoological results of a fifth expedition to East Africa. IV. Amphibians from Nyasaland and Tete. *Bull. Mus. comp. Zool. Harv.* 110: 325–406.

McCoid, M.J. & Fritts, T.H. (1980a). Observations of feral populations of *Xenopus laevis* (Pipidae) in southern California. *Bull. Sth. Calif. Acad. Sci.* 79: 82–86.

McCoid, M.J. & Fritts, T.H. (1980b). Notes on the diet of a feral population of *Xenopus laevis* (Pipidae) in California. *SWest. Nat.* 25: 272–275.

McCoid, M.J. & Fritts, T.H. (1989). Growth and fatbody cycles in feral populations of the African clawed frog, *Xenopus laevis* (Pipidae), in California with comments on reproduction. *SWest. Nat.* 34: 499–505.

Merkle, S. & Hanke, W. (1988). Long-term starvation in *Xenopus laevis* Daudin - I. Effects on general metabolism. *Comp. Biochem. Physiol.* 89B: 719–730.

Moreau, R.E. (1966). *The bird faunas of Africa and its islands*. Academic Press, New York.

Moron, Y. (1947). Observations sur *Xenopus tropicalis* (Gray). *Bull. Soc. zool. Fr.* 72: 128–133.

Munsey, L.D. (1972). Salinity tolerance of the African pipid frog, *Xenopus laevis*. *Copeia* 1972: 584–586.

Nieuwkoop, P.D. & Faber, J. (1956). *Normal table of* Xenopus laevis *(Daudin)*. North-Holland Pub. Co., Amsterdam.

Noble, G.K. (1924). Contributions to the herpetology of the Belgian Congo based on the collection of the American Museum Congo Expedition, 1909–1915. Part III. Amphibia. *Bull. Am. Mus. nat. Hist.* 49: 147–347.

Perret, J.-L. (1966). Les amphibiens du Cameroun. *Zool. Jb. (Syst.)* 93: 289–464.

Picker, M.D. (1985). Hybridization and habitat selection in *Xenopus gilli* and *Xenopus laevis* in the southwestern Cape Province. *Copeia* 1985: 574–580.

Poynton, J.C. (1964). The biotic divisions of Southern Africa, as shown by the Amphibia. *Monogr. biol.* 14: 206–218.

Poynton, J.C. & Broadley, D.G. (1991). Amphibia Zambesiaca 5. Zoogeography. *Ann. Natal Mus.* 32: 221–277.

Prinsloo, J.F., Schoonbee, H.J. & Nxiweni, J.G. (1981). Some observations on the biological and other control measures of the African clawed frog *Xenopus laevis* (Daudin) (Pipidae, Amphibia) in fish ponds in Transkei. *Water S A* 7: 88–96.

Rau, R.E. (1978). The development of *Xenopus gilli* Rose & Hewitt (Anura, Pipidae). *Ann. S. Afr. Mus.* 76: 247–263.

Romspert, A.P. (1976). Osmoregulation of the African clawed frog, *Xenopus laevis*, in hypersaline media. *Comp. Biochem. Physiol.* 54A: 207–210.

Rose, W. (1950). *The reptiles and amphibians of Southern Africa*. Maskew Miller, Capetown.

Rose, W. (1962). *The reptiles and amphibians of Southern Africa*. (2nd edn). Maskew Miller, Capetown.

Savage, R.M. (1971). The natural stimulus for spawning in *Xenopus laevis* (Amphibia). *J. Zool., Lond.* **165**: 245–260.

Schiøtz, A. (1963). The amphibians of Nigeria. *Vidensk. Medd. dansk. naturh. Foren.* **125**: 1–92.

Schmidt, K.P. & Inger, R.F. (1959). Amphibians exclusive of the genera *Afrixalus* and *Hyperolius*. *Explor. Parc natn. Upemba Miss. G.F. de Witte* No. 56: 1–264.

Schoonbee, H.J., Prinsloo, J.F. & Nxiweni, J.G. (1992). Observations on the feeding habits of larvae, juvenile and adult stages of the African clawed frog, *Xenopus laevis*, in impoundments in Transkei. *Water S A* **18**: 227–236.

Seymour, R.S. (1973). Energy metabolism of dormant spadefoot toads (*Scaphiopus*). *Copeia* 1973: 435–445.

Street, F.A. (1981). Tropical palaeoenvironments. *Prog. phys. Geogr.* **5**: 157–185.

Tinsley, R.C. (1973). Studies on the ecology and systematics of a new species of clawed toad, the genus *Xenopus*, from western Uganda. *J. Zool., Lond.* **169**: 1–27.

Tinsley, R.C. (1975). The morphology and distribution of *Xenopus vestitus* (Anura: Pipidae) in Central Africa. *J. Zool., Lond.* **175**: 473–492.

Tinsley, R.C. (1981). Interactions between *Xenopus* species (Anura Pipidae). *Monit. zool. ital. (N.S.) (Suppl.)* **15**: 133–150.

Tinsley, R.C. (1995). A new species of *Xenopus* (Anura: Pipidae) from the highlands of Ethiopia. *Amphibia-Reptilia* **16**: 375–388.

Tinsley, R.C., Kobel, H.R. & Fischberg, M. (1979). The biology and systematics of a new species of *Xenopus* (Anura: Pipidae) from the highlands of central Africa. *J. Zool., Lond.* **188**: 69–102.

Tinsley, R.C. & Sweeting, R.A. (1974). Studies on the biology and taxonomy of *Diplostomulum (Tylodelphylus) xenopodis* from the African clawed toad, *Xenopus laevis*. *J. Helminth.* **48**: 247–263.

Tymowska, J. & Fischberg, M. (1973). Chromosome complements of the genus *Xenopus*. *Chromosoma* **44**: 335–342.

Unsworth, B.R. & Crook, E.M. (1967). The effect of water shortage on the nitrogen metabolism of *Xenopus laevis*. *Comp. Biochem. Physiol.* **23**: 831–845.

Visser, S.A. (1962). Chemical investigations into a system of lakes, rivers and swamps in S.W. Kigezi, Uganda. *E. Afr. agric. For. J.* **28**: 81–86.

Witte, G.F. de (1941). Batraciens et reptiles. *Explor. Parc. natn. Albert Miss. G.F. de Witte* **33**: 1–261.

Wood, J.A. (1965). Some notes on *Xenopus laevis* (Daudin). (Amphibia, Pipidae). *Jl E. Africa nat. Hist. Soc.* **25**: 57–68.

Worthington, S. & Worthington, E.B. (1933). *Inland waters of Africa.* Macmillan, London.

Zielinski, W.J. & Barthalmus, G.T. (1989). African clawed frog skin compounds: antipredatory effects on African and North American water snakes. *Anim. Behav.* **38**: 1083–1086.

4 Natural hybridization between *Xenopus laevis laevis* and *X. gilli* in the south-western Cape Province, South Africa

M.D. PICKER, J.A. HARRISON and D. WALLACE

Synopsis

Hybridization occurs between the endangered *Xenopus gilli* and *X. laevis laevis* in the south-western Cape Province. *X. gilli* is adapted to the acidic blackwaters that are found in the unique fynbos vegetation biome (Cape Floral Kingdom). Under conditions of disturbance these ponds become accessible to *X. laevis*, which leads to hybridization. Phenotypically, and on features of the mating call and female release call, natural *X. gilli* × *X. laevis* hybrids fall into two distinct groupings which are *X. gilli*-like or *X. laevis*-like. The erythrocytic nucleus diameters of hybrids do not implicate a polyploid condition in any of the hybrids examined. This, and the near absence of hybrids that are phenotypically intermediate between the parental species, suggest that hybridization has led to the backcross of F_1 individuals with the parental species, without the involvement of polyploidy. The future survival of *X. gilli* thus appears to be threatened not only by habitat destruction, but also by subtle and extensive introgression with *X. laevis* genes.

Introduction

Xenopus gilli is a rare species, which was known only from approximately eight localities in South Africa before 1970. More intensive fieldwork has increased the number of localities to 35 in 1985 (Picker 1985; Picker & de Villiers 1988, 1989), extending the range of *X. gilli* eastwards. These localities are all situated in coastal lowlands, with the exception of two old inland collections made in Namaqualand. Apart from the latter, all localities fall within the Cape Floral Kingdom (fynbos biome), a vegetation type endemic to the south-western Cape Province of South Africa (Fig. 4.1). This biome is characterized by a high degree of endemism in its floral (Cowling, Holmes

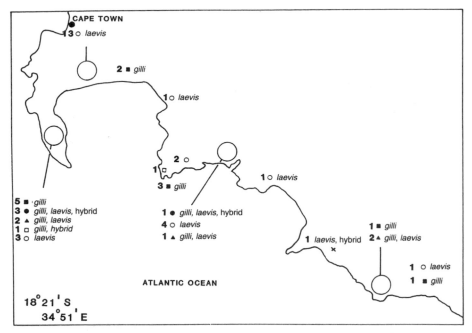

Fig. 4.1. Post-1980 distribution of *X. gilli, X. laevis* and hybrids in the south-western Cape Province (modified from Picker & De Villiers 1989). The number of waterbodies supporting various populations is given before the taxon.

& Rebelo 1992) and faunal elements (including Amphibia) (Passmore & Carruthers 1979). *X. laevis laevis* has a very wide distribution that extends from summer rainfall regions in the north, southwards from Malawi and Namibia until it overlaps with *X. gilli* in the winter rainfall region of the south-western Cape (Picker & De Villiers 1989). The two species are easily separated morphologically. *X. gilli* is much smaller, with distinctive dorsal and ventral markings. Additionally it differs from *X. laevis* in having fewer circumorbital plaques, no subocular tentacle, a pointed head and other morphometric differences (Poynton 1964).

At most localities some frogs display a phenotype that has features of both species. The possible hybrid nature of such individuals was first suggested by Rau (1978) and later confirmed genetically by Kobel, Du Pasquier & Tinsley (1981).

Observations

Population analysis

Frogs were either seine-netted or funnel-trapped in suitable bodies of water, and seven measurements (Fig. 4.2) were taken with vernier calipers from

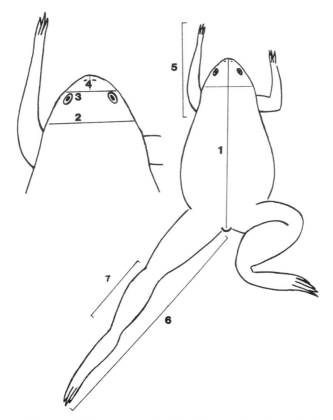

Fig. 4.2. Mensural features of *Xenopus* used in discriminating *X. gilli*, *X. laevis*, and hybrids between the two. 1, snout–vent length; 2, widest head width; 3, narrowest head width; 4, snout length; 5, forelimb length; 6, hindleg length; 7, tibial length.

chilled animals comprising: 27 male *X. laevis*, 53 female *X. laevis*, 12 male *X. laevis*-like hybrids, 45 female *X. laevis*-like hybrids, 20 *X. gilli* males, 45 *X. gilli* females, 15 male *X. gilli*-like hybrids and 24 female *X. gilli*-like hybrids. Allocation to a taxon was made on pattern alone (*X. gilli* differs from *X. laevis* in the presence of dorsal striping and ventral marbling). Hybrids were initially recognized by their incomplete expression of these characters, or by their very large size, combined with *X. gilli*-like features. Morphological data were subjected to an analysis of variance and a stepwise discriminant functions analysis.

Discriminant functions analysis

Within each species, the sexes separate from one another, with some minor overlap (Fig. 4.3). There is no overlap between species. The four hybrid groups are most closely associated with their own sex within either *X. gilli* or *X. laevis*

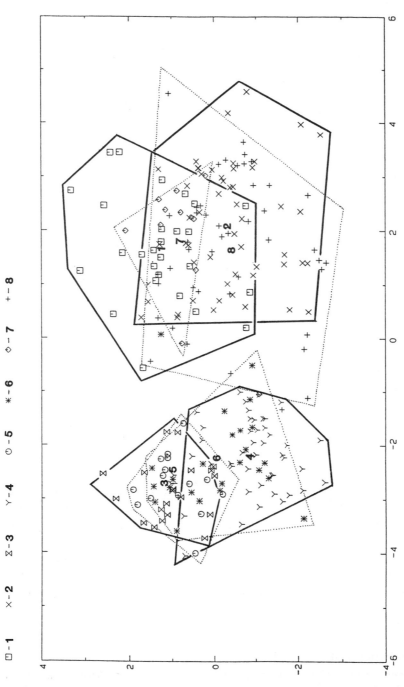

Fig. 4.3. Discriminant functions analysis plot, using transformed variables, and examining 241 cases. The eight clustered groups as indicated by the key are: 1, *X. laevis* males; 2, *X. laevis* females; 3, *X. gilli* females; 4, *X. gilli* males; 5, *X. gilli*-like hybrid females; 6, *X. gilli*-like hybrid males; 7, *X. laevis*-like hybrid males; and 8, *X. laevis*-like hybrid females. The first canonical variable is the linear combination of variables entered that best discriminate among the groups, the second canonical variable is the next best linear combination orthogonal to the first one.

groupings. In fact, they are so closely associated with the respective parental species that, were it not for the *a priori* separation on the basis of dorsal and ventral markings, they could be mistaken morphometrically for individuals of the parent species. There is a small amount of overlap between females of the hybrid groups. The *X. laevis*-like hybrids are the most distinct category, with the *X. gilli*-like hybrids closely resembling *X. gilli*. Surprisingly, the hybrids did not cluster between the parental species, as would be expected for F_1 hybrids such as those produced in the laboratory (H. Kobel pers. comm.). Thus natural hybrids appear to represent backcrosses to the parental species rather than F_1 crosses.

Vocalizations of hybrids

The separation of the hybrids into *X. gilli*-like and *X. laevis*-like hybrids is supported by the structure of the mating call and female release call of the two hybrid groups. The mating calls of the two species differ significantly in dominant frequency, note repetition rate and call duration. Hybrid mating calls are variable in structure: *X. laevis*-like hybrid calls are structurally intermediate between those of the parental species, but more similar to those of *X. laevis* (Fig. 4.4). The small sample examined may not, however, be representative of the group. The *X. gilli*-like hybrid calls closely resemble those of *X. gilli*, but have some of the features of the *X. laevis* call. The female release calls of *X. gilli* and *X. laevis* are also distinctive, differing in note repetition rate, dominant frequency and call duration. All hybrid female release calls differ significantly from those of both parental species in note repetition rate. In addition, the *X. gilli*-like hybrid calls differ from the *X. laevis* call in all parameters, but closely resemble the *X. gilli* call. However, the *X. laevis*-like hybrid call differs markedly from the *X. gilli* call in pulse repetition rate (Fig. 4.5).

Ploidy

Under experimental conditions, female hybrids are known to produce normal as well as some endoreduplicated eggs. Offspring resulting from the latter would be polyploid backcrosses (Kobel & Du Pasquier 1986; Kobel this volume pp. 73–80). It would be possible to detect such polyploids in nature by measuring the nuclear diameter of erythrocytes, as this is known to be proportional to the number of genomes present (Engel & Kobel 1983). However, none of the 47 hybrid frogs examined showed a nuclear diameter typical of triploidy or higher ploidy condition.

Habitat preferences

Approximately 75 localities within the known or probable distribution of *X. gilli* were sampled. The range of *X. gilli* was found to extend 150 km further eastward than previously known. However, extinction appears to have taken place at 60% of the original localities in the last 50 years. The results of this survey showed that *X. gilli* is restricted to seepages

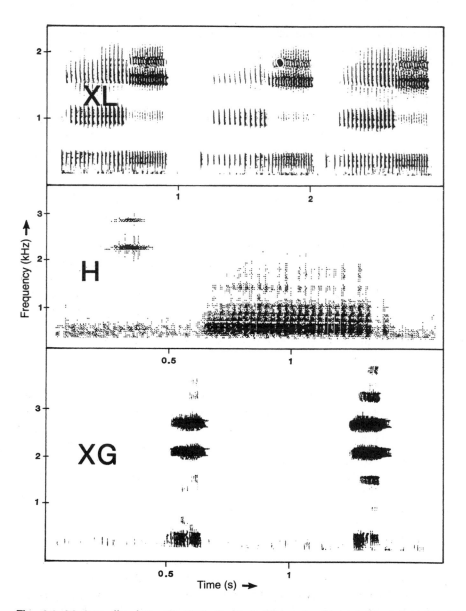

Fig. 4.4. Mating calls of *X. gilli* (XG), *X. laevis* (XL) and a *X. laevis*-like hybrid (H). The mating call of the hybrid more closely resembles that of *X. laevis*, but contains elements of the brief tonal call of *X. gilli* and resembles the *X. gilli* call in the frequency band used.

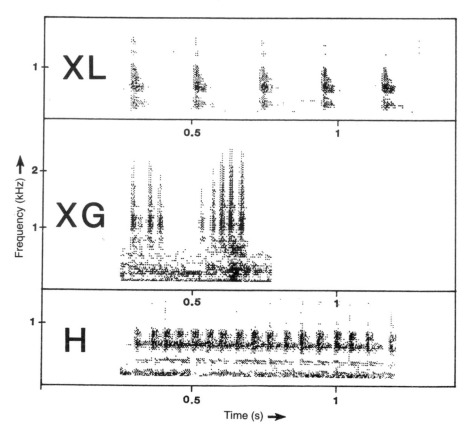

Fig. 4.5. Female release calls of *X. laevis* (XL), *X. gilli* (XG) and a *X. laevis*-like hybrid (H). The hybrid call illustrated is intermediate to that of the parental species, comprising groups of notes separated by short intervals of silence.

and ponds deeply stained with humic compounds. These result from the breakdown of phenolic-rich plant matter in the soil to a variety of organic acids such as humic, fulvic and hymenomelanic and also humin (Burges, Hurst & Walkden 1964). Such blackwaters are frequently very acidic, and are characteristic of the fynbos vegetation biome (Picker & de Villiers 1988, 1989). Hybrids also occur in these waterbodies; within the same general habitat, *X. laevis* often occurs in clear (mostly artificial) waterbodies that are interspersed between the blackwater ponds (Picker 1985). There does not appear to be any preference amongst any of the taxa examined for temporary versus permanent waterbodies (Table 4.1). However, whereas *X. laevis* and hybrids are frequently found in disturbed waterbodies, *X. gilli* rarely is. It would appear that any changes to the fynbos vegetation (such as replacement by alien vegetation) will reduce the input of humic compounds to the drainage system, resulting in clear waterbodies of elevated pH that are suitable for

Table 4.1. Habitat parameters for *X. gilli, X. laevis* and hybrid-containing ponds. The 47 waterbodies examined were all situated in the Cape of Good Hope Nature Reserve. Hybrid ponds also contained *X. laevis, X. gilli* or both, and therefore these ponds were included in the calculation of habitat associations of *X. laevis* and *X. gilli* where appropriate.

	X. gilli alone	*X. laevis* alone	Hybrids
Number of ponds occupied by the taxon	11	29	7
Percentage of the ponds occupied by a taxon associated with:			
Blackwater	100	64	100
Disturbed blackwater	17	87	86
Temporary waters	54	32	35

invasion and breeding by *X. laevis*. Such circumstances offer conditions that are tolerated by adults of both species, and favour hybridization. It has been shown (Picker, McKenzie & Fielding 1993) that if water is sufficiently acidic, particularly if concentrations of humic compounds are high, embryos of *X. gilli* will survive whereas those of *X. laevis* will not. But undisturbed blackwaters that provide these conditions and sustain a population of *X. gilli* uncontaminated by hybridization are now rare and to be found only in two ponds in the Cape of Good Hope Nature Reserve (Table 4.2). This consequence of an altered habitat, combined with the relative paucity of natural blackwaters in the fynbos, threatens the future survival of uncontaminated *X. gilli* populations.

Intensive agricultural practices, eutrophication and elimination of most natural fynbos vegetation have resulted in local extinction of the original *X. gilli* populations (Picker & de Villiers 1989). The blackwaters which are home to *X. gilli* are unusual for softwaters in that they have high levels of sodium chloride, no doubt derived from brine-laden winds (Loveridge 1980; Picker *et al.* 1993). In contrast to the exacting habitat requirements of *X. gilli, X. laevis* is both widespread and ecologically undemanding. It occurs in almost all kinds of waterbodies in southern Africa, but does not favour lentic situations or the acidic blackwaters of the south-western Cape (Tables 4.1 and 4.2). Its adaptability has enabled it to increase its range and density considerably by invading man-made dams and other waterbodies (Van Dijk 1977). Both species have a breeding peak in early spring (July to October) (Berk 1939; Rau 1978), a factor facilitating hybridization. The hybrids have been found in about a third of the known *X. gilli* localities (Picker & de Villiers 1989), generally in blackwater ponds that have suffered some habitat disturbance (Tables 4.1 and 4.2). Although the hybrids are also present in ponds that appear to be in a pristine condition, these waterbodies are situated near ponds that have been disturbed or are man-made. Contamination through such ponds seems highly probable.

Table 4.2. The pH values of various waterbodies containing populations of *X. gilli*, *X. laevis* and hybrids. Mixed species populations do occur in many of the ponds, but the analysis below is by species, not species combinations.

	Ponds containing		
	X. gilli	*X. laevis*	Hybrids
Southern Cape			
Number of waterbodies examined	11	29	6
Mean pH (SD)	6.6 (2.1)	6.8 (1.2)	5.5 (1.7)
Range	3.5–8.7	3.7–9.8	4.2–8.7
Cape Point (Cape of Good Hope Nature Reserve)			
Number of waterbodies examined	8	9	5
Mean pH (SD)	4.9 (1.3)	5.9 (1.7)	5.7 (1.7)
Range	3.5–7.1	4.3–8.7	4.2–8.7

Population densities

The population densities of *X. gilli* in three ponds in the Cape of Good Hope Nature Reserve have remained constant at approximately 200–400 individuals for a number of years, although in the last two years there has been a sudden drop in the populations in Gillidam, a pond which previously supported up to 400 animals. The reasons for this decline are not known, although the frogs are capable of overland migrations of at least 1.5 km from one pond to another during the wet winter.

Conclusions

Xenopus gilli is a very specialized frog that occurs only along a narrow coastal strip from Cape Point to Cape Agulhas. While it is possible that it occurs at two localities inland, it has to date not been rediscovered at these old collection sites. The populations on the Cape Flats have been virtually exterminated, owing to intensive agricultural and housing developments that have taken place in the region. In contrast, *X. laevis* has a very wide geographical distribution and, not unexpectedly, has very wide ecological tolerances. Today it occurs sympatrically with *X. gilli* at all *X. gilli* localities. It is likely that *X. laevis* always occurred in geographical sympatry with *X. gilli*, but ecological sympatry and hybridization may be a recent, human-induced event accompanying fynbos disturbance.

As in other anurans, information for species recognition is encoded in the mating call (Picker 1983). However, mismatings are likely to occur under conditions of crowding, especially if one species numerically dominates the other. Thus a gravid female approaching a vocalizing male of her own species may well come into close contact with a non-conspecific male *en route*, and amplexus may occur. Hybridization between the species can easily be demonstrated in the laboratory with experimental crosses (Kobel this volume pp. 73–80). Although such crosses give rise to polyploid offspring, this was not recorded in the 47 frogs we examined from a variety of locations. It would seem that the majority of these frogs represent multiple backcrosses to either of the parental species rather than F_1 individuals. The abundance of hybrids indicates that they have no apparent selective disadvantage.

The continued survival of *X. gilli* depends on the extent of future habitat degradation, and the importance and extent of introgression of *X. laevis* genes into the *X. gilli* gene pool. The only strongholds for *X. gilli* are ponds that are so acidic and have such high levels of humic compounds that they exclude *X. laevis*. Such ponds are rare, and only two exist in the Cape of Good Hope Nature Reserve. The exclusion appears to reside in the ability of *X. gilli* embryos to withstand the toxic effects of humic compounds and low pH better than *X. laevis* embryos (Picker *et al.* 1993). In blackwater, *X. laevis* embryos are only able to tolerate pH values of 4.2, whereas *X. gilli* survive at pH 3.6. At pH 4.5, *X. laevis* embryos die even at blackwater concentrations of 0.3, whereas *X. gilli* embryos survive in blackwater that has been concentrated five times. The future survival of this unique *Xenopus* may well depend on these adaptations; however, the very narrow habitat requirements of *X. gilli*, its limited geographical range and the increasing destruction of fynbos and blackwater ponds do not bode well for the continued existence of *X. gilli*.

Acknowledgements

We are grateful to the Director of the Cape of Good Hope Nature Reserve for permission to work on the *X. gilli* populations in the Reserve. Both Hansruedi Kobel and Rheinhold Rau enthusiastically supported this work, and Hansruedi Kobel kindly provided useful comment on the manuscript.

References

Berk, L. (1939). Studies in the reproduction of *Xenopus laevis*. III. The secondary sex characters of the male *Xenopus*: the pads. *S. Afr. J. med. Sci.* 3: 47–60.

Burges, W.A., Hurst, H.D. & Walkden, B. (1964). The phenolic constituents of humic acid and their relation to the lignin of plant cover. *Geochim. cosmochim. Acta* 28: 1547–1554.

Cowling, R.M., Holmes, P.M. & Rebelo, A.G. (1992). Plant diversity and endemism. In *The ecology of fynbos: nutrients, fire and diversity* 4: 62–112. (Ed. Cowling, R.M.). Oxford University Press, Oxford.

Engel, W. & Kobel, H.R. (1983). The Z-chromosome is involved in the regulation of H-W (H-Y) antigen gene expression in *Xenopus*. *Cytogenet. Cell Genet.* 35: 28–33.

Kobel, H.R. & Du Pasquier, L. (1986). Genetics of polyploid *Xenopus*. *Trends Genet.* 2(12): 310–315.

Kobel, H.R., Du Pasquier, L. & Tinsley, R.C. (1981). Natural hybridization and gene introgression between *Xenopus gilli* and *Xenopus laevis laevis* (Anura: Pipidae). *J. Zool., Lond.* **194**: 317–322.

Loveridge, J.P. (1980). *The habitat requirements of* X. gilli *in the Cape Point Nature Reserve.* Unpublished research report, Department of Nature and Environmental Conservation, Cape Town.

Passmore, N.I. & Carruthers, V.C. (1979). *South African frogs.* Witwatersrand University Press, Johannesburg.

Picker, M.D. (1983). Hormonal induction of the aquatic phonotactic response of *Xenopus. Behaviour* **84**: 74–90.

Picker, M.D. (1985). Hybridization and habitat selection in *Xenopus gilli* and *Xenopus laevis* in the south-western Cape Province. *Copeia* **1985**: 574–580.

Picker, M.D. & de Villiers, A.L. (1988). *Xenopus gilli*: species account. In *South African red data book—reptiles and amphibians.* (Ed. Branch, W.R.). *S. Afr. natn. Sci. Prog. Rep.* **151**: 25–28.

Picker, M.D. & de Villiers, A.L. (1989). The distribution and conservation status of *Xenopus gilli* (Anura: Pipidae). *Biol. Conserv.* **49**: 169–183.

Picker, M.D., McKenzie, C.J. & Fielding, P. (1993). Embryonic tolerance of *Xenopus* (Anura) to acidic blackwater. *Copeia* **1993**: 1072–1081.

Poynton, J.C. (1964). The Amphibia of southern Africa: a faunal study. *Ann. Natal Mus.* **17**: 1–334.

Rau, R. (1978). The development of *Xenopus gilli* Rose and Hewitt (Anura, Pipidae). *Ann. S. Afr. Mus.* **76** (6): 247–263.

Van Dijk, D.E. (1977). Habitats and dispersal of southern African Anura. *Zool. afr.* **12**: 169–181.

5 Reproductive capacity of experimental *Xenopus gilli* × *X. l. laevis* hybrids

HANS RUDOLF KOBEL

Synopsis

Hybridization between *X. l. laevis* and the rare *X. gilli* represents an additional threat to the latter species whose geographical range is restricted to a small coastal strip from the Cape Peninsula to Cape Agulhas, South Africa. Laboratory studies on the reproductive capacity of experimental hybrids between the two species revealed that male F_1 hybrids are sterile because of failure of meiotic chromosome pairing and subsequent univalent meiosis. Female F_1 hybrids produce two egg types: small eggs issue from normal oocytes and big eggs from polyploidized oocytes. While the former are aneuploid because of univalent meiosis and give lethal embryos, big endoreduplicated eggs contain complete chromosome sets of both parental species. After fertilization with spermatozoa of either species, they develop into triploid females that again may produce the two egg classes.

However, a single F_1 female in one of the hybrid families studied behaved differently in that its small eggs also showed normal viability, giving rise to diploid recombinant offspring. These F_2 backcross animals were fertile in both sexes and no hybrid breakdown of fertility was apparent. Thus, animals in which homoeologous chromosome pairing and gene exchange between *gilli* and *laevis* chromosomes do occur could be the source of gene introgression into both species, as observed in interbreeding populations, especially on the Cape Peninsula.

Introduction

The presence on the Cape Peninsula of hybrids between *X. gilli* and *X. l. laevis* was suspected first by Rau (1978), though the size differences between the two species seemed to preclude interbreeding. Subsequently, more specimens with a hybrid constitution were observed (Kobel, Du Pasquier & Tinsley 1981; Picker 1985; Picker & de Villiers 1989; Picker, Harrison & Wallace, this volume pp. 61–71), but most often the odd specimens show only a few hybrid traits and are difficult to classify. As results of the following experimental breeding experiments suggest, many of these specimens may represent cases of extensive gene introgression into both species. Crosses were made between *X. l. laevis* females and males of *X. gilli* by artificial fertilization of stripped eggs with spermatozoa suspension. The reciprocal cross could not be obtained because *X. gilli* females almost exclusively laid unmatured oocytes.

F$_1$ hybrids: males

Hybrid males are sterile. Testes appear normal in size. However, the majority of chromosomes in spermatocytes I remain as univalents. This failure to pair of homoeologous chromosomes of the two species leads to a random reduction in chromosome number during the first meiotic division and to aneuploid cells. In anaphases I, some chromosomes often lag behind the movement of others. Second meiotic divisions are relatively infrequent and result in spermatids of unequal nuclear size. Germinal cysts contain a reduced and irregular number of spermatids, most of which are pycnotic. Among the few differentiated spermatozoa, many are abnormal with double heads, multiple tails or misshapen head curvature.

F$_1$ hybrids: females

Females produce two egg classes, small eggs and big ones of about double the volume of the former. The frequency of big eggs depends on the female (Table 5.1) and a given female always lays a consistent proportion of the two

Table 5.1. Proportion and respective viability of the two egg classes in F$_1$ hybrids between *X. gilli* and *X. l. laevis*. F$_1$ female LG$_{15}$ is exceptional in the high viability of both egg types.

F$_1$ female	'Big eggs': triploid frogs		'Small eggs': diploid frogs	
	Blastula (n)	Feeding (%)	Blastula (n)	Feeding (%)
LG$_2$	134	55	27	4
LG$_3$	43	65	8	0
LG$_5$	230	57	195	6
LG$_7$	91	65	38	5
LG$_{13}$	77	44	184	2
LG$_{14}$	82	73	4	0
LG$_{17}$	116	84	5	0
Mean	773	63	461	4
LG$_{15}$	89	56	94	78

egg types. Analysis of lampbrush chromosomes of oocytes revealed that small eggs issue from diploid (36 chromosomes) oocytes containing many univalents (mean 23.4), whereas big eggs result from polyploidized oocytes having 36 bivalents in their nuclei (Müller 1977). It is presumed that in a fraction of the oocytes, possibly during the transition from oogonia to oocytes, all the chromosomes undergo an additional replication so that each chromosome is then composed of four identical chromatids. Such oocytes thus have a diploid number of bivalents whose four identical chromatids segregate normally during meiosis. The resulting eggs are therefore diploid and contain both parental genomes (diploid with regard to the parental species which are in fact both ancient tetraploids). This hypothesis has been shown to be correct, since it is possible to obtain isogenic *Xenopus* clones from these hybrid females by gynogenetic development of their endoreduplicated eggs (Kobel & Du Pasquier 1975).

The viability of the two egg types is very different (Table 5.1). Univalent meiosis in diploid oocytes leads to a random reduction in chromosome number; however, since meiotic divisions intervene only after the complete differentiation of eggs, small eggs are functional gametes that can be fertilized and progress into early development. However, owing to their aneuploid constitution, development is abortive and only occasionally does a 'small egg' embryo survive beyond metamorphosis.

Big endoreduplicated eggs, by contrast containing complete genomes of both parental species, develop normally into triploid backcross animals. According to the male used for the backcross, these specimens phenotypically resemble the respective backcross species.

In the most thoroughly studied hybrid family, a single female, LG_{15}, did not conform to this scheme in that small eggs also gave rise to fully viable offspring (Table 5.1). Small oocytes in gynogenetically-obtained isogenic clones of this particular female show normal synapsis between *gilli* and *laevis* chromosomes and normal numbers of chiasmata (W.P. Müller pers. comm.). In this case, the genetic control of homoeologous chromosome pairing fails in the diploid oocytes while it still operates in endoreduplicated polyploid oocytes of the same female. From wheat and other plant hybrids it is known that suppression of homoeologous chromosome pairing is probably regulated by genes whose function is to remove non-homologous pairing in the normal course of meiosis. One may suppose that synapsis control is a fundamental process which applies to *Xenopus* as well; at least LG_{15} demonstrates that chromosomes of the two species are sufficiently similar in structure to act as homologues.

Backcrosses with LG_{15} females thus produce two different kinds of offspring, triploids from big eggs and diploids from small eggs which are recombinant for *gilli* and *laevis* chromosomes. Again, following the backcross males, they closely resemble the backcross species. However, their phenotypes are very diverse, reminiscent of certain specimens found on the Cape Peninsula.

F_2 backcross hybrids

Triploid offspring resulting from fertilization of endoreduplicated eggs with spermatozoa of either parental species again produce small and big eggs. After fertilization, the latter develop into tetraploids of both sexes, a phenomenon discussed by Kobel elsewhere in this volume (pp. 391–401).

The small eggs issue from oocytes that have two chromosome sets of the backcross species and a third genome from the second species. While the two homologous chromosome sets form bivalents, the third alien one remains as univalents. Meiosis thus contributes to the egg a complete genome of the backcross species whereas the univalents are distributed at random. Zygotes resulting from these small eggs are hyperdiploid, and a certain percentage of them survive to adulthood (Kobel, Egens de Sasso & Zlotowski 1979). Such specimens could theoretically introduce so-called supernumerary or B chromosomes into these species, as is known to occur in a number of Amphibia (Green 1991). Since no *Xenopus* specimen with an irregular chromosome number has been found so far, hyperdiploids do not seem to be competitive in nature.

The case of LG_{15}

The normal viability of the small eggs of LG_{15} clone females poses several questions:

Does homoeologous chromosome pairing occur again in the next, F_2, generation or are there indications of hybrid breakdown?

Does endoreduplication also occur in recombinant diploid F_2 females?

Are recombinant F_2 males fertile?

The viability of small eggs appeared to be normal: 83.4% surviving stage 50 tadpoles from 3856 blastulae produced by the 32 recombinant F_2 females listed in Table 5.2. Apparently no problems with synapsis nor signs of hybrid breakdown are present in diploid recombinant F_2 females.

The incidence of endoreduplicated big eggs is listed in Table 5.2 for diploid L(G) and (L)G females which issued from three different backcrosses with both parental species (the recombinant maternally-derived genome is set in brackets). The two backcrosses with X. *laevis* behaved very differently; while several L(G) females of the first backcross family did produce big eggs at high frequency, only a single female among 12 tested of the second backcross laid a few endoreduplicated eggs. The backcross with X. *gilli* also showed a high incidence of endoreduplicated eggs.

Diploid recombinant F_2 males can be fertile (Table 5.3); whether fertility is restored in all these males is difficult to judge. Cytological preparations of testes from the infertile males of Table 5.3, made at the same time as the spermatozoa suspensions used for fertilization, showed a normal aspect

Table 5.2. Incidence (%) of endoreduplicated big eggs in diploid recombinant F_2 backcross families

($\female$ LG_{15} × $\male$ L_1)			($\female$ LG_{15} × $\male$ L_2)			($\female$ LG_{15} × $\male$ G_1)		
$\female$ L(G) No.	%	n	$\female$ L(G) No.	%	n	$\female$ (L)G No.	%	n
1	96	209	1	0	200	2	65	130
2	90	200	2	0	200	3	86	454
4	0	200	3	0	200	4	54	200
5	93	137	4	0	200	5	0	200
6	1	200	5	0	38	7	0	53
8	0	200	6	0	200	8	3	290
			7	0	169	10	93	41
			8	0	210	11	1	134
			9	0	28	13	0	200
			10	0	200	14	0	84
			11	1.5	203	15	0	200
			14	0	200	17	1	201
						18	5	210
						19	1	201

n = total number of fertilized eggs (blastulae).

but for the fact that few spermatozoa were present. A cyclic production of spermatozoa seems fairly normal under laboratory conditions and these males could have been in a pre- or post-reproductive phase as well.

Another question concerns whether endoreduplication can also occur in hybrid males. No polyploid spermatocyte I could be detected in testis preparations, which suggests that it would be, at best, a rare event.

The results of these breeding experiments permit the following predictions (Fig. 5.1)

1. The interbreeding *X. gilli* and *X. l. laevis* populations may contain not only F_1 hybrids but also triploid backcross animals and even recombinant triploids of further backcross generations.

2. Depending on the incidence of 'LG_{15} type' females, diploid recombinant F_2 backcross animals may be present as well. Such animals could cause extensive gene introgression into both species, especially as the males are also fertile.

Indeed, this latter possibility can explain the presence of the many odd specimens in certain Cape populations (Picker *et al.* this volume pp. 61–71).

Examples of interbreeding in the zone of sympatric overlap of two species are known for several other Anura, e.g. *Bombina* in Europe. Generally, hybrid zones form rather steep clines of gene introgression and do not necessarily lead to a blending of the two species concerned. The situation may be dramatically different for *X. gilli* since this species lives only on the coast while *X. l. laevis* occupies all the inland territory. It has been supposed that the latter species has invaded the coastal region relatively recently. However, Pleistocene sea levels were substantially lower, perhaps by as much as 120 m, during the last glacial

Table 5.3. Fertility of diploid recombinant F_2 backcross males

Male[a]	Female[a]								Cytology
	Haploid eggs				Endoreduplicated diploid eggs				
	LL$_1$		LL$_2$		LG$_{112}$		LG$_{113}$		
	% develop.	n[b]	% develop.	n[b]	% develop.	n[b]	% develop.	n[b]	
Control of egg quality									
X. l. laevis	86	232	75	264	64	163	88	214	Many spermatozoa
F$_2$-backcross males									
(L)G$_1$	80	41	75	226	45	157	84	153	Many spermatozoa
(L)G$_2$	0	143	0	38	–	–	–	–	Few spermatozoa; mainly spermatocytes I
(L)G$_3$	–	–	0	176	–	–	–	–	Few spermatozoa; mainly spermatocytes
(L)G$_4$	93	312	–	–	61	74	75	145	Many spermatozoa
L(G)$_1$	0	90	0	172	–	–	–	–	Few spermatozoa
L(G)$_2$	0	158	0	43	–	–	–	–	–
L(G)$_3$	0	123	–	–	–	–	–	–	No spermatozoa; mainly spermatocytes I
L(G)$_4$	78	304	–	–	88	59	91	74	Many spermatozoa

[a](L)G: from cross female LG$_{15}$ × male X. gilli; L(G): female LG$_{15}$ × male X. l. laevis; LL: X. l. laevis from the Cape; LG$_{112}$ and LG$_{113}$: F$_1$ hybrids.
[b]n = number of blastulae = 100%.

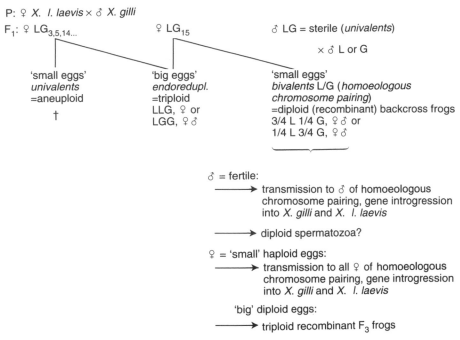

Fig. 5.1. Summary of fertility characteristics in female *Xenopus l. laevis* × male *X. gilli* hybrids and in F_2 backcrosses.

maximum (Hendey 1983), exposing vast areas of the continental shelf. Such a territory could have corresponded with the present habitat requirements or tolerances of *X. gilli* (including temporary ponds of high salinity and low pH) (Picker 1985). However, such habitats are now restricted to a narrow coastal zone, and interbreeding with the omnipresent *X. l. laevis* threatens to eliminate this loveliest of all *Xenopus* species.

Acknowledgements

I am grateful to the Department of Nature and Environmental Conservation of the Cape Province, South Africa, for permission to work with the protected *X. gilli*, to the George & Antoine Claraz Foundation for financial support, and to R. Rau and M.D. Picker for hospitality and help during field work.

References

Green, D.M. (1991). Supernumerary chromosomes in amphibians. In *Amphibian cytogenetics and evolution*: 333–357. (Eds Green, D.M. & Sessions, S.K.). Academic Press, San Diego.

Hendey, Q.B. (1983). Cenozoic geology and palaeogeography of the Fynbos region. In *Fynbos palaeoecology: a preliminary synthesis*: 35–60. (Eds Deacon, H.J., Hendey, Q.B. & Lambrechts, J.J.N.). Council for Scientific and Industrial Research, Pretoria.

Kobel, H.R. & Du Pasquier, L. (1975). Production of large clones of histocompatible, fully identical Clawed Toads (*Xenopus*). *Immunogenetics* 2: 87–91.

Kobel, H.R., Du Pasquier, L. & Tinsley, R.C. (1981). Natural hybridization and gene introgression between *Xenopus gilli* and *Xenopus laevis laevis* (Anura: Pipidae). *J. Zool., Lond.* 194: 317–322.

Kobel, H.R., Egens de Sasso, M. & Zlotowski, Ch. (1979). Developmental capacity of aneuploid *Xenopus* species hybrids. *Differentiation* 14: 51–58.

Müller, W.P. (1977). Diplotene chromosomes of *Xenopus* hybrid oocytes. *Chromosoma* 59: 273–282.

Picker, M.D. (1985). Hybridization and habitat selection in *Xenopus gilli* and *Xenopus laevis* in the south-western Cape Province. *Copeia* 1985: 574–580.

Picker, M.D. & de Villiers, A.L. (1989). The distribution and conservation status of *Xenopus gilli* (Anura: Pipidae). *Biol. Conserv.* 49: 169–183.

Rau, R. (1978). The development of *Xenopus gilli* Rose & Hewitt (Anura, Pipidae). *Ann. S. Afr. Mus.* 76: 247–263.

6 Feral populations of *Xenopus* outside Africa

R.C. TINSLEY and M.J. McCOID

Synopsis

Although now limited naturally to Africa south of the Sahara, the clawed frogs actually have the capacity for a wide geographical distribution. During the past 30 years or more, *Xenopus* has been introduced into a range of temperate regions, in North and South America, Europe and some isolated localities such as Ascension Island. All these introductions appear to have been of one species, *Xenopus laevis* from the Cape, South Africa. A critical review shows that many reports have not been substantiated recently. Our data are based on populations in the USA and the UK. Long-term records of colonies in the UK, where *X. laevis* are individually dye-marked for recapture studies, show relatively mobile and long-lived individuals, some being caught regularly for up to nine years.

In the UK, population growth is limited by temperature and periodic drying-out of habitats. Conditions favouring successful recruitment (warm, wet summers) occur irregularly. In California, environmental conditions appear to be optimal, maturation may take only eight months, and enormous populations have built up in a series of independent drainage systems between the US–Mexico border and northern Los Angeles. These populations have defied attempted eradication schemes and now appear to be uncontrollable.

The success of *Xenopus* in disturbed habitats, where prey resources for the adult frogs (mainly macroinvertebrates) are often absent, may be explained by the advantages of cannibalism. The high reproductive rate of *Xenopus* leads to large populations of phytoplankton-feeding tadpoles which exploit seasonal algal blooms. Field data confirm that adults may rely on their own offspring as a food source, enabling older individuals to survive periods of food shortage: the nutrient resource represented by algal populations is exploited by the adult frogs via their young. This strategy enables rapid population growth under favourable conditions for reproduction and dispersal (as in California); however, where reproduction is an irregular event and populations are localized, cannibalism may be partly responsible for poor annual recruitment (as in the UK).

Introduction

The widespread use of the African clawed frog, *Xenopus laevis*, in biological research, including physiology, biochemistry, genetics and developmental biology, and in human pregnancy diagnosis, has relied upon a series of attributes of this 'laboratory animal'. These advantages include ease of maintenance in the laboratory (the simplicity of husbandry for an aquatic animal, relative freedom from infection and disease, an undemanding diet), responsiveness to induction of reproduction and ability for repeated and relatively prolific breeding under aseasonal laboratory conditions. The ease of maintenance in captivity also made *Xenopus* attractive to the pet trade. Comparatively rapidly in the 1950s and 1960s, *X. laevis* became established in aquaria in many areas of the world. It is therefore not surprising that, as in the case of many other experimental animals and pets, loss of interest, end of an experiment, misguided ethics or curiosity occasionally resulted in the release of captives. The 'hardiness' which had made *Xenopus* ideal for laboratory maintenance then proved to be a considerable advantage for adaptation to new environments. Feral populations were first recorded in the 1960s and a long series of new reports continues to the present. Some of the initial records coincide with the end of the use of *Xenopus* in human pregnancy diagnosis. So far, there have been reports of feral *Xenopus* in Europe (including the United Kingdom), Chile, Ascension Island (South Atlantic) and the United States. In the USA, populations have been reported from 11 States, although the present status of many of these populations is unknown. All reports involve a single species, *Xenopus laevis*, almost certainly the southern *X. l. laevis* exported from the Cape. The common thread linking all these extralimital populations is the occupation of disturbed or artificial habitats. Known sites are generally man-made ponds, flooded excavations or canalized water courses, or, in the case of natural waterbodies, are subject to high environmental variability.

In this review, we summarize records from the literature, from correspondence and from our own fieldwork (in the UK, Virginia, Arizona and California by RCT, and in California by MJM). We outline the principal features of the ecology of these feral populations and evaluate the factors influencing the prospects of permanent establishment of introduced populations.

Extant populations

United Kingdom: Isle of Wight

Populations were established in about 1962 following deliberate introduction and animals were subsequently transferred to several nearby sites. The habitats are ephemeral ponds adjacent to the sea on cliffs of unstable clay which are subject to continuous erosion by rain and wave action, especially during winter storms. Subsidence causes some ponds to disappear but new ponds form continuously in recently collapsed areas. Pond areas range from 1 to 100 m^2, and water depth varies from 10 cm to 1 m. Summer droughts result in the ponds drying, but the *Xenopus* presumably withdraw into fissures in the clay substrate until the ponds refill. This locality has been monitored since 1980. Recent surveys of the ponds and nearby streams have not produced any specimens and it is possible that the population has declined or disappeared.

United Kingdom: South Wales

Populations in two adjacent watercourses were first noted in 1979 but were already well established at this time. The area is relatively well drained with underlying limestone geology. One population occurs in a river (3 to 4 m wide, 10 cm to 1 m deep) that may dry up during the summer but is fast-flowing in winter. The *Xenopus* apparently move freely between several man-made ponds and the river. The second population occupies the grounds of an abandoned castle and country house. Underground water-storage cisterns constructed beneath the castle provide a permanent habitat for the *Xenopus* which also occur in adjacent streams and in a series of ornamental weirs constructed to form alternating pools and waterfalls. This site, like that on the Isle of Wight, is close to the sea and experiences significant salt spray in storms. The two drainages are separated across a watershed of less than 1 km, so it is possible that both populations result from a single introduction. A total of over 350 adult *Xenopus* have been individually dye-marked, released and periodically recaptured in a 12-year study of these populations.

United Kingdom: south-east and south-west England

There are other reliable reports of *Xenopus* in the UK, including large populations of tadpoles in a pond near London in 1990, and direct observations of adults and tadpoles in ornamental ponds in Kent in 1987. Established populations have not yet been located and the animals may not have survived (R.C. Tinsley unpublished). River Authority personnel involved with monitoring fish populations have commented on occasional but consistent captures of adult *Xenopus* in waterways in the south-west of England. So far, these reports suggest survival of isolated individuals (released or escaped): there is no information on whether populations have been established by successful breeding.

Mainland Europe

There is fragmentary evidence that *X. laevis* has been released and has survived for some years in several countries, including Germany (area of Hamburg) and the Netherlands. There has been no recent confirmation of long-term survival although the experience of the studies in the UK is that *Xenopus* populations may remain largely undetected. In the Netherlands, about 100 *X. laevis* tadpoles were collected in a canalized ditch near Utrecht in 1979, and a large adult was caught in a funnelnet in a river near Gorichem in 1974 (M.S. Hoogmoed pers. comm.).

Chile

A collection of live post-metamorphic frogs was received by RCT in 1985 from a reliable source who reported that they originated from a 'wild population' in

Chile. The locality is unknown but Pefaur (1994) referred to the introduction of X. *laevis* into South America, including Chile, as a potential threat to native amphibians.

Ascension Island

Loveridge (1959) reported X. *laevis* collected in 1958 from a small man-made pond near the summit of Green Mountain. He mentioned that the first specimen from Ascension Island was collected in 1944 and *Xenopus* were probably imported from South Africa during World War II for diagnostic purposes. A naturalist visiting the Island in the early 1980s also observed *Xenopus* in a mountain-top pond, probably the same population.

United States: Arizona

Xenopus laevis was introduced into man-made ponds at the Arthur Pack Golf Course in Tucson, Arizona, in the 1960s. Studies in 1988, 1989, 1990 and 1991 (by RCT) confirmed that substantial populations occur here. The individual responsible for this introduction also released *Xenopus* at the same time into other man-made waterbodies in southern Arizona. Recent surveys by University of Arizona personnel have failed to document the persistence of *Xenopus* in Arizona except in Tucson.

United States: California (Los Angeles County)

In 1974, X. *laevis* was recorded in the Placerita Canyon area of northern Los Angeles County (Zacuto 1975). Concern was expressed by State of California officials that these animals might prejudice survival of an endangered fish (*Gasterosteus aculeatus williamsoni*). Studies and control procedures were initiated at that time. A survey of Placerita Canyon in 1990 (by RCT) documented enormous populations in man-made ponds and intervening streams. X. *laevis* in now known to occupy ponds and streams in Soledad, Agua Dulce and Placerita Canyons. It is possible that these populations resulted from deliberate introductions, as the areas of occupation have unlimited public access and are either State or County parks. An additional population mentioned by St. Amant, Hoover & Stewart (1973) in Palmdale, Los Angeles County, has not subsequently been surveyed.

United States: California (Orange County)

First records of X. *laevis* in Greater Los Angeles were documented by St. Amant & Hoover (1969) and St. Amant *et al.* (1973). Specimens were collected in 1968 in channels and ponds built to confine watercourses of the Santa Ana River. St. Amant *et al.* linked this introduction with escapes from commercial pet dealers in the early 1960s. Sampling in the same areas in 1986 (by RCT) did not yield any *Xenopus*, but there had been substantial building developments in the original sites and it is likely that animals may persist in suitable neighbouring habitats. A different introduction in Irvine (southern Orange County), adjacent to the University of California, was investigated by RCT in July 1989: large numbers of metamorphs were found dead in a recently drained pond but live adults were caught in adjacent permanent lakes.

United States: California (Riverside County)

A small series of *Xenopus* was collected in southern Riverside County (Arroyo Seco Creek, Santa Margarita River drainage) in 1974 (and subsequent data on population structure suggested that this was the year of initial establishment). Studies were conducted in the mid 1970s on two man-made ponds, approximately 0.8 km apart, built to confine waters of the Arroyo Seco Creek (McCoid & Fritts 1980a,b, 1989; McCoid 1985). Both ponds were in a private park which attracted visitors primarily from the Greater Los Angeles area. It appears likely that this population resulted from the release of pets. In the mid 1970s, one pond (108 m^2) contained a post-metamorphic population estimated at about 500 (McCoid & Fritts 1980a). The second pond was considerably larger and deeper, but apparently contained a smaller population. Surveys of one pond in 1989 (by RCT) and the other in 1992 (see McCoid, Pregill & Sullivan 1993) did not yield any *Xenopus*, but it is unlikely that they have been eliminated from both sites. Additional collections (by RCT) in 1986 in the City of Riverside—man-made ponds isolated from other known populations—yielded a small number of adults and large numbers (tens of thousands) of tadpoles.

United States: California (San Diego County)

The first reports of *Xenopus* in San Diego County were from the Sweetwater River drainage in 1971 (Mahrdt & Knefler 1972). However, it is possible that populations in this drainage were established in the mid 1960s by escapees from a commercial pet dealer (McCoid & Fritts 1980a). Populations are now known from virtually every permanent and semi-permanent body of water in western San Diego County (Tecolote, San Diego, Sweetwater, Otay and Tijuana Rivers). Studies were conducted on populations in the Sweetwater River drainage in the mid 1970s (McCoid & Fritts 1980a,b, 1989). Surveys of some known sites (Sweetwater drainage) were conducted by RCT in 1983 and 1989 and yielded large numbers of *Xenopus*. Surveys of these same sites and the Tecolote, Tijuana and San Diego Rivers in 1990 and 1992 (McCoid *et al.* 1993) did not reveal any specimens. Between 1987 and 1991, southern California experienced an extended drought which probably prevented successful recruitment (McCoid *et al.* 1993). However, despite the lack of recently collected specimens, it is unlikely that *Xenopus* has disappeared from many sites in San Diego County. The enormous populations of *X. laevis* observed in the Tijuana River on the US–Mexico border in the mid 1970s probably indicate that the Municipality of Tijuana, Baja California Norte, Mexico, also supports a population.

Populations now probably extinct

There are a number of unverified, anecdotal reports of X. *laevis* occurring in widely-scattered locations in the United States (Louisiana, Nevada, New Mexico and Wyoming). At present, there is no way to evaluate these reports and such populations should be considered tentatively as extinct. For some other populations, there is reliable information to confirm former occurrence and, in several cases, to document their extinction.

United States: California (Yolo County)

In 1973, free-living X. *laevis* were reported in a drainage system on the University of California Davis campus. In 1975, a number of adult frogs escaped from university holding facilities into this same waterway (Zacuto 1975). The California Department of Fish and Game embarked on a poisoning programme that was apparently successful (Zacuto 1975). Conversations with biologists at U.C.–Davis in 1992 indicated that X. *laevis* had not recently been sighted in the locality.

United States: Colorado

In 1990, a single X. *laevis* was collected in a beaver pond (elev. 3150 m) near the town of Keystone (Bacchus, Ricter & Moler 1993). Surveys at the same locality in 1991 did not reveal any additional specimens. It is possible that this individual was a released pet and no population has been established.

United States: Florida

A single specimen was found near Tampa, but no other information or date of collection is recorded: the status of the population is unknown.

United States: North Carolina

The exact locality of this population is not known nor the dates that it persisted. However, a population existed for a number of years in ponds of a fish hatchery. Eradication was accomplished by drainage of the ponds in autumn which allowed complete freezing during the winter.

United States: Virginia

This population (south of Washington, DC) was first recorded in 1982 (Zell 1986) and was sampled by RCT in 1987. The habitat was an artificial pond (c. 300 m²) in a nature reserve. Systematic trapping and removal by local conservation personnel of many hundreds of adult and juvenile *Xenopus*, coupled with severe weather conditions, probably resulted in elimination by 1988. However, the possibility that *Xenopus* might have migrated from this pond into neighbouring habitats was not followed up.

United States: Wisconsin

During the summer of 1972 (June to August), a large number of late-stage larvae (51–54; Nieuwkoop & Faber 1967) were collected in an artificial pond in Greenfield Park, Milwaukee, Wisconsin. No further specimens were obtained in subsequent years, which indicates that this population probably did not survive winter conditions.

Biology of introduced populations

Characteristics

Xenopus is highly distinctive ecologically, both in comparison with other anuran amphibians and with fishes which may potentially share a similar lifestyle. The aquatic *Xenopus* species can survive prolonged periods without water and can migrate overland, which enables them to exploit temporary habitats inaccessible to fish. Unlike most other amphibians, they feed underwater and take a wide range of live prey. They also scavenge for dead prey, detecting it by olfaction. Adults can exploit the productivity of plankton blooms in recently-flooded habitats by eating their filter-feeding tadpoles. *Xenopus* is also very resilient, able to survive prolonged periods without food, with a maximum longevity exceeding 20 years, a wide temperature tolerance, a comparatively short generation time (eight months under optimum conditions), and an extended breeding season leading to prolific reproduction (see Tinsley, Loumont & Kobel, this volume pp. 35–59). These characteristics, which contributed to the attractiveness of *Xenopus* for laboratory research and for the pet trade, have become important advantages for the survival of populations introduced into new environments.

Feeding studies

McCoid & Fritts (1980b) analysed stomach contents from a population in California and summarized other unpublished feeding reports. They concluded that even though vertebrate prey were present seasonally, slow-moving macroinvertebrates formed the bulk of the diet. Data from other populations in the US and the UK are similar. A major concern of State conservation agencies in the US has been prompted by anecdotal reports that *X. laevis* represents a threat to endangered fishes and native amphibians. These reports are unsubstantiated and not supported by the experimental data of Avila & Frye (1978). McCoid & Fritts (1980a,b) did record an introduced fish as a minor component of the stomach contents of Californian *X. laevis*, but considered that predation occurred only under exceptional circumstances.

Cannibalism is an attribute of *Xenopus* which is important for the success of feral (and native) populations. In California, cannibalism apparently allows colonization of newly created habitats in the absence of an established macroinvertebrate prey base (McCoid & Fritts 1980a, 1993). Exploitation of newly created habitats also enables *Xenopus* to multiply rapidly before predators and competitors become established. Cannibalism may also allow survival of long-established populations in the UK but, because most offspring may be consumed, could inhibit recruitment and ultimately affect long-term population stability.

The ability of *Xenopus* to tolerate considerable periods of total starvation is also an important factor aiding colonization of new habitats and survival through adverse seasonal conditions. Studies on *X. laevis* in Arizona have consistently revealed that, in July, all individuals in large population samples have empty stomachs, reflecting the depletion of prey resources in midsummer (R.C. Tinsley unpublished).

Reproduction

In California, reproduction is both asynchronous and opportunistic: cues for spawning are associated with water temperatures of at least 20 °C and these occur during much of the year. Successful reproduction (in terms of larval growth and metamorphosis) may be closely tied to the availability of phytoplankton and this is optimal between April and August. However, larvae have been found during a 10-month span of one year (McCoid & Fritts 1989). Reproduction occurs virtually every year under normal conditions in California, but during the late 1980s and early 1990s successful recruitment has probably been inhibited by insufficient rainfall (McCoid *et al.* 1993).

Reproduction in the UK populations may be determined primarily by a critical balance between temperature and water availability. The relatively high temperatures necessary for stimulating spawning are usually associated with dry weather, which leads to disappearance of the ponds where *Xenopus* occurs (in both Wales and the Isle of Wight), so that spawning and/or tadpole survival become impossible. Wet summers, when waterbodies remain adequate for spawning, are generally also cool, and temperatures may not reach the levels required to promote breeding. In the UK, the habitats of *Xenopus* only occasionally receive adequate rainfall during a warm summer, so recruitment is likely to occur only intermittently. Two further characteristics of UK habitats reduce the chances of successful reproduction. First, the summer season is relatively short so that metamorphosis may not be possible before the onset of winter. Second, suitable habitats are relatively confined and this may prevent emigration and dispersal of parents after spawning. If adults and offspring remain together during the relatively long developmental period of the tadpoles, all or a majority of the young may be eaten.

The reproductive potential of *X. laevis* is high. Studies in California show that large females (with a snout–vent length of >104 mm) may contain as many as 17 000 secondary follicles and females at first ovulation (65 mm SV length) may contain 1000 secondary follicles (McCoid & Fritts 1989). Field and laboratory studies indicated that large females may release only a few thousand eggs per reproductive session, but there may be multiple spawnings per individual during the extended reproductive season in California, leading to huge numbers (hundreds of thousands) of larvae of various stages in some ponds.

Fatbody cycles in California are related to size, sex, reproductive condition and time of year (McCoid & Fritts 1989). Male and female fat weights were roughly equivalent until the first reproduction. After this, female fatbody weight decreased to 29–33% of adjusted male weight. At the onset of reproduction in the spring, both sexes experienced parallel seasonal decreases in fatbody weights. Sexual differences in recovery of the fatbodies after reproduction is probably attributable to virtually continual mobilization of fat reserves for ovulation in females.

Temperature relations

Xenopus laevis from the Cape region of South Africa is adapted to a Mediterranean climate and the most successful extralimital populations occur in this ecological zone elsewhere, most notably California. Further information is needed on the *Xenopus* in the equivalent Mediterranean zone in Chile. Populations in the Cape region of South Africa experience winter water temperatures below 10 °C and summer temperatures around 25 °C. Populations in southern California experience a similar but somewhat warmer regime, and may not be limited to the same extent by low winter temperatures. Populations in the UK experience much less favourable conditions, with water temperatures routinely less than 10 °C for prolonged periods in winter and only occasionally over 20 °C in summer. The *Xenopus* in South Wales which occur in underground cisterns experience a buffered environment with year-round temperatures of 10–15 °C (adequate for survival but not warm enough for reproduction). The ability of adults to tolerate low temperature is demonstrated by observations on the Virginia population: *Xenopus* could be caught in midwinter after cutting through 10 cm of ice and lowering baited traps (G.A. Zell pers. comm.). In addition to confirming the survival ability of *X. laevis*, this indicates that animals are active and continue to respond to the presence of food at extreme low temperatures. The record from Colorado also confirms the occurrence of *Xenopus* in water at freezing point.

Maximum temperature tolerance is illustrated by data for the Arizona population. *X. laevis* was collected in summer from ponds that reach 30 °C: after capture, transfer to water only 1 or 2 °C warmer resulted in death within a few minutes. Observations in California show that *Xenopus* construct pits 30–40 cm deep in mud substrates in evaporating ponds during the summer. Temperatures at the surface of these pits were 30 °C or more, but animals remained submerged at the bottom of the pits where daily temperatures were near 20 °C.

When ponds and rivers dry completely during summer drought, *X. laevis* is able to aestivate. While this phenomenon has not been observed directly in

California, laboratory studies indicated an ability to survive in moist conditions for as long as three months (McCoid & Fritts 1980a). Both the principal UK sites are subject to drying each summer and the *Xenopus* must move into underground fissures until the arrival of autumn rains.

Age and growth

Estimates of growth, size at maturity and age of large specimens have been made for certain California populations (McCoid & Fritts 1989). Age and size at maturity were estimated to be six months after metamorphosis and approximately 65 mm SVL for both sexes. Subsequently, females grew more rapidly than males. Large males (up to 80 mm SVL) were estimated to be between two and four years of age and large females (up to 119 mm SVL) between 4.5 and 15 years of age. Mark–recapture investigations of the UK South Wales population have recorded the same individuals for up to nine years, which indicates that adults can be long-lived in nature. A mark–recapture study was conducted for a much shorter period of time in California (a five-month period during winter) and data indicated that mean growth of adults was 1.2 mm/month (at 8–15 °C). Macroinvertebrate prey resources did not appear to be limiting.

Dispersal and movements

Data based on marked *X. laevis* in the UK populations indicate that adults move freely between bodies of water, often undertaking overland migration. Some individuals were repeatedly recaptured in the same small pond (which suggests a 'home base') but were found to migrate in late spring to a much larger pond about 0.2 km away which served as a spawning site. Most of the route was overland, first within woodland and then over open grassland, but the *Xenopus* must also have crossed a relatively fast-flowing river (3–4 m wide) and a road to reach the spawning site. Significantly, one return journey to the 'home base' was found to be accomplished rapidly (between recaptures 48 h apart), suggesting more or less purposeful navigation (rather than a random wandering until a suitable habitat was encountered). These types of studies have not so far been undertaken in California but, in populations that included recently metamorphosed and adult *Xenopus*, juvenile frogs were observed to disperse out of ponds, using existing water courses or sheet-flooding during heavy rains (McCoid & Fritts 1980a).

Defences

In extralimital situations, where co-evolved predators do not occur, it can be assumed that predatory pressure would be less than that in the native range, and this may have contributed to the success of *X. laevis* in California. McCoid & Fritts (1993) reviewed anti-predator mechanisms which include synchronized aerial breathing by adults, escape responses involving rapid reverse movements, selection of breeding sites and production of slippery skin secretions which make capture difficult and induce a 'gaping response' in predatory snakes. Other skin secretions have anti-microbial properties and reduce the risk of infectious disease (see Kreil, this volume pp. 263–277).

Legal considerations

Since the establishment of *X. laevis* in the USA and the UK, conservation agencies in these countries have taken action to reduce the risk of further introductions. Much of the interest in the USA has been based on the unsubstantiated reports that *X. laevis* may have a negative impact, through predation, on native aquatic vertebrates. In the USA, unless a detrimental species is deemed to have broad or national interest, decisions regarding introduced species are left to local State agencies. State conservation agencies, in the southern tier of States from California east to North Carolina, were surveyed and only California, Arizona, Louisiana and Florida banned the possession or release of *Xenopus* spp. Other states (New Mexico, Texas, Mississippi, Alabama, Georgia, South Carolina and North Carolina) have no prohibitions against the importation or possession of *Xenopus*. In the UK, release of *Xenopus* has been prohibited by the Wildlife and Countryside Act, 1981. In the known UK localities, *X. laevis* shares habitats (including breeding sites) with five of the six species of native amphibians (*Rana temporaria, Bufo bufo, Triturus vulgaris, T. helveticus* and the endangered *T. cristatus*), but the ecological interactions are largely undocumented.

Conclusions

The ecological information relating to the series of feral populations is limited but it indicates that introduced populations of *X. laevis* have become established in a relatively diverse range of environmental regimes. Conditions in California are probably near optimal for the success of *X. laevis* and may surpass those experienced in native situations. Populations in California can reach enormous densities and have been established for over 30 years. At many localities, populations have continued to expand despite intensive eradication efforts which have involved habitat destruction and treatment with a range of poisons. Populations in other regions have been less successful and provide some indication of the environmental limits of introduced *X. laevis*. In Arizona, summer temperatures are close to the upper thermal limits of the species, but the established populations are probably also confined by lack of alternative habitats. In Wisconsin, North Carolina and Virginia, the animals experience prolonged periods under ice during winter (although low temperature alone does not prejudice survival). Populations in the UK emphasize the hardiness and persistence of feral populations. Despite only periodic temperature and rainfall regimes that might allow successful reproduction, populations have survived in the UK, in low densities, for over 25 years. Thus, in extralimital environments, X. *laevis* appears to be constrained primarily by habitat availability and temperature conditions, although biological factors (e.g. cannibalism) may play a role. As mentioned above, populations in the UK may be self-regulating and may suffer low recruitment because of cannibalism. On the other hand, under conditions of high productivity, efficient recruitment and freedom to disperse (in California), cannibalism is a major advantage in the exploitation of disturbed habitats.

The recent considerable expansion of the geographical range of *Xenopus*, which has been a product of man's activities, has actually led to the re-introduction of *Xenopus* into some areas of former occurrence. Species closely resembling present-day representatives occurred in parts of South America in the Cretaceous (Baez, this volume pp. 329–347): it is intriguing that the genus became extinct here, given its success in ecologically similar regions of Africa. With its adaptability for colonization, *Xenopus* shares many of the features of other 'commensals' of man which have become established around the world. Its range could therefore become much greater, particularly with the most recent exploitation of this laboratory animal—or at least its oocytes—for protein synthesis. The capacity of *Xenopus* for future colonization could potentially include many other world regions where it has been employed in laboratory research: possibilities include the Mediterranean areas of Europe and the Middle East (irrigated areas in Israel, for instance), Japan and Australia, where it would not be surprising if intensive laboratory use of *Xenopus* had occasionally resulted in accidental escapes. The evidence of this review would suggest that *Xenopus* might be less effective in invading established aquatic communities where predators and competitors are already integrated (lakes and swamplands, for instance); however, introduced *Xenopus* may be very successful in disturbed habitats, especially those created by human activities (artificial waterbodies accompanying mining excavations, irrigation schemes, etc). Indeed, these are exactly the circumstances in which *Xenopus* thrives in Africa (see Tinsley, Loumont & Kobel, this volume pp. 35–59).

All feral populations so far identified have comprised X. *laevis*: this can be related to the fact that most laboratory animals and pets have originated from Cape suppliers. In addition, X. *tropicalis* and X. *epitropicalis* are regularly exported to Europe and North America with tropical fish from West Africa. These species would be unlikely to survive and breed in temperate areas: they require relatively constant high temperatures (25 °C and above), are intolerant of cool seasons, and show rather poor laboratory survival even under controlled conditions. On the other hand, they might represent effective colonizers in other tropical regions. *Xenopus borealis*, exported from Kenya, should have significant potential for establishment where conditions correspond to those of the East African highland habitats, but there are, so far, no known records. *Xenopus muelleri* would be more heat-tolerant, and X. *clivii* and certain montane species would be more cold-tolerant, than X.

laevis, and thus potentially capable of successful establishment in desert environments and in cool temperate regions respectively. However, these species are rarely exported from Africa by commercial suppliers.

Acknowledgements

Many people contributed information for this review: in particular, we thank Paul Moler, Hobart Smith, Charles Painter, Keith Guyse, Robert Hansen, Randy Wilson and Dave Cook. Rebecca Hensley reviewed a draft of the manuscript. Many of the ideas formulated by the junior author were in collaboration with Thomas Fritts. RCT is particularly grateful for fieldwork help from Mark Simmonds, Joe Jackson, Ree Simovich, Greg Zell, John Wright and John Measey, and for research grant support from the Natural Environment Research Council (including GR3/6661).

References

Avila, V.L. & Frye, P.G. (1978). Feeding behavior of the African clawed frog (*Xenopus laevis* Daudin): (Amphibia, Anura, Pipidae): effect of prey type. *J. Herpet.* **12**: 391–396.

Bacchus, S., Ricter, K. & Moler, P. (1993). Geographic distribution. *Xenopus laevis*. *Herpet. Rev.* **24**: 65.

Loveridge, A. (1959). Notes on the present herpetofauna of Ascension Island. *Copeia* **1959**: 69–70.

Mahrdt, C.R. & Knefler, F.T. (1972). Pet or pest? *Environ. Southwest* **446**: 2–5.

McCoid, M.J. (1985). An observation of reproductive behavior in a wild population of African clawed frogs, *Xenopus laevis*, in California. *Calif. Fish Game* **71**: 245–246.

McCoid, M.J. & Fritts, T.H. (1980a). Observations of feral populations of *Xenopus laevis* (Pipidae) in southern California. *Bull. Sth Calif. Acad. Sci.* **79**: 82–86.

McCoid, M.J. & Fritts, T.H. (1980b). Notes on the diet of a feral population of *Xenopus laevis* (Pipidae) in California. *SWest. Nat.* **25**: 272–275.

McCoid, M.J. & Fritts, T.H. (1989). Growth and fatbody cycles in feral populations of the African clawed frog, *Xenopus laevis* (Pipidae), in California with comments on reproduction. *SWest. Nat.* **34**: 499–505.

McCoid, M.J. & Fritts, T.H. (1993). Speculations on colonizing success of the African clawed frog, *Xenopus laevis* (Pipidae), in California. *S. Afr. J. Zool.* **28**: 59–61.

McCoid, M.J., Pregill, G.K. & Sullivan, R.M. (1993). Possible decline of *Xenopus* populations in southern California. *Herpet. Rev.* **24**: 29–30.

Nieuwkoop, P.D. & Faber, J. (1967). *Normal Table of* Xenopus laevis *(Daudin)*. North-Holland Pub. Co., Amsterdam.

Pefaur, J. (1994). III CLAH: a synopsis. *Froglog* (Newsletter of IUCN, SSC and Declining Amphibian Populations Taskforce) No. 9: 4.

St. Amant, J.A. & Hoover, F.G. (1969). Addition of *Misgurnus anguillicaudatus* (Cantor) to the California fauna. *Calif. Fish Game* **55**: 330–331.

St. Amant, J.A., Hoover, F.G. & Stewart, G.R. (1973). African clawed frog, *Xenopus laevis laevis* (Daudin), established in California. *Calif. Fish Game* **59**: 151–153.

Zacuto, B.J. (1975). *The status of the African clawed frog* Xenopus laevis *in Agua Dulce and Soledad Canyons*. Unpublished report, California Department of Fish and Game.

Zell, G.A. (1986). The clawed frog: an exotic from South Africa invades Virginia. *Va Wildl.* February **1986**: 28–29.

Behaviour, sensory perception and development

7 Sensory perception and the lateral line system in the clawed frog, *Xenopus*

ANDREAS ELEPFANDT

Synopsis

Xenopus lives in stagnant and often turbid waters and is nocturnal. Correspondingly, the organization of its sensory systems differs from that in terrestrial frogs. *Xenopus* retains its lateral line system after metamorphosis. By means of this system, *Xenopus* can analyse single waves as well as waves impinging simultaneously on it.

In the adult *Xenopus*, refraction of the eyes is for vision in air. The dorsal and slightly frontal orientation of the eyes allows the frog, when floating at the water surface, to see the whole hemisphere above water binocularly. A fovea is absent. Several features indicate specialization for nocturnal vision. Behavioural tests show good discrimination of form and strong, irreversible preference of some forms to others. It is suggested that a major function of vision is analysis of the environment above water.

In the absence of vision in water during darkness, touch and chemoreception might become more important for orientation. For touch, no specialization has so far been found in *Xenopus*. For smell, the adult *Xenopus* possesses, in addition to the principal cavity for smelling in air and the vomeronasal cavity of terrestrial frogs, an additional rostral cavity for smelling in water. A dermal fold between the principal and the rostral cavity closes either cavity according to whether the animal is submerged or not. *Xenopus* has taste buds on the palate and the floor of the mouth. They are smaller than in other frogs, but not less sensitive. *Xenopus* is aroused by the odour of meat in water, an effect that can also be elicited by some amino acids.

Attempts to condition frogs generally fail. *Xenopus*, however, shows good learning abilities for wave and sound stimuli, and to some degree for visual and chemical stimuli. It is the only frog so far in which long-term memory and complex learning have been shown. This demonstrates that *Xenopus* can associate sensory inputs with variable meanings and learn complex relations in its environment.

Introduction

Sensory organs provide an animal with information necessary to survive in its environment. Consequently, one might expect different sensory equipment in animals of contrasting ecology. Considerable divergence may occur between related animals living in water or on land, such as the fully aquatic *Xenopus* and terrestrial frogs, because in these two media the signals of equal modality differ in their properties and in the information they convey. The sensory differences can concern three aspects: the structure and occurrence of organs, sensory abilities, and the biological relevance of the sensory input.

In his pioneering study, Kramer (1933) demonstrated that in *Xenopus* there has been a major change from terrestrial frogs in the relative importance of information from vision and from the lateral line. In *Xenopus*, the lateral line system, which serves for detection and analysis of water movements around the body, is retained throughout life and has become a central sensory system for orientation. Vision, on the other hand, has been reduced. It is of little use at night and in turbid water, which is the predominant environment of *Xenopus*. Consequences of this interchange are seen even in gross structures of the brain. The tectum opticum, which typically is a major structure in frogs and bulges out dorsally and at the sides, is small in *Xenopus* and hardly exceeds the dimensions of the brain. On the other hand, the main processing station for the lateral line system in the midbrain, the torus semicircularis, is enlarged in *Xenopus* and expands into the third ventricle. In the absence of vision, other sensory modalities also become more important, such as chemoreception, tactile orientation and possibly hearing. In terrestrial frogs non-visual modalities are often subordinate to vision, in that their major function is to bring objects into visual focus, so that decisions about further procedure are based on visual inspection. This does not hold for *Xenopus*. In the laboratory, for instance, blind individuals can be maintained without disadvantage in the same aquarium with sighted individuals even though the water is clear and the food is given during the day, that is, when vision could be used. Even for sighted animals, food lying on the bottom of the tank is not found by vision. Rather, the animals are aroused by the odour and then swim around in search of the food until they touch it: it is a common experience to see a *Xenopus* pass very close to food and miss it unless it makes contact with the object. In feeding arousal, objects of acceptable size are taken into the mouth without visual inspection, and only there is the decision about palatability made.

Today, knowledge of sensory systems and stimulus processing in *Xenopus* has increased considerably. *Xenopus* has proved to be useful for studies of neuronal circuitry and development because it can be maintained and reared easily in the laboratory, and it is hardy and survives many operative manipulations. Most studies, because they are prompted by interest in general mechanisms of neuronal organization, for which *Xenopus* is merely used as a model, neglect the specific biological function of the sensory system in this animal whose ecology differs so dramatically from that of common frogs. In contrast, the present article focuses on the sensory biology of *Xenopus* in its environment. Omitting general amphibian features that are documented elsewhere in the literature, specific characters of organ structures and sensory abilities in *Xenopus* are described. Particular attention is given to wave analysis with the lateral line system because of the system's importance for *Xenopus* and its absence in terrestrial frogs. Next, vision and chemoreception are described, and finally studies of learning, that is, of the ability of *Xenopus* to evaluate sensations on the basis of previous experience. Hearing and touch will not be considered. Hearing is dealt with in a separate chapter (Elepfandt, this volume pp. 177–193), and no specializations for touch have been found so far in *Xenopus* (Fox 1994).

The lateral line system

In correspondence with its fully aquatic life, *Xenopus* retains its lateral line system as an adult. The lateral line system is found in fish and tadpoles, but in only some adult amphibians. Among anurans, retention of the lateral line system after metamorphosis was known in all pipids and a few discoglossids (Escher 1925), but has recently been found in various other families (Fritzsch 1989; Elepfandt 1989a; Hellmann, Topel & Elepfandt 1994). All these species are characterized by an aquatic lifestyle. Thus, retention of the lateral line system appears to be a function of ecology rather than of systematics. Since the occurrence of an anuran-type ear in pipids suggests that these frogs had terrestrial ancestors (Elepfandt, this volume pp. 177–193), the lateral line system in adults may be a neotenic character.

The lateral line organs
Arrangement and structure of the organs
An adult *Xenopus* has approximately 180 lateral line organs distributed over its head and trunk. They are visible on the skin as small elongated structures that may measure up to 2×0.5 mm in fully grown animals. The number, location and orientation of the organs are somewhat variable both between individuals and between species (Vigny 1977). The majority of the organs are arranged in three lines on both sides of the animal's trunk (Fig. 7.1a). In the medial line the organs are oriented perpendicular to the line, whereas in the dorsal and ventral lines they are oriented in parallel. A small group of organs is located behind the head. The remaining quarter to third of the organs are located dorsally and ventrally on the head in various small lines, most numerous in a circle around each eye (Shelton 1970).

As in all amphibians, the lateral line organs of *Xenopus* are superficial neuromast organs that stand free on the skin so that water impinges directly on them. From each organ, four to 12 cupulae extend for fractions of a millimetre into the water (Fig. 7.1b). Their orientation is perpendicular to the organ's axis. Water passing over them deflects the cupulae. At the base of each cupula are inserted the sensory hairs of 40–80 sensory hair cells which are stimulated by the deflection (Fig. 7.1c, d). The morphology and

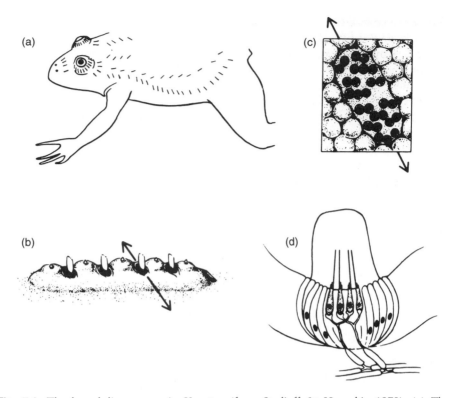

Fig. 7.1. The lateral line organs in *Xenopus* (from Strelioff & Honrubia 1978). (a) The arrangement of the organs on the body of *Xenopus*. (b) A single lateral line organ with its flag-shaped cupulae. Arrows: direction of maximal sensitivity of the sensory hair cells to water movements. (c) Horizontal section through the skin under a cupula, showing the sensory hair cells. Arrows as in Fig. 1b. (d) Drawing of a cupula and its underlying base, showing the insertion of the sensory hairs into the cupula and the connections of the hair cells to the afferent fibres.

physiology of the hair cells are the same as in the inner ear (Hudspeth 1983). Hair deflection toward the kinocilium depolarizes the cell, opposite deflection hyperpolarizes it, and deflection in other directions elicits responses according to the cosine with the axis of maximal sensitivity. In *Xenopus*, the hair cells in each lateral line organ are oriented so that maximal stimulation is produced by water movement along the side of the flag-shaped cupula. Since the force of water running alongside the cupula is friction, which is proportional to velocity, the lateral line organs in *Xenopus* are sensors for water velocity. Each organ's hair cells are arranged in two groups of opposite polarity so that deflection of the cupula in one direction will excite the cells of one group and hyperpolarize the others, and vice versa. Each group synapses to one of the two afferent nerve fibres of the organ. As a result, the two afferent neurons from a lateral line organ show opposite directional sensitivity with regard to local water flow (Görner 1963; Russell 1976).

Response characteristics of the organs

The response of the lateral line organs to wave stimuli has been studied in *X. laevis* by electrophysiological recording from the afferent neurons. The afferent neurons are spontaneously active. The response to a wave consists of a modulation of the discharge frequency so that spikes are synchronized toward the peaks of the wave. The frequency and intensity of the modulation encode the frequency and amplitude of the wave. All afferent neurons possess equal tuning curves: the response range is 0.1–40 Hz, with maximal sensitivity at 25 Hz (Kroese, van der Zalm & van den Bercken 1978). At this frequency, wave amplitudes of 0.2 µm can be detected (Görner 1963; Elepfandt & Wiedemer 1987). Wave amplitudes are encoded over at least 80 dB, that is, a range of 1:10 000 (Elepfandt & Wiedemer 1987).

The directional response characteristic of the afferent lateral line neuron is sinusoidal with regard to local water flow at the organ (Görner 1963). However, in the intact animal and with regard to waves originating at some distance—that is, under natural conditions—the directionality is determined strongly by the animal's wave shadow. When the frog is floating at the water surface, with its head slightly protruding, surface waves will easily impinge on ipsilateral organs but will have to go around the animal in order to reach contralateral organs. The result is a cardioid directional sensitivity (Elepfandt & Wiedemer 1987).

Most if not all organs are innervated by inhibitory efferent neurons as well. Whereas each lateral line organ has its own two afferent neurons, there are only six to eight efferent neurons on each side, each of which projects to several lateral line organs. They are only activated when *Xenopus* is moving (Russell 1971). Whether they block or only reduce the sensitivity of the lateral line organ during the movement is still controversial (Roberts & Meredith 1989). During night-time field studies, I observed a *Xenopus* chasing a rapidly fleeing small fish in midwater: since this activity requires functioning of the lateral line organs in order to detect lateral escape movements of the fish, the sensitivity of the organs was obviously not blocked during this action.

Wave analysis

Nature of stimuli

There are three major sources of water movements around the frog: surface waves produced by objects moving on the water, waves coming from objects moving within the water, and water flow that results from the animal's own movement or is met in fast-flowing streams or rivers. Each of these produces characteristic flow patterns around the animal.

Analysis of water-surface waves has been studied most. *Xenopus* analyses these waves to catch prey. An insect that flounders on the water and tries to escape produces surface waves that expand around it. By turning and swimming toward the origin of the impinging wave *Xenopus* will find it. Wave localization is achieved by means of the lateral line system: Kramer (1933) showed that blinding *Xenopus* does not impair its wave localization whereas its responses to waves cease after its lateral line organs are also destroyed. The turn response of *Xenopus* is elicited easily under laboratory conditions, which make surface waves a favourite stimulus for analysis of the lateral line system. If not mentioned otherwise all behavioural tests described below were carried out with surface waves.

Although they appear simple, surface waves possess complicated physical characteristics. Both the decrease of amplitude with distance and the velocity of wave propagation depend on wave frequency. Waves of higher frequency decay faster than waves of lower frequency and, above 13 Hz, they also propagate faster. Thus, when the water surface is hit at a particular point by the jerky movement of an insect or by a water droplet falling on it, the resulting frequency mixture at that point (equivalent to noise in sound) disperses in the course of expansion. As a consequence, when the wave has travelled some distance, it arrives as a frequency-modulated wave sweeping from higher to lower frequencies, and the rapidity of the sweep decreases with the distance from the origin. Below 13 Hz, on the other hand, wave propagation accelerates with decrease in frequency, so that the propagation is lowest for 13 Hz. The general speed of surface waves is slow: it is 28 cm/s at 6.5 Hz, 23 cm/s at 13 Hz and 25 cm/s at 26 Hz. The respective wave lengths are 4.3 cm, 1.7 cm and 0.95 cm. The wave length of the lowest frequency of these three approximates to the size of small froglets. Waves of lower frequencies therefore may be difficult for small frogs to detect, because they will shake the animal rather than move relative to it. Fortunately, low frequencies are produced mainly by abiotic wave sources such as wind, whilst biogenic wave sources usually produce frequencies above 10 Hz (Lang 1980). Water-surface waves also extend into lower levels, but diminish rapidly with depth. At the depth of one wavelength, motion amplitude has fallen off to 0.02%, which implies a frequency dependence of the amplitude decrement on the metric scale. Further information on water-surface waves is found in Bleckmann (1985).

As a consequence of these wave properties, a natural frequency-modulated wave produces a complicated pattern of stimulation on the frog. Not only does the wave reach the proximal organs earlier than the more distant ones, the proximal organs are also stimulated by a stronger wave with faster frequency sweep and with larger *relative* amplitudes of its higher-frequency components. In addition, the high-frequency components are detected only by the organs close to the water surface, while deeper organs are stimulated only by the lower-frequency components, a character reminiscent of the hydromechanic frequency dispersal in the cochlea (Oed & Elepfandt 1993). Finally, the animal's wave shadow induces reflections and deformations of the wave.

Considerably less is known about the form and analysis of water movements produced by motions *within* the water. *Xenopus* locates balls vibrating at low frequency in the water (Kramer 1933), and the effective stimulus of ball vibration in water can be calculated (Kalmijn 1988). However, such continuous vibration may not be a usual stimulus of biological origin. More relevant and probably running further in the water appear to be turbulences and eddies produced by movements of animals. They may result, for example, from the beat of the hind legs of a swimming *Xenopus*, from a fish's tail beat, or a crayfish's escape tailflip (Bleckmann *et al.* 1991). The properties of such turbulences of biological origin are little known as yet. In our laboratory, water disturbances have been analysed that are produced by small fish approximately 5 cm long: these fish produce vortices of about 5 cm diameter that run at 3–4 cm/s over distances of up to 20 cm (T. Breithaupt unpubl.). From these observations one may guess that *Xenopus* can detect each other at a distance of approximately 20 cm. The distance for the detection of a source depends, of course, on the size of the animal, whether it is a jerking worm or a large mammal drinking at the water surface.

It is not known whether *Xenopus* analyses the flow resulting from its own motion or from moving water. Such flow is considerably greater than that produced by other animals at a distance. The inhibition of the lateral line organs during motion in *Xenopus* (see Russell 1971) may compensate for the resulting strong deflection of the cupulae. Water flow during the animal's motion also differs qualitatively from the waves discussed above. In wave stimulation, the frog remains stationary in relation to the bulk of the water around it, and it is the differential movement of water at different regions of the body that leads to stimulation of lateral line organs. When the frog moves, however, a common flow direction is produced along the body. The self-induced flow is different when the animal moves in open water from when it passes a nearby object. Such differences are detected by, for instance, some fish (Dijkgraaf 1963), but nothing is known of *Xenopus* in this respect.

Single wave analysis

The most obvious ability for wave analysis in *Xenopus* is wave localization, shown by the frog's turn toward the origin of surface waves that pass along its body, as mentioned above. Wave localization has been examined in *X. laevis* and *X. muelleri* (Elepfandt 1982; Görner, Moller & Weber 1984). Waves from all directions can be localized at an accuracy of ± 5°. No difference was found between the two species.

The ability of *X. laevis* to assess the distance of a frontal wave source has been studied by Buschmann & Görner (1990). Though *Xenopus* hardly ever approaches a wave source in a single move, the length of approach movements increases slightly with the distance of the source. The cue for this change in behaviour seems to be the curvature of the wave.

In addition to surface wave localization, *Xenopus* can localize midwater vibrations (Kramer 1933) and also has some vertical wave localization (Elepfandt 1984).

Wave frequency discrimination has been shown in *X. laevis* by go/no-go conditioning. Discrimination was found within the range 5–30 Hz (Elepfandt, Seiler & Aicher 1985). At the optimum around 14 Hz, the relative limen was 4%. This means that 14 Hz is distinguished from 14.5 Hz, even though each stimulus

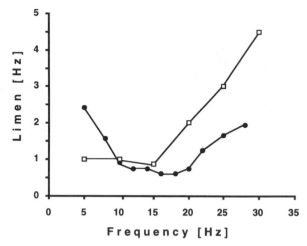

Fig. 7.2. Discrimination limen for water surface waves in *Xenopus laevis*. Open squares indicate the limen for presentation of single waves. Filled circles indicate the limen for superimposed waves. For frequencies above 10 Hz, discrimination is better in wave superimposition.

lasted only 5 s and interstimulus intervals were at least 30 s. Discrimination deteriorated rapidly toward higher frequency: at one octave above 14 Hz the limen exceeded 10% (Fig. 7.2). The recognition of the wave is of absolute pitch quality: after three weeks without stimulation the recognition accuracy for a previously rewarded frequency was as good as after one day (Elepfandt 1986a).

Dual wave analysis
In a natural environment, a wave may be disturbed by ripples produced by wind blowing on the water or by waves coming from other sources. Therefore, *X. laevis* was examined by presenting the animal with two waves simultaneously so that they overlapped at the frog. Under these conditions, *Xenopus* still localized the wave sources by orienting towards them (Fig. 7.3) (Elepfandt 1986b, 1989a). Localization even occurred when the two waves were of equal frequency. For orienting turns the interstimulus angle had to be at least 60°. At smaller angles intermediate orientations appeared: it remains to be elucidated, however, whether the intermediate orientations are due to sensory limits or a behavioural strategy.

When wave frequency discrimination was tested with two waves presented simultaneously from 45° left and right, discrimination of frequencies above 10 Hz was even better than it had been in the tests with single wave presentation (Fig. 7.2). Discrimination limen was below 1 Hz at most frequencies (Elepfandt & Lebrecht 1994).

Neural organization of wave analysis
While identification and localization of single waves can conceivably be carried out by comparing the discharge volleys from the lateral line organs, this does

Fig. 7.3. Turn response of *Xenopus* to two waves presented simultaneously. The two waves overlap at the position of the frog. Nevertheless, the frog responds by turning towards one of the wave sources (Elepfandt 1989a).

not suffice for analysis of superimposed waves. When two waves are presented simultaneously, the afferent discharge pattern from the organ reflects the sequence and amplitude of peaks resulting from the local superimposition of the two waves at its position. In frequency discrimination at the discrimination limen, for instance, this superimposition results in a beat of the two frequencies at 1 Hz or less. Wave analysis under such conditions thus requires computation of the component waves by comparing the local superimpositions at several lateral line organs, taking into consideration the relative location of the respective organs and the distance between them. Such analysis has been described so far only in electroreception in weakly electric fish, where the ability to analyse beats is shown in the animal's jamming avoidance response. Electroreception has evolved from the mechanoreceptive lateral line system (Bullock & Heiligenberg 1986). Thus, the ability for beat analysis in electroreception might be inherited from the mechanoreceptive lateral line system.

Why does *Xenopus* have so many lateral line organs? Small groups of lateral line organs are entirely sufficient to localize single waves from any direction with undiminished accuracy (Elepfandt 1982, 1984). They are also sufficient for localization when two waves are presented simultaneously. It seems that the lateral line organs form an array in which small groups of organs are capable of full analysis of the water movements at their respective place on the animal's body. The high number of organs may then provide for simultaneous analysis of water movements at various locations on the body.

Although, in the peripheral nervous system, small groups of lateral line organs suffice for localization of waves from *any* direction, the midbrain is organized *topologically* with regard to wave directions: each side of the midbrain is essential for localization of contralateral waves, and small areas within each side are essential for localization of waves from specific directions (Elepfandt 1988a,b; Zittlau, Claas & Münz 1986). In the forebrain, a small area in the caudal septum has been found to be important for the responsiveness to contralateral waves (Traub & Elepfandt 1990). The details of the effects resemble 'sensory neglect', a phenomenon known in mammals and associated with attentional processes up to—in humans—conscious perception.

The visual system

In turbid water and at night, which includes the main sphere of activity of *Xenopus*, vision seems to be of little use. As mentioned in the Introduction, the eye is hardly used by adult *Xenopus* even in clear water during the day. A function often seen in the laboratory is protective arousal: when a large object moves rapidly over the aquarium, frogs that have been hanging at the water surface dive to the bottom to hide. This reaction is useful against predatory birds that feed on *Xenopus*. Interestingly, it is restricted to objects above the animal. Objects passing by the side of the aquarium rarely elicit escape reactions. The anatomical data given below indicate that the eye of adult *Xenopus* is constructed for seeing in air and specialized for scotopic vision. Thus, the functional area of vision in adult *Xenopus* seems to be above water rather than within it.

The eyes

In contrast to conditions in air, there is no refraction of the light at the cornea in water. In tadpoles, therefore, all refraction has to be achieved by the lens alone. Isolated lenses from *Xenopus* tadpoles produce well-resolved images of grids under water (Chung, Stirling & Gaze 1975). During metamorphosis, the lens flattens in terrestrial frogs in order to adapt the refraction of the eye for conditions in air. In *Xenopus*, the lens remains roughly spherical (Chung *et al.* 1975). Nevertheless, the refraction also changes at metamorphosis. At late larval stage the eye becomes hyperopic in water (Chung *et al.* 1975), that is, the focus lies behind the retina. In the fully grown adult, the image of objects in air lies exactly at the level of the retina, whereas for stimuli in water the eye is strongly hyperopic. That is, the refraction of the eye of adult *Xenopus* is matched to vision in air. Behavioural tests, on the other hand, show that *Xenopus* can identify objects in air when it is submerged (see below). This would suggest some mechanism for optical compensation, which has not so far been found.

A characteristic feature of the eyes in adult *Xenopus* is their more or less upright orientation. The interocular angle is only 50°, the elevation of the eyes 50–60° (Schneider 1957; Grant & Keating 1986). For each eye, the visual field is nearly a hemisphere in water: in air it is somewhat greater, owing to the refraction at the cornea (Fig. 7.4) (Schneider 1957). As a consequence of the upright orientation of the eyes practically the whole area above the frog is seen binocularly, whereas a large area ventral to the frog is not seen. In the frog's normal floating and slightly perpendicular posture at the water surface, the binocular visual field of the frog coincides roughly with the area above the water. This match, together with the eye's refraction for vision in air, suggests that binocular control of the area on and above the water surface is a major reason for the eyes' vertical orientation.

The dorsal orientation of the eyes develops only at and after metamorphosis. At larval stage 58, the eyes are still oriented completely to the sides. In the juvenile, the interocular angle has been reduced to approximately 70°, and the last 20° of convergence develop during adulthood (Grobstein & Comer 1977; Grant & Keating 1986). Of interest with respect to neuronal processing is that, from stage 66 which is late metamorphosis, the reorientation of the retina as a whole is achieved by selective addition of retinal fibres at the ventral lateral rim, so that the individual retinal neurons retain a constant orientation of their receptive field in space (Grant & Keating 1986).

The general neuronal circuitry of the retina in *Xenopus* conforms to the usual pattern in frogs. The eyes show, however, specializations for scotopic vision. The usual protective mechanisms against bright light are absent: the iris does not contract in response to light (Weale 1956), and the epithelial pigment of the retina does not move in response to illumination but remains in a fixed position distal to the receptors (Saxen 1954). The number of ganglion cells and visual afferent fibres is only 59 000 in the adult *Xenopus* (Wilson 1971), which is just 15% of the corresponding number in ranid frogs (Maturana 1959). In mammals, a low number of ganglion cells and a spherical lens are characteristic of nocturnal species, because they enhance light collection for the individual cell (Walls 1942; Hughes 1977). The corresponding features in *Xenopus* might have the same function. The density distribution of ganglion cells in the retina of *Xenopus* is very even: only in the dorsal third of the retina—corresponding to horizontal vision—can a faint indication of a visual streak be found (Graydon & Giorgi 1984) but the cell density is increased just twofold. This is about the weakest increase found in a vertebrate retina, which suggests that there is no direction of best vision in *Xenopus*. Rather, the whole hemifield above the water is inspected with panoramic vision, with possibly a minimal improvement for horizontal directions.

Like other frogs, *Xenopus* possesses two rods and two cones. Electrophysiological evidence strongly indicates that *Xenopus* has typical amphibian dichromatic colour vision (Donner 1965). Two minor differences may affect spectral vision. First, the rod/cone ratio in *Xenopus* is only 1.06 as compared to 1.24 in *Rana temporaria* (Saxen 1954). Second, the red rods absorb maximally at 523 nm, in contrast to 500 nm in terrestrial frogs. In amphibian tadpoles, two kinds of visual pigments are found in this rod, namely rhodopsin, absorbing

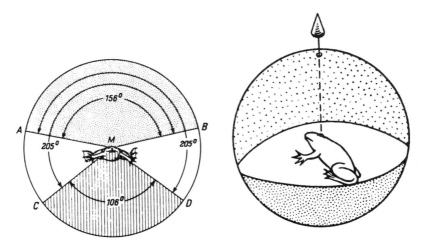

Fig. 7.4. The visual field of *Xenopus* under water (Schneider 1957). Left: vertical visual field. The figure shows the binocular (dotted) and monocular (white) visual fields and the area not seen (hatched). In air, the visual field of each eye is increased by 15–20° in all directions so that the binocular field becomes a full hemisphere. Right: spherical projection of the visual field. This figure, drawn originally for a sitting specimen, has been turned to show that the eyes look straight up in *Xenopus*' normal floating position. The binocular field is indicated with widely spaced dots, the area not seen with close dots.

maximally at about 500 nm, and porphyrhodopsin, with maximal absorption at 523 nm (Crescitelli 1973). In adults only rhodopsin is found in terrestrial frogs, whereas in *Xenopus* only porphyrhodopsin is retained (Crescitelli 1973). Correspondingly, Donner (1965) found, by recording from the optic nerve, that scotopic spectral sensitivity was maximal near 520 μm, with a secondary peak around 460–470 μm which he attributed to the green rod.

The visual afferent fibres project to the same nuclei as in ranid frogs, but with slight differences in their relative distribution to the nuclei and in the interconnections (Levine 1980). The difference is not surprising in view of the different orientation of the eyes and the comparatively large extent of the binocular visual field, but its functional implications remain to be elucidated.

Studies of vision

There are few behavioural tests of vision in *Xenopus*. Usually these tests examine vision in air with the animal submerged.

Schimmler (1952) carried out a detailed study of form vision in *X. laevis* using figures presented above water level. Since her thesis has not been published, it is referred to here in some detail.

Schimmler tested adult *X. laevis* individually in tanks of 30×20×15 cm filled with water 8 cm deep. The tests were conducted at room temperature; light intensity was 35–100 lx. The paradigm was a two-choice discrimination between two figures cut out of cardboard and suspended horizontally 3 cm above the water. The size of the figures was 1.5–6 cm², with the centres of the figures 5 cm apart from each other. They were presented simultaneously slightly in front of the animal on its left and right sides. The reactions assessed were: approach to one figure, snapping towards it, or catching behaviour directed at it. They were initially induced by food suspended just beneath the figures, but the tests were carried out without reinforcement. Until the approach, *Xenopus* was totally submerged. The objects were held motionless, but *Xenopus* often swam around beneath the objects so that a motion of the object on the retina might have resulted.

In a first series, shape vision was tested. When black figures of equal size were presented, *Xenopus* spontaneously preferred the figure with less profile. That is, the sequence of preferences was circle over square, square over triangle, triangle over tripod (⚚). The preference could not be reversed by training. The animals also distinguished between a rectangle (axis ratio 1:2) and an ellipse with the same lengths of the axes. When the form of the ellipse was made more similar to the rectangle, discrimination became uncertain or vanished. Presentation of white figures on a black background rather than black on white yielded the same sequence of form preferences; the preference also remained unchanged when only contours of the figures were presented. Tests such as presentation of a black rectangle with an inserted white circle versus a black circle with an inserted white rectangle demonstrated that the peripheral profile is the determining part of the figure. If the same figure was presented black on white and white on black, *Xenopus* strongly preferred the black figure. The preference for the dark figure could counteract the effect of contour: a black filled triangle was preferred to a white filled circle. The brightness of the background did not affect the preferences.

Next, the effect of stimulus orientation was tested. A bar with left–right orientation was strongly preferred to a bar with longitudinal orientation. When bars were presented simultaneously with crosses, bars with left–right orientation were preferred to the cross, those with longitudinal orientation were not. When two triangles were presented so that one pointed with its tip caudally and the other frontally, the triangle pointing caudally was clearly preferred. In tests with vertical presentation of the figures (rather than parallel to water surface), *Xenopus* preferred the horizontally extending bar to the upright one and the triangle standing on its base to that standing on its tip. From these and related experiments, Schimmler concluded that the left–right preference is most pronounced for contours located frontally close to the snout and that it may be related to the ease with which the object may be ingested.

Finally, Schimmler tested for the effect of size. *Xenopus* preferred larger figures. If the larger figure was at least twice the size of the smaller one, this preference overcame the shape preference. If, however, the size difference was less, choice was made according to shape. Also, a small black figure was preferred to a large white one of the same shape up to a size ratio of 1:6.

Schimmler's data, though lacking some critical controls, demonstrate form vision in *Xenopus* under low photopic conditions.

In the same laboratory and with essentially the same procedure as Schimmler, Ronde (1950) tested colour vision and brightness vision in *X. laevis* using Ostwald's colour and grey series. The animals could easily identify red, blue, yellow and green, with best discrimination between shades of yellow and red. Discrimination was constant if the illumination was at least 1 lx; at lower intensities it became uncertain. Controls to determine whether the colours could always be distinguished from shades of grey were, however, insufficient. In a further test, Ronde found correct distinctions between a black and a white surface down to an illumination of 0.05–0.1 lx.

Burgers (1952) used the optomotor reaction of *X. laevis* to rotations of a vertically striped cylinder to study its visual abilities. He found that the animal detects even minimal differences in brightness. Illumination was provided by a 150 W lamp hanging 30 cm above the beaker.

Chemoreception

In the absence of visual cues, olfaction becomes the third sense that *Xenopus* can use over distance, the other two being wave detection with lateral lines, and hearing. Intraspecific communication by means of odorants is well known in the mating of urodeles. In anurans, similar evidence has been reported only in *Pipa pipa*: water previously occupied by receptive females excites the males (Rabb & Rabb 1963). A second potential use of olfaction is in prey detection. This function is easily observed in *Xenopus*. If meat is introduced without disturbance into an aquarium, as the smell disperses, *Xenopus* becomes aroused and swims around with feeding motions of its arms (Kramer 1933). Apparently there is little searching along an intensity gradient: the frog can closely approach the source of the smell, but unless it makes contact it may move away. *Xenopus* may also be capable of assessing the quality of the water it is living in. In the laboratory, replacement of old water in an aquarium by new water seems to arouse the frogs. In South Africa, *Xenopus* accumulates in fish ponds into which fowl manure has been added, and it has been suggested that the frogs find the ponds by olfaction (Du Plessis 1966). Thus there are several functions in *Xenopus* in which olfaction is, or may be, involved. If potential food cannot be seen, the importance of tasting is increased. This, too, can easily be seen in the laboratory. *Xenopus* in feeding mood will put nearly anything of appropriate size into its mouth, and only there is the decision made between further ingestion or spitting out.

The organs of chemoreception

In terrestrial frogs, metamorphosis involves a shift in olfaction from sensing chemicals in water to detecting airborne odours. *Xenopus* remains aquatic and the structure of the olfactory organ and its changes during metamorphosis have therefore been the object of considerable interest (Föske 1934; Paterson 1939; Altner 1962; Weiß 1986). The larval olfactory organ conforms to the normal anuran pattern. It consists of a principal cavity that is connected to a more ventrally located smaller cavity, the vomeronasal organ or Jacobson's organ. The sensory epithelia of these two cavities are separated from each other by undifferentiated epithelium. Both cavities show normal changes and enlargement during metamorphosis. As part of the process, the sensory epithelium in the principal cavity undergoes modifications, which include the development of Bowman glands for smelling substances in air. Only the eminentia olfactoria, which is a small prominence in the principal cavity of terrestrial frogs and is densely occupied by sensory epithelium, does not develop in *Xenopus*. In adult *Xenopus*, the principal cavity is always filled with air, even when the frog is submerged (Altner 1962). The vomeronasal organ, on the other hand, is filled with fluid throughout life. In addition to these cavities, a third cavity not found in terrestrial frogs develops in *Xenopus* rostrolateral to the principal cavity. Beginning at premetamorphosis, this rostral cavity grows and differentiates into various lobi and develops its own sensory epithelium (Föske 1934). The cavity is filled with water and its sensory epithelium has a structure that is characteristic for detection of odours in water (Weiß 1986). This pattern is found in all pipid frogs (Paterson 1951). Pipids are thus the only vertebrates that possess three olfactory sensory epithelia separated by undifferentiated epithelium.

Between the principal cavity and the rostral cavity, dermal folds develop at the external nares during metamorphosis. They are used to close either of the two cavities. When *Xenopus* is breathing through its choanes at the water surface, the rostral cavity is closed. When *Xenopus* dives it closes the principal cavity, retaining the air volume.

The afferent projection has been studied by Weiß (1986). As in all amphibians, the nerve fibres from the principal cavity project to the main olfactory bulb, whereas the nerve from the vomeronasal organ runs to the accessory olfactory bulb. When the sensory epithelium of the rostral cavity develops, its nerve fibres begin to occupy the ventral olfactory bulb, whereas the projection from the principal cavity becomes restricted to the dorsal olfactory bulb.

Taste buds in frogs are generally concentrated on the tongue and on the walls of the mouth. *Xenopus* lacks a tongue, but taste buds are found on the palate and floor of the mouth, with some concentration on the front and sides (Kramer 1933). The buds are smaller than in other frogs (M. Witt pers. comm.) but have normal electronmicroscopical appearance (Toyoshima & Shimamura 1982). Electrophysiological investigation (Yoshii *et al.* 1982) showed that the morphological reduction does not entail a decrease in sensitivity.

Smelling and taste

Despite a considerable number of studies of the molecular mechanisms of frog olfaction, knowledge about the sensory abilities is still very scanty: 'Amphibia are among the best studied species as far as olfaction is concerned ... but nobody really knows what frogs smell in their ... life' (Schild 1991). The crucial problem seems to be to find not only substances that affect the olfactory system, but odorous stimuli that have some biological meaning in the animals' natural environment.

In *Xenopus*, virtually nothing is known about detection of odours from the air. When *Xenopus* is floating at the water surface, with its nostrils out of the water, the rostral water-filled cavity is closed by the dermal folds and air can be inhaled into the lungs and on the way it also passes into the principal cavity. Altner (1962) failed to elicit responses of *X. laevis* in such a situation by presenting the animal with the odour of *Tubifex* juice. When presented under water this stimulus reliably elicits a feeding reaction from *Xenopus*. *Xenopus* is reported to be able to find its way to new ponds over distances of several kilometres. The only available cue seems to be olfaction. This is supported by Du Plessis (1966) who suggested that one function of smelling airborne odours may be in locating pond habitats.

In order to achieve olfaction in water, *Xenopus* has developed a special mechanism for ventilation of its rostral cavity, which is a blind sac. The principal cavity is closed under water by the dermal fold. From the rostral cavity, water is ejected as a small upward jet for approximately 2 cm (Kramer 1933; Altner 1962). The subsequent inflow occurs from water that is level with the nares or lower, so that a real change of water in the cavity takes place. Exchange of water is repeated at irregular intervals, generally every few seconds. It is limited to the nose: there is no support by buccal movements, and no water passes from the mouth to the nose or vice versa (Altner 1962). When *Xenopus* is stimulated by a smell this may lead to a change in the rhythm of pulsation by the nostrils and, when the excitement increases, to greater activity. This activity includes swimming around on the bottom, together with feeding movements of the arms.

Altner (1962) tested the sensitivity of *X. laevis* to solutions of β-phenyl-ethylalcohol, γ-phenylpropylalcohol, citral and β-ionon by conditioning. The thresholds were 5×10^{-7} M to 5×10^{-8} M, suggesting that *Xenopus*, like trout or humans, is a species that is not very sensitive to smell. Terpineol was also sensed by *Xenopus* but the response was more one of fright and no threshold could be determined (Altner 1962).

Kramer (1933) found that extracts of *Tubifex* or mealworms are very effective in eliciting the feeding reaction of *Xenopus*. Hemmer & Köhler (1975) showed that histidin alone can elicit the feeding response. Kruzhalov (1983b) tested the efficacy of 20 amino acids by adding a 0.4 ml portion of a 10^{-2} M solution of individual acids to a volume of 2 l of water in which a *X. laevis* was located. He found that nine amino acids are effective in eliciting significantly increased feeding activity. Except for alanin, all are essential amino acids in humans. On the other hand, tryptophan, which is essential in humans, was not effective. The most effective acids are histidin, methionin, isoleucin and phenylalanin. The basic amino acids were most effective, hydrophobic amino acids had a high or medium effect and hydrophilic amino acids had little or no effect. The effectiveness of L- and D-isomers could differ (Kruzhalov & Zhukova 1988). GABA is also detected by *X. laevis* (Kruzhalov 1984). Even the most effective substances, however, were less effective than meat juice, which is a combination of these stimuli.

The responses of *X. laevis* tadpoles to some chemicals were tested by Kiseleva (1989). An infusion of nettle tea strongly increased the tadpoles' general activity, whereas betain, glutamine and amylacetate did not affect activity. When the latter substances were given only to one side of an aquarium tank the tadpoles tended to stay on the side where betain had been given but to avoid the side with glutamine. Amylacetate had no effect. The effect of glutamine differs from that on tadpoles of *Rana temporaria* and *R. esculenta*, in which glutamine had elicited a pronounced feeding response (Kiseleva & Manteifel 1982; Kiseleva 1986). The reaction to glutamine changes at metamorphosis: young froglets of *Xenopus* show a positive reaction to glutamine, as do adult frogs.

Since these substances are dissolved in water, their detection is possible by olfaction or gustation, or both. By testing animals that had been made anosmic it could be shown that histidin and methionin, at least, are detected through olfactory reception (Kruzhalov 1982). Further evidence might be gained by electrophysiological analysis but studies of *Xenopus* are lacking. However, in *R. temporaria*, essential amino acids are more effective in eliciting neural responses than non-essential amino acids (Kruzhalov 1983a), which corresponds to the behavioural data in *Xenopus*. The correlation was 0.65, with $P < 0.01$ (Kruzhalov & Zhukova 1988).

Electrophysiological recordings from the gustatory nerve of *X. laevis* showed high sensitivity to amino acids and bitter substances, thresholds being 200–20 000 times lower than in the bullfrog, which uses sight in feeding (Yoshii *et al.* 1982). The basic and the neutral amino acids were effective whereas the acidic acids were not. Comparison of the effectiveness of hydrophobic amino acids in the gustatory nerve and in behavioural tests gave a correlation of 0.77, $P < 0.01$ (Kruzhalov & Zhukova 1988), which suggests that the amino acids might be detected by gustation. Further studies are necessary to distinguish between the two chemoreceptory modalities.

Interspecific differences can be considerable even within the *Xenopus* group. Yamashita, Kohriyama & Kiyohara (1989) found 10 to 100 times lower thresholds to L-tyrosine, L-tryptophane and L-phenylalanine in *X. borealis* than in *X. laevis*. Compared with *X. laevis*, gustatory receptors in *X. borealis* were more sensitive to adenosine and strychnine but less to quinine.

Learning

Learning, that is, association of sensations with functional meanings, is an important ability for orientation in the environment. Whereas learning has been studied to a considerable degree in fish, reptiles, birds and mammals, attempts to show learning in frogs under controlled conditions have almost completely failed: 'the available data are insufficient to decide whether either classical or instrumental associations (or both) are to be found in amphibians' (Macphail 1982).

Xenopus has become an exception to these limitations. Reports have been given of several forms of visual conditioning: Ronde (1950) conditioned *Xenopus* to colour discrimination, and Schimmler (1952) tested form discrimination by means of conditioning. Altner (1962) conditioned *Xenopus* to odours in water. Karplus, Algom & Samuel (1981) conditioned tadpoles of *Xenopus* to dark avoidance and succeeded in demonstrating retention of the learned task beyond metamorphosis. This was the first demonstration of long-term memory in a frog.

Learning has been studied most thoroughly in conditioning *X. laevis* to water waves (Elepfandt 1985). Operant learning was demonstrated by presenting the frog with monofrequent waves. When each turn response toward the wave source was rewarded by food, the initially low response rate increased up to 95% or more within a few days. When turn responses were not rewarded any more, the turn rate decreased slowly within two weeks to 20% or less. Wave frequency discrimination was tested by both go/no-go and two-choice procedures. Within 4–8 training days, the initially equal rate of response to

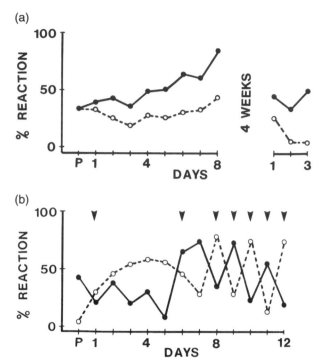

Fig. 7.5. Learning of wave frequency discrimination in *X. laevis* (Elepfandt 1985). The continuous line indicates the response rate to 20 Hz, the dotted line responses to 12 Hz. (a) Discrimination learning and long-term memory. Starting from equal response rates at pretraining day P, response rates to the two waves separate during training. After four weeks of rest without stimulation, discrimination reappears immediately when tests are resumed. (b) Complex learning in serial discrimination reversal. After initial discrimination training up to day P, the reinforcement for responses to the two waves is reversed. The frog learns the new discrimination within 3–4 days. After subsequent reversals (arrowheads) the new discriminations are learned faster and faster.

the two frequencies differentiated, resulting in a significantly higher rate of response to the rewarded frequency than to the other (Fig. 7.5). Subsequently, long-term memory was examined by retesting the animal after a rest of some weeks: the previously learned discrimination was performed correctly from the first day (Fig. 7.5a), indicating long-term memory of wave frequency. *Xenopus* could be trained to respond either to the higher or to the lower frequency; there was no innate preference.

Complex learning was tested by serial discrimination reversal for mono-frequent waves. When the animal has learned the first discrimination, the reinforcement scheme is reversed. When the new discrimination has been learned, the rewarding scheme is reversed again, and so forth. In this reversal paradigm, *Xenopus* improved rapidly, so that after three or four reversals, it would perform the new discrimination correctly at the first day of presentation

(Fig. 7.5b) (Elepfandt 1985). Thus, *Xenopus* had learned the general scheme for frequency discrimination, that is, a learning set. Learning set development was also found when the frequencies were changed whenever the frog had learned the discrimination of the given pair (Elepfandt 1989b).

In another test, the reward for the turn response to the water wave was made contingent on stimuli from another modality: the frogs had to learn to distinguish a wave that was presented simultaneously with a tone from a wave presented without tone or, in another task, to discriminate between two waves presented with different tones. In both cases, conditioning was successful, and the paradigms have been used to determine the hearing threshold and the auditory frequency discrimination in *X. laevis* (see Elepfandt this volume, pp. 177–193).

These data show that *Xenopus* is able to learn various tasks. They also demonstrate long-term memory and several forms of complex learning. Thus, *Xenopus* is capable of learning rules in its environment. The form of acquisition of the learning tasks by *Xenopus* conforms to that known from other vertebrates.

Acknowledgement

The author's studies on the lateral line system and on learning have been supported by grants from DFG (El 75). Thanks are due to R. Tinsley for revising the English text of the manuscript.

References

Altner, H. (1962). Untersuchungen über Leistungen und Bau der Nase des südafrikanischen Krallenfrosches *Xenopus laevis* (Daudin, 1803). *Z. vergl. Physiol.* **45**: 272–306.

Bleckmann, H. (1985). Perception of water surface waves: how surface waves are used for prey identification, prey localization, and intraspecific communication. *Prog. sens. Physiol.* **5**: 147–166.

Bleckmann, H., Breithaupt, T., Blickhan, R. & Tautz, J. (1991). The time course and frequency content of hydrodynamic events caused by moving fish, frogs, and crustaceans. *J. comp. Physiol. (A)* **168**: 749–757.

Bullock, T.H. & Heiligenberg, W. (1986). *Electroreception*. Wiley, New York.

Burgers, A.C.J. (1952). Optomotor reactions of *Xenopus laevis. Physiol. comp. Oecol.* *('s Grav)* **2**: 272–281.

Buschmann, P. & Görner, P. (1990). Distance localization of the center of a surface wave in the clawed toad *Xenopus laevis* Daudin. In *Brain—perception—cognition. Proceedings of the 18th Göttingen neurobiology conference*: 165. (Eds Elsner, N. & Roth, G.). Georg Thieme Verlag, Stuttgart.

Chung, S.H., Stirling, R.V. & Gaze, R.M. (1975). The structural and functional development of the retina in larval *Xenopus. J. Embryol. exp. Morphol.* **33**: 915–940.

Crescitelli, F. (1973). The visual pigment system of *Xenopus laevis*: tadpoles and adults. *Vision Res.* **13**: 855–865.

Dijkgraaf, S. (1963). The functioning and significance of the lateral-line organs. *Biol. Rev.* **38**: 51–105.

Donner, K.O. (1965). The scotopic spectral sensitivity of the clawed toad (*Xenopus laevis*). *Commentat. biol. Soc. Sci fenn.* **28**(4): 1–8.

Du Plessis, S.S. (1966). Stimulation of spawning in *Xenopus laevis* by fowl manure. *Nature, Lond.* **211**: 1092.

Elepfandt, A. (1982). Accuracy of taxis response to water waves in the clawed toad (*Xenopus laevis* Daudin) with intact or with lesioned lateral line system. *J. comp. Physiol.* **148**: 535–545.

Elepfandt, A. (1984). The role of ventral lateral line organs in water wave localization in the clawed toad (*Xenopus laevis* Daudin). *J. comp. Physiol. (A)* **154**: 773–780.

Elepfandt, A. (1985). Naturalistic conditioning reveals good learning in a frog (*Xenopus laevis*). *Naturwissenschaften* **72**: 492–493.

Elepfandt, A. (1986a). Wave frequency recognition and absolute pitch for water waves in the clawed frog, *Xenopus laevis. J. comp. Physiol. (A)* **158**: 235–238.

Elepfandt, A. (1986b). Detection of individual waves in an interference pattern by the clawed frog, *Xenopus laevis* Daudin. *Neurosci. Lett. Suppl.* **26**: S380.

Elepfandt, A. (1988a). Central organization of water wave localization in the clawed frog, *Xenopus*. I. Involvement and bilateral organization of the midbrain. *Brain Behav. Evol.* **31**: 349–357.

Elepfandt, A. (1988b). Central organization of water wave localization in the clawed frog, *Xenopus*. II. Midbrain topology for wave directions. *Brain Behav. Evol.* **31**: 358–368.

Elepfandt, A. (1989a). Wave analysis by amphibians. In *The mechanosensory lateral line: neurobiology and evolution*: 527–541. (Eds Coombs, S., Görner, P. & Münz, H.). Springer-Verlag, New York.

Elepfandt, A. (1989b). Transfer between two complex learning tasks in the aquatic clawed frog, *Xenopus laevis. Behav. Brain Res.* **33**: 322.

Elepfandt, A. & Lebrecht, S. (1994). Wave frequency discrimination in wave super-position by the clawed frog, *Xenopus laevis*. In *Sensory transduction. Proceedings of the 22nd Göttingen neurobiology conference*: 403. (Eds Elsner, N. & Breer, H.). Georg Thieme Verlag, Stuttgart.

Elepfandt, A., Seiler, B. & Aicher, B. (1985). Water wave frequency discrimination in the clawed frog, *Xenopus laevis. J. comp. Physiol. (A)* **157**: 255–261.

Elepfandt, A. & Wiedemer, L. (1987). Lateral-line responses to water surface waves in the clawed frog, *Xenopus laevis. J. comp. Physiol. (A)* **160**: 667–682.

Escher, K. (1925). Das Verhalten der Seitenorgane der Wirbeltiere und ihrer Nerven beim Übergang zum Landleben. *Acta zool., Stockh.* **6**: 307–414.

Föske, H. (1934). Das Geruchsorgan von *Xenopus laevis. Z. Anat. EntwGesch.* **103**: 519–550.

Fox, H. (1994). The structure of the integument. In *Amphibian biology* **1**: 1–32. (Eds Heatmole, H. & Barthalamus, G.T.). Surrey Beatty and Sons, Chipping Norton, NSW.

Fritzsch, B. (1989). Diversity and regression in the amphibian lateral line and electrosensory system. In *The mechanosensory lateral line: neurobiology and evolution*: 99–114. (Eds Coombs, S., Görner, P. & Münz, H.). Springer-Verlag, New York.

Görner, P. (1963). Untersuchungen zur Morphologie und Elektrophysiologie des Seitenlinienorgans vom Krallenfrosch (*Xenopus laevis* Daudin). *Z. vergl. Physiol.* 47: 316–338.

Görner, P., Moller, P. & Weber, W. (1984). Lateral-line input and stimulus localization in the African clawed toad *Xenopus* sp. *J. exp. Biol.* 108: 315–328.

Grant, S. & Keating, M.J. (1986). Ocular migration and the metamorphic and postmetamorphic maturation of the retinotectal system in *Xenopus laevis*: an autoradiographic and morphometric study. *J. Embryol. exp. Morph.* 92: 43–69.

Graydon, M.L. & Giorgi, P.P. (1984). Topography of the retinal ganglion cell layer of *Xenopus*. *J. Anat.* 139: 145–157.

Grobstein, P. & Comer, C. (1977). Post-metamorphic eye migration in *Rana* and *Xenopus*. *Nature, Lond.* 269: 54–56.

Hellmann, B., Topel, U. & Elepfandt, A. (1994). Adult *Rana ehrenbergii* retain a mechanosensory lateral-line system. In *Sensory transduction. Proceedings of the 22nd Göttingen neurobiology conference*: 404. (Eds Elsner, N. & Breer, H.). Georg Thieme Verlag, Stattgart.

Hemmer, H. & Köhler, S. (1975). Rhythmische Änderungen der Reaktion auf verschiedene olfaktorische Reize beim Krallenfrosch (*Xenopus laevis*). *Experientia* 31: 449–450.

Hudspeth, A.J. (1983). Mechanoelectrical transduction by hair cells in the acoustico-lateralis sensory system. *A. Rev. Neurosci.* 6: 187–215.

Hughes, A. (1977). The topography of vision in mammals of contrasting life style: comparative optics and retinal organization. In *Handbook of sensory physiology VII. 5. The visual system in vertebrates*: 613–756. (Ed. Crescitelli, F.). Springer-Verlag, New York.

Kalmijn, A.J. (1988). Hydrodynamics and acoustic field detection. In *Sensory biology of aquatic animals*: 83–130. (Eds Atema, J., Fay, R.R., Popper, A.N. & Tavolga, W.N.). Springer-Verlag, Berlin.

Karplus, I., Algom, D. & Samuel, D. (1981). Acquisition and retention of dark avoidance by the toad, *Xenopus laevis* (Daudin). *Anim. Learn. Behav.* 9: 45–49.

Kiseleva, E.I. (1986). [Behaviour of tadpoles of *Rana esculenta* at exposure to chemical stimuli.] *Zool. Zh.* 65: 1199–1206. [In Russian with English abstract.]

Kiseleva, E.I. (1989). [Behavioural reactions of *Xenopus laevis* tadpoles and juveniles to chemical stimuli and the effect of previous chemosensory experience.] *Zool. Zh.* 68: 85–92. [In Russian with English abstract.]

Kiseleva, E.I. & Manteifel, Yu.B. (1982). [The behavioural reactions of common toad and common frog tadpoles to chemical stimuli.] *Zool. Zh.* 61: 1669–1681. [In Russian with English abstract.]

Kramer, G. (1933). Untersuchungen über die Sinnesleistungen und das Orientie-rungsverhalten von *Xenopus laevis* Daud. *Zool. Jb. (Physiol.)* 52: 629–676.

Kroese, A.B.A., van der Zalm, J.M. & van den Bercken, J. (1978). Frequency response of the lateral line organ of *Xenopus laevis*. *Pflügers Arch. Eur. J. Physiol.* 375: 167–175.

Kruzhalov, N.B. (1982). [Chemoreception of amino acids in frogs.] In *Chemical signals of animals*: 214–218. Nauka, Moskva. [In Russian.]

Kruzhalov, N.B. (1983a). [Amino acids as olfactory stimuli in the frog *Rana temporaria*.] *Zh. evol. Biokhim. Fiziol.* 19: 275–281. [In Russian; translation in *J. evol. Biochem. Physiol.* 19: 205–211.]

Kruzhalov, N.B. (1983b). [Essential and non-essential amino acids as chemical stimuli in the clawed frog *Xenopus laevis*.] *Zh. evol. Biokhim. Fiziol.* 19: 503–506. [In Russian; translation in *J. evol. Biochem. Physiol.* 19: 369–371.]

Kruzhalov, N.B. (1984). [Behavioural reactions of the clawed frog *Xenopus laevis* to amino acids.] *Zool. Zh.* 63: 1828–1834. [In Russian with English abstract.]

Kruzhalov, N.B. & Zhukova, V.M. (1988). [Behavioural reactions of the clawed frog *Xenopus laevis* to L- and D- amino acid isomers.] *Zh. evol. Biokhim. Fiziol.* 24: 451–454. [In Russian; translation in *J. evol. Biochem. Physiol.* 24: 350–353.]

Lang, H.H. (1980). Surface wave discrimination between prey and nonprey by the backswimmer *Notonecta glauca* L. (Hemiptera, Heteroptera). *Behav. Ecol. Sociobiol.* 6: 233–246.

Levine, R.L. (1980). An autoradiographic study of the retinal projection in *Xenopus laevis* with comparison to *Rana*. *J. comp. Neurol.* 189: 1–29.

Macphail, E.M. (1982). *Brain and intelligence in vertebrates*. Clarendon Press, Oxford.

Maturana, H.R. (1959). Number of fibres in the optic nerve and the number of ganglion cells in the retina of anurans. *Nature, Lond.* 183: 1406–1407.

Oed, K. & Elepfandt, A. (1993). Place principle for frequency analysis in the lateral line system: evolutionary basis of sound frequency analysis? In *Gene—brain—behaviour. Proceedings of the 21st Göttingen neurobiology conference*: 83. (Eds Elsner, M. & Heisenberg, M.). Georg Thieme Verlag, Stuttgart.

Paterson, N.F. (1939). The olfactory organ and tentacles of *Xenopus laevis*. *S. Afr. J. Sci.* 36: 390–404.

Paterson, N.F. (1951). The nasal cavities of the toad *Hemipipa carvalhoi* Mir.-Rib. and other Pipidae. *Proc. zool. Soc. Lond.* 121: 381–415.

Rabb, G.B. & Rabb, M.S. (1963). Additional observations on breeding behavior of the Surinam toad, *Pipa pipa*. *Copeia* 1963: 636–642.

Roberts, B. & Meredith, G.E. (1989). The efferent system. In *The mechanosensory lateral line: neurobiology and evolution*: 445–459. (Eds Coombs, S., Görner, P. & Münz, H.). Springer-Verlag, New York.

Ronde, G. (1950). *Über den Farbensinn des afrikanischen Krallenfrosches* (Xenopus laevis *Daudin*). Doctoral thesis: Universität München.

Russell, I.J. (1971). The role of the lateral-line efferent system in *Xenopus laevis*. *J. exp. Biol.* 54: 621–642.

Russell, I.J. (1976). Amphibian lateral line receptors. In *Frog neurobiology: a handbook*: 513–550. (Eds Llinás, R. & Precht, W.). Springer-Verlag, Berlin, Heidelberg, New York.

Saxen, L. (1954). The development of the visual cells. Embryological and physiological investigations on Amphibia. *Ann. Acad. Sci. fenn.* (4) 23: 1–93.

Schild, D. (1991). Olfactory information processing. *Encycl. Human Biol.* 5: 523–531.

Schimmler, F. (1952). *Untersuchungen über das Formensehen von* Xenopus laevis *Daudin*. Doctoral thesis: Universität München.

Schneider, D. (1957). Die Gesichtsfelder von *Bombina variegata*, *Discoglossus pictus* und *Xenopus laevis*. *Z. vergl. Physiol.* 39: 524–530.

Shelton, P.M.J. (1970). The lateral line system at metamorphosis in *Xenopus laevis* (Daudin). *J. Embryol. exp. Morphol.* 24: 511–524.

Strelioff, D. & Honrubia, V. (1978). Neural transduction in *Xenopus laevis* lateral line system. *J. Neurophysiol.* 41: 432–444.

Toyoshima, K. & Shimamura, A. (1982). Comparative study of ultrastructures of the lateral-line organs and the palatal taste organs in the African clawed toad, *Xenopus laevis*. *Anat. Rec.* **204**: 371–381.

Traub, B. & Elepfandt, A. (1990). Sensory neglect in a frog: evidence for early evolution of attentional processes in vertebrates. *Brain Res.* **530**: 104–106.

Vigny, C. (1977). *Étude comparée de 12 espèces et sous-espèces du genre* Xenopus. Doctoral thesis: Université de Genève.

Walls, G.L. (1942). *The vertebrate eye and its adaptive radiation*. Cranbrook Institute of Science, Bloomfield Hills.

Weale, R.A. (1956). Observations on the direct effect of light on the irides of *Rana temporaria* and *Xenopus laevis*. *J. Physiol.* **132**: 257–266.

Weiß, G. (1986). *Die Struktur des Geruchsorgans und des Telencephalons beim südafrikanischen Krallenfrosch* Xenopus laevis *(Daudin) und ihre Veränderungen während der Metamorphose*. Doctoral thesis: Universität Regensburg.

Wilson, M.A. (1971). Optic nerve fibre counts and retinal ganglion cell counts during development of *Xenopus laevis* (Daudin). *Q.J. exp. Physiol.* **56**: 83–91.

Yamashita, S., Kohriyama, H. & Kiyohara, S. (1989). Subspecies differences in neural gustatory responses of *Xenopus*. *Zool. Sci. (Tokyo)* **6**: 1079.

Yoshii, K., Yoshii, C., Kobatake, Y. & Kurihara, K. (1982). High sensitivity of *Xenopus* gustatory receptors to amino acids and bitter substances. *Am. J. Physiol.* **243**: R42–R48.

Zittlau, K.E., Claas, B. & Münz, H. (1986). Directional sensitivity of lateral line units in the clawed toad *Xenopus laevis* Daudin. *J. comp. Physiol. (A)* **158**: 469–477.

8 Sound production and acoustic communication in *Xenopus borealis*

DAVID D. YAGER

Synopsis

Xenopus has an underwater acoustic communication system based on simple clicks arranged in different temporal patterns.

Focusing on *X. borealis*, I have found a sound production mechanism highly adapted to the underwater environment—in contrast to terrestrial anurans, these frogs do not use a moving air column to produce their vocalizations. The larynx of the male is enormously enlarged to form an internal vocal sac. The arytenoid cartilages are modified to form two discs whose smooth medial surfaces are normally tightly apposed. When contraction of bipennate laryngeal muscles causes the discs to separate with high acceleration, a single click results. The click is most probably produced by an 'implosion' as air rushes into the cleft between the discs. The frequency spectrum of the click is determined by the vibrational characteristics of the laryngeal walls.

Laboratory and field studies show that *X. borealis* arranges its clicks into three discrete, context-dependent temporal patterns. At night, isolated males produce an advertisement call consisting of single clicks at 2–4/s. This call is highly attractive to females. When chasing or clasping another frog, males produce an 'approach call' with clicks at 10/s. Males and non-receptive females produce a 'release call' (clicks at 20/s) when clasped. However, males also use this click pattern when fighting and in situations suggesting a territorial function.

While each *Xenopus* species has a distinctive advertisement call, the role of the call in maintaining species integrity in zones of sympatry is not clear. Transects across a narrow sympatry zone between *X. borealis* and *X. laevis victorianus* in Kenya revealed strong ecological separation between the species. In the few ponds where both species occurred, hybrids (8% of the animals collected) were also present. The presence of hybrids was indicated by the occurrence of aberrant calls and by physical measurements (on field-collected and laboratory-reared animals) that showed intermediate body forms.

Introduction

As befits a genus of aquatic animals that are active primarily at night and inhabit murky, eutrophic ponds, *Xenopus* has a well-developed acoustic communication system. Males of every species examined produce vocalizations consisting of clicks arranged in species-specific temporal patterns (Vigny 1979). Female sound production has been documented only in *X. l. laevis* (Russell 1954; Hannigan & Kelley 1986) and *X. borealis* (Yager 1992b), but is likely also to be present across the genus.

Until recently, the literature on *Xenopus* acoustic communication has consisted almost exclusively of observations, often anecdotal, on a single species, *X. l. laevis* (see Picker 1980; Yager 1992a for references). Males of this species produce an 'advertisement call' consisting of a complex, continuous trill that Picker (1983) has shown in two-choice phonotaxis experiments to effectively, and selectively, attract conspecific females. Kelley and her colleagues (Hannigan & Kelley 1986; Kelley 1986; Kelley & Tobias 1989) have extensively studied a 'release call'—single clicks produced at about two per second—produced by females. Unreceptive females produce this vocalization when clasped by a male, and this 'ticking' is effective in causing the male to release the female. Receptive females do not 'tick' when clasped. Both Picker and Kelley have also noted a low-amplitude male 'release call'.

In her survey of *Xenopus* calls, Vigny (1979) noted that several other species have at least two call types. In fact, our current knowledge of the *X. l. laevis* (Picker 1980, 1983; Kelley 1986) and *X. borealis* (Yager 1992a) repertoires suggests that most, if not all, species may have 3–6 call types, defined by acoustic characteristics and relevant behavioural context. However, as demonstrated by the comparison below of *X. l. laevis* and *X. borealis* female release calls, specific call types may vary considerably in importance among species.

The acoustic environment *Xenopus* experiences differs considerably from that of its terrestrial counterparts (Kinsler & Frey 1962; Fine & Lenhardt 1983). Sound wavelengths are almost five times longer underwater, attenuation with distance is minimal, reflections from the bottom and the surface complicate the sound field, and water depth can have a profound impact on sound transmission, especially in shallow water and at low frequencies. We might expect that animals communicating under these conditions would diverge significantly from terrestrial communication patterns by, for instance, using much higher-frequency signals and lower source intensities. Since *Xenopus* is most probably secondarily aquatic (L. Trueb pers. comm.), it provides an excellent opportunity to contrast terrestrial and aquatic communication systems.

Whereas the advertisement call of *X. l. laevis* is the most complex in the genus (Vigny 1979), that of *X. borealis* is the simplest, consisting of single clicks repeated at 2–4 per second. Thus, this species has offered an attractive opportunity to examine (1) the sound production mechanism (Yager 1992b), an aspect of *Xenopus'* sound communication likely to be strongly affected by the move from terrestrial to aquatic environment; (2) the call repertoire and associated behaviours (Yager 1992a); and (3) *Xenopus'* acoustic ecology in the field, especially with regard to hybridization. These topics will be reviewed here in detail for *X. borealis* and can be compared and contrasted with *X. l. laevis* (Kelley this volume pp. 143–176) and augmented by Elepfandt's discussion of *Xenopus* hearing (also this volume, pp. 177–193).

The sound production mechanism of *X. borealis*

Xenopus borealis males show no motion of any part of their body while calling, nor do they release bubbles or show any indication of buoyancy changes. Yet a male can produce more than 1500 clicks over a period of 15 min between trips to the surface to breathe. Clearly, these frogs must have a sound production mechanism markedly different from that of terrestrial anurans and must somehow be emancipated from the use of a moving, vibrating air column as a sound source.

While the overall structure of the *Xenopus* respiratory tract is like that of terrestrial anurans, the larynx is highly specialized (Ridewood 1898; Yager 1992b; Figs 8.1, 8.2). There are two distinct suites of structural modifications, and each suite has clear functional correlates. The larynx of *X. borealis* shows strong sexual dimorphism (Fig. 8.1; also see Kelley this volume pp. 143–176). I will focus on the males since they are the primary call producers and their larynx is the most highly developed.

The arytenoid cartilages and click production

In most anurans, the arytenoid cartilages comprise two shell-like structures surrounding the glottis (Trewavas 1933). Intrinsic laryngeal musculature controls airflow (and sound production) by opening and closing the glottis. In *X. borealis*, the glottal cartilage and musculature retain their terrestrial forms. However, the posterior portions of the arytenoid cartilages have become greatly elaborated to form two stout, calcified rods terminating posteriorly in a pair of discs (Fig. 8.2). The very smooth medial faces of the discs are normally held tightly apposed in the midline by surrounding tissue rich in elastin as well as by the adhesive forces of fluid between them. Besides the glottal portions of the arytenoids, these discs are the only movable structures in the larynx. There are no vocal cords.

Movement of the discs is controlled by a pair of very large, intrinsic muscles originating over much of the posterior surface of the larynx (Fig. 8.2). The muscle fibres are arranged predominantly in a bipennate pattern with a broad central tendon directed anteriorly. Each tendon enters the larynx from the side by turning 90° over a cartilaginous 'pulley', and inserts onto the lateral aspect of the discs. Contraction of this pair of muscles pulls the two arytenoid discs apart.

The anatomy of the females' arytenoid cartilages is similar, but the structures are much smaller, even given the larger body size of the females. In particular, the bipennate muscles are much less developed.

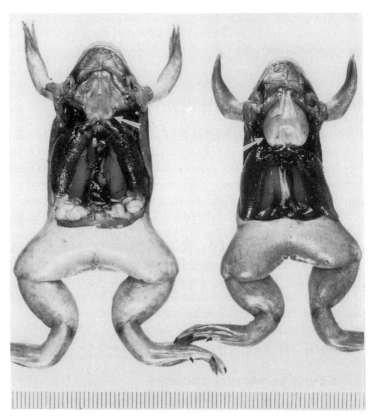

Fig. 8.1. Male (on right) and female *Xenopus borealis* to show the size, location and extreme sexual dimorphism of the larynx (arrows). The pectoral girdles and all viscera except the lungs have been removed. The lateral quarter on each side of the male larynx is covered by the bipennate muscle. Snout–vent lengths for adult males of this species are typically 40–60 mm and for females 40–80 mm. Scale is in millimetres. From Yager (1992b).

In *X. l. laevis* males, the anatomy of the arytenoids is like that of *X. borealis* males, but the rods are distinctly shorter compared to the disc diameter (Ridewood 1898; Kelley 1986).

The modifications of the arytenoid cartilages and the development of the bipennate muscles are the basis for the production of the clicks. In surgical preparations, disconnecting the bipennate muscles from the larynx at their origins makes it possible to pull on both muscles simultaneously in their normal direction of action, thus mimicking the effects of contraction, without other confounding effects such as compression of the cricoid box. At first, increasing tension on the muscles produces no motion of the discs. When sufficient force is applied, however, the discs suddenly separate and a click is produced. Their return to the rest position in the midline is silent. Electrical stimulation of the muscles through

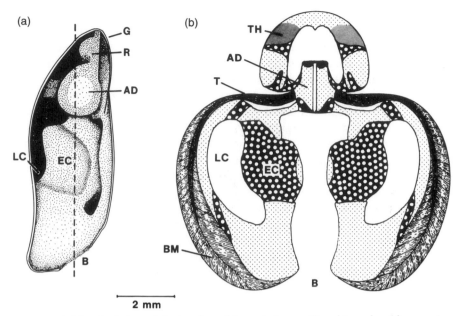

Fig. 8.2. (a) Mid-sagittal view of the interior of the male larynx. Dorsal is to the right; anterior to the top. G: glottis; R: arytenoid rod; AD: arytenoid disc; EC: the elastic cartilage block; LC: the lateral air chamber; B: bronchial opening. The dotted line indicates the plane of the horizontal section in (b). (b) Horizontal section through the male larynx. TH: thyrohyals; BM: bipennate muscle; T: tendon of the bipennate as it turns 90° to insert onto the arytenoid disc. Other abbreviations as in (a). From Yager (1992b).

implanted wires gives an identical result. Pulling or stimulating only one muscle simply pulls both discs to one side and is not accompanied by sound production. Acoustical analysis confirms that laboratory-produced clicks are almost identical to clicks from the advertisement call of a normal male recorded in a pond under approximately free-field acoustic conditions (Fig. 8.3a).

Any manipulation that interferes with the motion of the discs silences previously vocal males. Effective muting procedures[1] include wedging the discs apart with tiny glass beads, gluing them together with tissue cement, cutting the surrounding elastic tissue so that the discs do not meet in the midline, and cutting the tendons of the bipennate muscles. After recovering from surgery, males placed with unreceptive females engaged in the same behaviours as the sham-operated frogs, but, whereas the control animals vocalized, the males with laryngeal manipulations were silent.

These experiments demonstrated that movement of the discs—specifically, their rapid separation—is both necessary and sufficient for click production. No moving air column is involved. An attractive hypothesis for the physical

[1] Anaesthesia and care protocols for all experiments met or exceeded prevailing state and federal requirements for humane care of experimental animals.

mechanism creating the sound is: (1) the smooth medial faces of the discs are normally held tightly together by elasticity of the surrounding tissue, but more importantly, by fluid adhesive forces; (2) contraction of the bipennates is initially isometric, building up tension; (3) when the tension exceeds the adhesive forces, the discs suddenly pull away from each other with very high acceleration; (4) this motion creates a vacuum between the discs; (5) turbulent air rushing into the interdisc space (an 'implosion') produces a click. The broad-frequency-band sound generated by such implosions is well-known to physicists and engineers (Albers 1965); a practical example is the troublesome—and noisy—creation and collapse of cavitation bubbles by ship's propellers. It also has precedent in zoology: tongue clicking, well-documented in fruit bats (Kulzer 1960) and humans, uses the same principle.

Clicks produced by the arytenoid mechanism when the laryngeal box is largely removed are very weak and have a broad frequency spectrum (Fig. 8.3b), as predicted from the physics. The typical *X. borealis* click, however, is intense (up to 115 dB sound pressure level (SPL)—141 dB re: 1μ Pa—at 1m) and tuned to a relatively narrow band centred on 2.7 kHz. A second suite of laryngeal modifications effects the transformation of the clicks.

The laryngeal box, tuning and sound radiation

The posterior two-thirds of the male *X. borealis* larynx forms an immense, hollow box that has no direct equivalent in non-pipid anurans (Figs 8.1, 8.2).

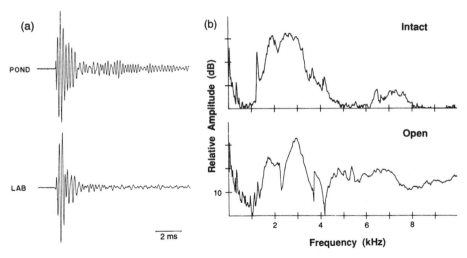

Fig. 8.3. (a) Oscillograms of a click produced at the centre of a 30 m diameter pond (upper) and of a click produced in the laboratory by electrical stimulation of the bipennate muscles (lower). (b) Power spectra demonstrating the effects of removing a rectangular portion of the ventral laryngeal wall. Clicks were produced by bipennate tendon-pull. The anterior end of the main air chamber was almost completely exposed by the surgery. Spectra are the average of >10 clicks. The 10 dB mark on the abscissa applies to both spectra. From Yager (1992 b).

The cricoid cartilages have expanded to form the large dorsal and ventral surfaces (plates). The air space within the box is divided into three longitudinal chambers. The central chamber is confluent anteriorly with the subglottal space and posteriorly with the bronchi.

Two large blocks of tissue separate the central from the lateral chambers (Fig. 8.2). This tissue is a highly cellular elastic cartilage not previously described in anuran larynges. It also fills the seams between the cricoid plates and surrounds the arytenoid discs. The small amount of elastin found in the female larynx is not organized into this tissue type, but rather forms aligned fibre bundles in the tissue around the discs. The function of the elastic cartilage blocks is unknown.

Clicks artificially produced in the laboratory by an intact male larynx have the same frequency spectrum as the clicks males produce when calling normally (Fig. 8.3a). However, any alteration of the cricoid (laryngeal) box broadens the frequency spectrum. Specifically, progressively removing pieces of the ventral cricoid plate while stimulating the bipennate muscles progressively detunes the clicks and reduces their amplitude (Fig. 8.3b). Opening the box by disconnecting the bronchi is particularly effective in broadening the spectrum; occluding the bronchi without disconnecting them, however, indicated that the lungs themselves are not involved in tuning the clicks. As further confirmation, a tiny spark gap was positioned in the central chamber near the discs. Outside the larynx the spark generated a click with a flat frequency spectrum from 1 to 10 kHz. In contrast, spark/clicks triggered inside the larynx had the same frequency profile as the arytenoid clicks.

These experiments demonstrate that the cricoid box determines the frequency spectrum of the clicks. In theory, the tuning of the clicks could come about by resonance of the air within the larynx as in a Helmholtz resonator, or by vibration of the walls of the cricoid box. I induced males to call underwater after breathing helium. Filling a Helmholtz resonator with helium would raise the resonant frequency by a factor of 2.85, but the helium-filled males called with the normal 2.7 kHz clicks. Thus, the tuning of the *X. borealis* males' clicks must be a function of the stiffness and mass of the walls of the cricoid box. Both the stiffness and the mass may be influenced by the connections of the walls to interior components such as the elastic tissue blocks.

Finally, having produced and tuned the clicks, the males must be able to radiate them effectively into the surrounding water to produce calls of adequate intensity. For animals the size of anurans calling at less than 5 kHz, the larger the sound source, the more sound energy will be radiated. This explains the vocal sacs common among terrestrial anurans. The same rule holds for *Xenopus* underwater. Since the body of a frog underwater is transparent to sound, the huge cricoid box can serve as an internal vocal sac, in this case both tuning and radiating the clicks.

Section summary

This unique sound production mechanism, with its independence from respiration and a moving air column, is unquestionably an adaptation to *X. borealis'* aquatic existence. Other members of the genus probably use the same basic mechanism, though with variations or elaborations to achieve the high click rates seen in some species. Among the other pipids, *Pipa* (Rabb 1960) and *Hymenochirus* (Ridewood 1899) appear to use a similar mechanism, but *Pseudhymenochirus* (Rabb 1969; D.D. Yager unpubl. obs.) seems to retain a more conventional sound production mechanism based on a moving air column.

The nature of *Xenopus'* sound production mechanism dictates that a male cannot vary the structure of the individual clicks; the frequency spectrum for the clicks of an individual male should remain constant. Any call repertoire must, therefore, be based on variations in the temporal patterning of the clicks.

The *X. borealis* call repertoire

Acoustic communication in terrestrial anurans concerns reproduction, either directly as in attracting mates or indirectly as in defending breeding territories (Wells 1977). While this is probably true for *Xenopus* as well, there has been no systematic effort to document the range of call types and their associated behaviours and an almost total lack of relevant data from the field.

To define the call repertoire for *X. borealis*, I recorded calling in a variety of behavioural contexts (Yager 1992a). Males were paired with females or other males; the reproductive state of each frog was controlled with hormone injections. A video image of animals interacting in an 80 l aquarium was combined with the image of the simultaneous oscillogram of the calls to form a split-screen video record of each encounter. This allowed unambiguous association of call type with behaviour even in complex acoustic interactions. Since there is no motion of any part of the frog when it calls, it is often impossible to determine which animal of the interacting pair is calling. However, by repeating selected experiments substituting a muted male, it was always possible to ascribe a particular call type to the correct animal.

The click

The fundamental unit of the *X. borealis* call is the click, and, as predicted from the sound production mechanism, its structure is the same in all of the calls in the males' repertoire (Fig. 8.4a). When recorded under approximately acoustic free-field conditions, the clicks are 2–3 ms in duration. The frequency spectrum extends from 1–4 kHz with a peak at 2.7 kHz. Individual males, however, have slightly different dominant frequency peaks ranging from 2500 to 2900 kHz.

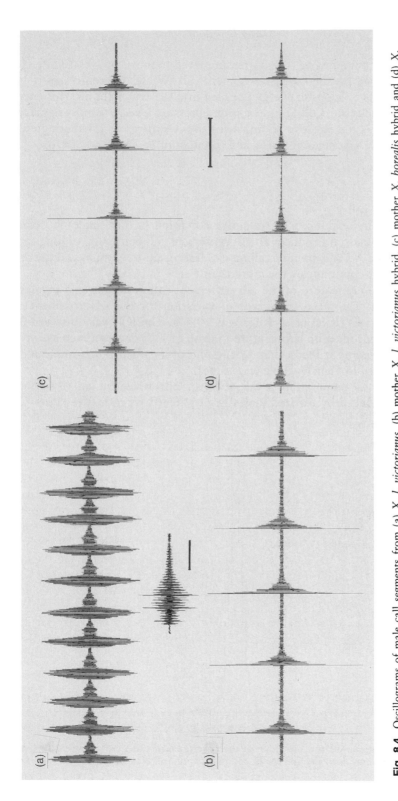

Fig. 8.4. Oscillograms of male call segments from (a) *X. l. victorianus*, (b) mother *X. l. victorianus* hybrid, (c) mother *X. borealis* hybrid and (d) *X. borealis*. The inset shows an expanded view of a single 'pulse' from an *X. l. victorianus* call to show that it is actually a series of many short clicks. In contrast, the calls of *X. borealis* and both hybrids consist only of single clicks. All recordings made at 24.5–26.0 °C. Scale bar for (a)–(d): 200 ms; bar for inset: 30 ms.

The clicks reach impressive sound pressure levels: a mean of 109 dB SPL (135 dB re: 1 μPa) at 1m for the advertisement call. While this represents only a modest amount of acoustic energy coupled into the water (in air, this energy would yield clicks of 73 dB SPL), the sound pressure levels are most significant since pressure is the effective stimulus for the *Xenopus* ear (Hetherington & Lombard 1982; Christensen-Dalsgaard, Breithaupt & Elepfandt 1990).

The repertoire

Advertisement call

When a male in heightened reproductive condition (a 'hot' male) is isolated from other animals, it produces an advertisement call consisting of long series of single clicks. A single bout of calling can last for up to 15 min, and the male can call almost continuously for more than 6 h.

A distinguishing feature of the advertisement call is its extreme regularity (Figs 8.4, 8.5). The mean coefficient of variation for the advertisement call interclick intervals (ICIs) of 27 frogs was 10.4%. The ICIs vary between 300 and 600 ms depending on temperature (a drop of 1 °C causes a increase in ICI of 19.8 ms). Except at the beginning and end of a calling bout, amplitude of the clicks is also extremely stable (Fig. 8.4a).

Justification for naming this vocalization an 'advertisement call' is twofold. First, it is produced by isolated males in heightened reproductive state—and

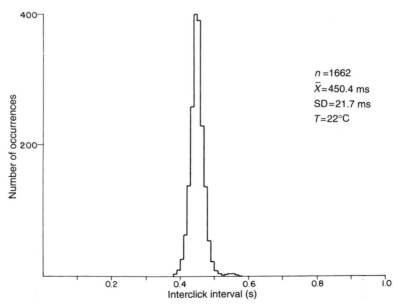

Fig. 8.5. An interclick interval histogram of the advertisement call from a single calling bout of an isolated *X. borealis* male. The coefficient of variation is 4.8%. Bin width is 10 ms. From Yager (1992a).

is the only call type produced in this situation. Second, in single-choice phonotaxis experiments in a 2 × 1.3 m concrete tank, females were strongly attracted to a loudspeaker broadcasting the call. The same females showed less response to the advertisement call of *X. l. victorianus*, and did not respond at all to white noise. Phonotaxis in *X. borealis* is very similar to that described in *X. l. laevis* by Picker (1980, 1983).

Approach call

A hot male responds to the presence of another animal (most probably detected by lateral line input) first by orienting and then by swimming rapidly towards it. When the male reorients it continues the advertisement call, but may slightly increase the repetition rate and amplitude; these changes are graded and highly variable. As the male swims towards the frog and clasps it, however, it produces a second, discrete call type, the 'approach call' (Fig. 8.6). ICIs for the approach call are 100–120 ms with CVs of 15–20%. This is the most intense call in the *X. borealis* repertoire. The male repeats the approach call frequently while clasping the second animal, often preceding it with several clicks of advertisement call, in effect creating a two-part signal. The behaviour of the hot male during the first portion of the encounter does not vary with the sex or reproductive state of the second animal. If the second animal is an unreceptive female, the male continues the approach call for the duration of his clasp. If the female is receptive, calling ceases except for rare bursts of the two-part call during egg laying.

Female release call

A 'cold' female communicates her lack of receptivity to the *X. borealis* male in several ways. Once clasped, the female assumes a rigid posture with fore and hind legs fully extended. (In contrast, a hot female keeps her hind legs flexed allowing apposition of the male and female cloacae.) There are also tremors of the unreceptive female's body. If the male continues in his clasp—which he generally does—the female begins kicking him in the belly with one of her hind legs. The male rarely persists more than 30 s after the female starts kicking. During the time she is clasped, the female may or may not produce a very weak release call consisting of trains of clicks with ICIs of 40–50 ms (similar to the male release/agonistic call in Fig. 8.6). The female click has a slightly lower dominant frequency than the male's, but is otherwise similar. In the face of the intense tactile signals the cold female gives the male, it is unlikely that the weak release call plays an important role in terminating the clasp.

The contrast of female behaviour between *X. borealis* and *X. l. laevis* is striking. Kelley and her colleagues (Kelley 1982; Hannigan & Kelley 1986; Kelley & Tobias 1989) have shown that unreceptive *X. l. laevis* females always produce a strong release call when clasped. While they do assume the same rigid posture as *X. borealis* females, they do not kick the males. Since strong tactile signals are always part of female unreceptive behaviour, the exact role

of the acoustic signals is not clear. Nevertheless, they are likely to be much
more important in X. *l. laevis* than in X. *borealis*.

Male release/agonistic call

A male clasped by another male produces long trains of moderately intense
clicks with ICIs of 40–50 ms (Fig. 8.6). The clasped male shows none of the
rejection behaviours of the cold female but, rather, swims about in an active
and jerky manner. However, two aspects of the experiments in the 80-l aquaria
suggest a more complex function than the traditional release call. First, the
call sometimes continues long after the clasp has broken off, something rarely
heard with a cold female. Second, in many instances both males produce the
vocalization simultaneously. This is particularly evident when the clicks of the

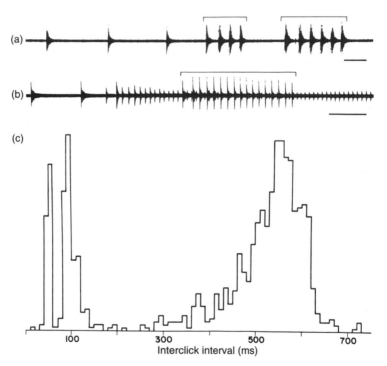

Fig. 8.6. The acoustic interaction between a 'hot' (see text) male and a 'cold' female. Note
that the time scales are different for the two oscillograms. (a) Clicks of advertisement call
precede two bursts of approach call (within bracket). Temperature 22 °C. Scale bar: 200 ms.
(b) The male produces three clicks of advertisement call and an unusually long burst of approach
call (within bracket) while clasping the female. She produces a long train of lower amplitude
clicks with short ICIs—the female release call. Temperature 23 °C. Scale bar: 400 ms. (c) The
cumulative ICI histogram for several periods of activity within a single behavioural experiment
run at 23 °C showing that there is no overlap in ICIs for the three call types present. The means
± SE for the three call types are: advertisement—542.8 ± 2.96, n = 793; approach—112.0 ± 1.6,
n = 181; release—54.3 ± 0.7, n = 86. Note that the advertisement call peak is broader than for
an isolated male. Bin width is 10 ms. From Yager (1992a).

males have slightly different dominant frequencies, and can also be seen on oscillograms as overlapping click trains.

Several experiments with males in a 2×1.3 m concrete tank revealed more complex behaviours than were seen in the 80 l aquaria. When a second male was introduced into the large tank at a point more than a metre distant from the 'resident', both males initially produced brief bouts of advertisement call. They then swam rapidly towards each other, sometimes producing release/agonistic calls when they were some distance apart. The two males then repeatedly tried to clasp one another, and both continued to produce the release/agonistic call. In several instances, the males grabbed one another around the shoulders and 'wrestled' violently. These vocal and physical interactions lasted for up to 2 min, after which one of the males—the loser?—swam away. From that point onward, only one male in the tank called; any attempt by the other to call was immediately met with advertisement call of increased intensity and rate which silenced the second frog. The silenced male would, however, resume normal calling after the other male was removed from the tank.

These observations suggest complex male–male interactions in *X. borealis* and raise the possibility of territoriality. While not previously known in *Xenopus*, robust male–male agonistic encounters and clear indications of territoriality have been documented in *Hymenochirus* (Österdahl & Olsson 1963; Rabb & Rabb 1963), possibly because the territory size for these small pipids is better encompassed by the aquaria generally used for observation. *Pipa* males also show some agonistic interactions (Weygoldt 1976).

Section summary

The *X. borealis* acoustic communication system comprises four discrete call types defined by interclick interval and behavioural context. While the significance—the 'meaning'—of the advertisement and female release calls seems evident, the function of the approach call is not as clear. That it has a higher sound pressure level and click rate than the advertisement call suggests simply that it is a signal of higher behavioural intensity. Finally, the release/agonistic call of the male plays a complex role that awaits further study.

The *X. borealis* call repertoire does not at all distinguish it from its terrestrial counterparts (Wells 1977). Judging from its acoustic communication system, we would expect the reproductive behaviour in the field to be much like that of many terrestrial species despite the tightly circumscribed three-dimensional world of the pond with its complex acoustics.

Field observations and hybridization studies

The following information derives from five months studying *X. borealis* calling behaviour in Kenya. The field work has been complemented by laboratory studies and recordings of captive animals in ponds in upstate New York.

General observations

Xenopus borealis is native to the highland regions of central and western Kenya (Vigny 1977). Its range is bordered on the north by that of *X. clivii*, on the west by that of *X. l. victorianus*, and on the east and south by *X. muelleri*. However, the only previously documented area of actual range overlap is between *X. borealis* and *X. l. victorianus* in the vicinity of Kaimosi and Kitale, Kenya (Vigny 1977).

While *X. borealis* can live in virtually any body of water, it is most likely to be found in small ponds and lakes and is rarely seen in moving water. The species much prefers foul, eutrophic water; some of my best collection sites were sewage treatment ponds. I found ample evidence that these animals can move overland more than a kilometre from pond to pond. Four of the collection sites (estimated frog populations of 20–150) were 1–3 km from the nearest neighbouring pond or stream. One site with >80 frogs was located on a ridge 2.5 km distant from and several hundred feet above the nearest possible *Xenopus* habitat. Reports from local naturalists indicate that these treks occur at night, primarily during the rainy season.

Males appear to spend the daylight hours in the deeper parts of the pond, and then move to shallow water (<1 m deep) from which they call at night. Localization using a movable hydrophone showed that in many cases males arranged themselves around the pond perimeter. In other cases, calling males were concentrated in underwater vegetation some distance from shore. They do not call from the surface. These calling site preferences can lead to loose aggregations of males. I did not, however, hear any indication of chorusing or co-ordinated calling. The typical distance between calling males varies considerably with population density.

Examination of museum specimens (looking for males with nuptial pads and for tadpoles) as well as anecdotal reports from Kenyans on the occurrence of tadpoles indicate that *X. borealis* breeds throughout the year in permanent bodies of water. In my collections (made in August–December), it was common to find a few males at each site with heavy nuptial pads while many others did not appear to be in reproductive condition. I recorded advertisement calls throughout my stay in Kenya.

Around-the-clock monitoring of ambient light levels and *X. borealis* calling for two days at each of four sites established that males call only at night. Calling rates start to increase as light levels begin to drop shortly before sundown. Calling reaches a maximum between 22:00 and 01:00, and then declines sharply, ceasing just before sunrise. The pattern for *X. l. victorianus* is identical.

Hybridization

Species-specific anuran advertisement calls are commonly thought to function in preserving species integrity by minimizing heterospecific matings (Wells

1977; Duellman & Trueb 1986). *X. borealis* and *X. l. victorianus* are similar in size and heterospecific matings are easily obtained in the laboratory. However, the two species have strikingly different advertisement calls (see below) that might serve in the field to help to limit hybridization. Vigny (1977) reported them to be sympatric in at least a small area in western Kenya. To begin understanding whether the advertisement calls of these species help to maintain species integrity, I have tried to define more exactly their zone of sympatry and to determine whether hybrids actually occur inside that zone.

In Kenya, I studied *Xenopus* in ponds at 17 locations along east–west transects crossing the suspected sympatry zone (Fig. 8.7). Frogs were collected at all but one of the sites, and at 11 of these sites I made tape recordings of underwater calling activity for one or more nights. Most animals were preserved in formalin, but some individuals of each species from sites well away from the overlap region were brought back to Cornell University alive.

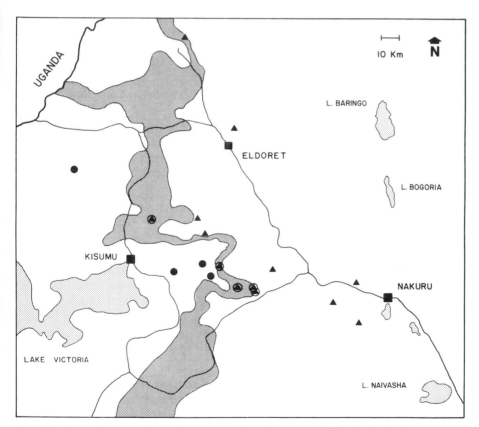

Fig. 8.7. Map of western Kenya showing the 17 collection sites. The darkly shaded region lies between 1500 and 1800 m elevation. Only *X. borealis* (triangles) was collected above 1800 m (to the east of the shaded region); only *X. l. victorianus* (circles) was collected below 1500 m (to the west). The sites where both species were heard and/or collected are indicated with circled triangles.

Using field-collected animals, I obtained laboratory-reared hybrids of both reciprocal crosses between *X. l. victorianus* and *X. borealis*. Blackler & Gecking (1972) and Brun & Kobel (1977) documented a fertilization block between *X. borealis* and *X. l. laevis* based on the jelly coat of the *X. borealis* eggs. *X. borealis* spermatozoa readily fertilize *X. l. laevis* eggs yielding almost normal numbers of offspring, but the reverse cross produces fewer than 5% viable offspring. A similar situation exists with *X. l. victorianus* and *X. borealis*. However, the very large number of eggs laid by the females provided >20 adult, mother-*borealis* hybrids for study.

Parental species

To define anatomical differences between *X. borealis* and *X. l. victorianus*, I made 16 physical measurements on each frog. These and several derived variables, e.g. ratio of head width at the eye to maximum head width, were chosen to match the variables used by Tinsley (1973, 1975) for his extensive measurements of several species. In summary, *X. borealis* is slightly larger than *X. l. victorianus*. It has longer limbs and fingers relative to body length, a broader head and less pointed snout, slightly larger eyes and longer subocular tentacles. The most convenient and reliable differentiating character, however, is the metatarsal tubercle: it is long and prominent in *X. borealis* and virtually absent in *X. l. victorianus*. Colour differences exist, but were not used in these analyses.

Laboratory-reared hybrid frogs have anatomical characteristics intermediate between the two parental species. Figure 8.8a shows a principal components analysis carried out on 20 laboratory-reared hybrids (the reciprocal crosses did not differ and are combined), 24 'pure' *X. borealis* (from the 2310 m elevation site; see below), and 17 'pure' *X. l. victorianus* (from the 1320 m sites). I identified a suite of variables describing head shape, eye size and limb and digit length (all independent of body length; metatarsal tubercle was excluded) which very effectively distinguished the two parent species (as independently defined by collection site and metatarsal tubercle length; see below). In the graphic analysis, the known hybrids generally occupy positions intermediate between the parental species. Considering only the metatarsal tubercle, the hybrids show an intermediate length: *X. l. victorianus*—0.19±0.02; hybrids —0.84±0.05; *X. borealis*—1.46±0.07 (mean ± SE in millimetres; a different measurement technique yielded slightly higher values for all animals compared to the full study below).

The advertisement calls of the two species differ in three major parameters (Fig. 8.4): (1) whereas *X. borealis* produces single clicks, *X. l. victorianus* produces short bursts of click-like pulses (the interpulse interval is 2–3 ms); (2) the interclick interval of *X. borealis* (400–600 ms) is twice the interburst interval of *X. l. victorianus* (150–250 ms); and (3) the *X. borealis* clicks have a broad frequency spectrum from 1000–4000 Hz with a peak at 2700 Hz while the spectrum of the *X. l. victorianus* lies between 1900 and 2800 Hz. The advertisement calls of the hybrids from the two reciprocal

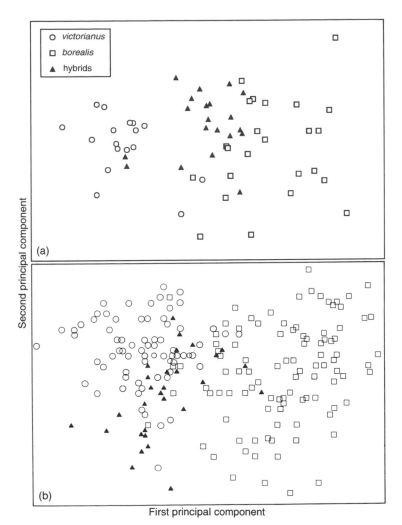

Fig. 8.8. Principal components analysis to distinguish parental species and hybrids using a suite of body size-independent variables, excluding the metatarsal tubercle. See text for details on group assignments. (a) The data set contained 'pure' *X. borealis* (squares), 'pure' *X. l. victorianus* (circles), and laboratory-reared hybrids (triangles). (b) The full data set, containing 'pure' *X. borealis* (squares), 'pure' *X. l. victorianus* (circles), and field-collected, suspected hybrids (triangles).

crosses are not intermediate between the parent calls but, rather, are both almost indistinguishable from the *X. borealis* advertisement call in their temporal parameters. However, hybrid frogs call more sporadically, and display distinctive irregularities in the click pattern at the beginning and end of calling bouts. The approach calls of the hybrids are aberrant, comprising a complex mix of *X. borealis* and *X. l. victorianus* characteristics.

In agreement with reports from other parts of Africa (reviewed in Tinsley

1975), *X. l. victorianus* in Kenya appeared only in collections from below 1800 m. In contrast, *X. borealis* occurred only above 1500 m, thus defining a strong ecological separation between the species. Any ponds containing both species would be expected to lie between the overlapping elevation extremes.

Sympatry and hybrids

Out of the 17 sites studied, five yielded both *X. borealis* and *X. l. victorianus*, as determined by anatomy and/or advertisement calls. The elevations of the five sites were 1570, 1630, 1720, 1750 and 1780 m. Tadpoles (species not determined) were present at three of the five sites, confirming active reproduction. Defining a sympatry zone between the two species as a band between 1500 and 1800 m along the eastern slope of the Rift Valley in Kenya includes not only my sites with both species (Fig. 8.7), but also those cited by Vigny (1977).

Since hybrid advertisement calls are like *X. borealis* calls, identifying hybrid calls in underwater recordings from the sympatry zone proved difficult. However, in two ponds with both species present, I did record some markedly aberrant calls. In a few cases, these were short bursts of hybrid approach, rather than advertisement, calling. In other cases, the distinctive irregularities at the onset of advertisement calls were clearly heard. In yet other cases, the vocalizations were unlike anything previously encountered. The presence of aberrant vocalizations in these ponds—and absence in ponds at lower or higher elevations—suggests that hybrid frogs were present.

The simplest analysis of the anatomy of frogs in the sympatry zone relies solely on the metatarsal tubercle. In the collections from the four sites yielding the longest mean tubercle lengths ('pure' *X. borealis*, at elevations of 1820, 2000, 2120 and 2310 m), 98% of the animals had lengths ≥ 0.6 mm; in the three collections giving the shortest mean lengths ('pure' *X. l. victorianus*, at 1320, 1320 and 1350 m), 89% had lengths ≤ 0.3 mm. At Kaimosi (1750 m), my largest collection in the sympatry zone, both species were actively calling. All tubercle lengths from 0 to >1.0 mm occurred (with roughly equal numbers at the extremes) and 21% were in the intermediate range between 0.3 and 0.6 mm, suggesting the presence of hybrids. The number of intermediate tubercle lengths in the 'pure' *X. l. victorianus* collections was somewhat higher than expected (11%). This may be because the 'pure' collection sites were < 30 km from the sympatry zone, and there are no topographic barriers to migration between sites. In contrast, the 'pure' *X. borealis* sites were generally more distant and a steep escarpment (the edge of the Rift Valley) blocks migration of *X. l. victorianus* toward *X. borealis* habitats.

Principal components analysis provided a more comprehensive evaluation of the anatomy. Figure 8.8b shows the graphical results of such an analysis using the same variables as in Fig. 8.8a. This analysis uses all the parental individuals from the tubercle analysis above (*n* = 182). When I added animals collected in the sympatry zone that had intermediate tubercle lengths to a data set containing both parental species, the resulting principal components analysis placed the putative hybrids at positions between the parent groups. In other words, these animals were intermediate in a range of distinguishing variables as well as in tubercle length. On the basis of both tubercle length and the principal components analysis, 23 of the animals from the mixed population ponds were suspected hybrids; this is 8% of the total collected.

Section summary

Xenopus borealis and *X. l. victorianus* call at the same times of year, the same times of day, and in the same ponds in a sympatry zone lying between 1500 and 1800 m along the eastern side of the Rift Valley in Kenya. Thus, there is ample opportunity for heterospecific mating, and both acoustic and anatomical data suggest that hybrids do occur.

Broad regions of sympatry between *Xenopus* species are rare; the narrow zone of sympatry shown by *X. borealis* and *X. l. victorianus* is much more typical (Tinsley 1981). A particularly well-documented sympatry occurs between *X. l. laevis* and *X. gilli* at the southern tip of South Africa (Kobel, Du Pasquier & Tinsley 1981; Simmonds 1985). Like *X. borealis* and *X. l. victorianus*, these two species have strikingly different advertisement calls (Vigny 1979; D.D. Yager unpubl. obs.), and *X. l. laevis/X. gilli* hybrids have been found (Kobel *et al.* 1981).

While *X. borealis/X. l. victorianus* hybrids apparently do occur, it is important to note that they are not common—less than 10% of the animals collected in the sympatry zone. Part of the explanation for this may be the fertilization block that largely prevents viable hybrids from *X. borealis*—but not *X. l. victorianus*—eggs. Nevertheless, this low value is remarkable considering the very high population density of frogs of both species in some of the ponds studied. The pond at Kaimosi, for instance, measured 34 × 56 m and an estimate based on drag-netting the entire pond put the adult population at 1000 frogs (plus 3000–4000 tadpoles). It would not be unreasonable to expect that even advertisement calls normally very effective in promoting conspecific mating would be less effective under such conditions.

Acknowledgements

Much of this work was carried out while in the laboratory of Dr Robert Capranica (Cornell University) who provided advice and support throughout the project. Antonie Blackler, Richard Tinsley and Garry Harned also provided important advice and help. My sincere thanks go to the Republic of Kenya (Mr E.K. Ruchiami, Office of the President) for permission to carry out my field work. None of the field studies would have been possible without the help of the many Kenyans who very generously let me work in their ponds. This research was supported by NIH Grant #NSO 9244=11 to R.R. Capranica, as well as an NSF Doctoral Dissertation Improvement Grant (#BNS-7911236), NIH Integrative Neurobiology Traineeship (#T32-MH15793), and Sigma Xi Grants-in-Aid to DDY.

References

Albers, V.M. (1965). *Underwater acoustics handbook II*. Penn State University Press, University Park.

Blackler, A.W. & Gecking, C.A. (1972). Transmission of sex cells of one species through the body of a second species in the genus *Xenopus*. II. Interspecific matings. *Devl Biol.* 27: 385–394.

Brun, R. & Kobel, H.R. (1977). Observations on the fertilization block between *Xenopus borealis* and *Xenopus laevis laevis*. *J. exp. Zool.* 201: 135–138.

Christensen-Dalsgaard, J., Breithaupt, T. & Elepfandt, A. (1990). Underwater hearing in the clawed frog, *Xenopus laevis*: tympanic motion studied with laser vibrometry. *Naturwissenschaften* 77: 135–137.

Duellman, W.E. & Trueb, L. (1986). *Biology of amphibians*. McGraw-Hill, New York.

Fine, M.L. & Lenhardt, M.L. (1983). Shallow-water propagation of the toadfish mating call. *Comp. Biochem. Physiol. (A)* 76: 225–231.

Hannigan, P. & Kelley, D.B. (1986). Androgen induced alterations in vocalizations of female *Xenopus laevis*: modifiability and constraints. *J. comp. Physiol. (A)* 158: 517–527.

Hetherington, T.E. & Lombard, R.E. (1982). Biophysics of underwater hearing in anuran amphibians. *J. exp. Biol.* 98: 49–66.

Kelley, D.B. (1982). Female sex behaviors in the South African clawed frog *Xenopus laevis*: gonadotropin-releasing, gonadotropic and steroid hormones. *Horm. Behav.* 16: 158–174.

Kelley, D.B. (1986). Neuroeffectors for vocalization in *Xenopus laevis*: hormonal regulation of sexual dimorphism. *J. Neurobiol.* 17: 231–248.

Kelley, D.B. & Tobias, M.L. (1989). The genesis of courtship song: cellular and molecular control of a sexually differentiated behavior. In *Perspectives in neural systems and behavior*: 175–194. (Eds Carew, T. & Kelley, D.B.). Alan R. Liss, Inc., New York.

Kinsler, L.E. & Frey, A.R. (1962). *Fundamentals of acoustics*. John Wiley & Sons, New York.

Kobel, H.R., Du Pasquier, L. & Tinsley, R.C. (1981). Natural hybridization and gene introgression between *Xenopus gilli* and *Xenopus laevis laevis* (Anura: Pipidae). *J. Zool., Lond.* 194: 317–322.

Kulzer, E. (1960). Physiologische und morphologische Untersuchungen über die Erzeugung der Orientierungslaute von Flughunden der Gattung *Rousettus*. *Z. vergl. Physiol.* 43: 231–268.

Österdahl, L. & Olsson, R. (1963). The sexual behavior of *Hymenochirus boettgeri*. *Oikos* 14: 35–43.

Picker, M.D. (1980). *Xenopus laevis* (Anura: Pipidae) mating systems—a preliminary synthesis with some data on the female phonoresponse. *S. Afr. J. Zool.* 15: 150–158.

Picker, M.D. (1983). Hormonal induction of the aquatic phonotactic response of *Xenopus*. *Behaviour* 84: 74–90.

Rabb, G.B. (1960). On the unique sound production of the Surinam toad, *Pipa pipa*. *Copeia* 1960: 368–369.

Rabb, G.B. (1969). Fighting frogs. *Brookfield Bandarlog* No. 37: 4–5.

Rabb, G.B. & Rabb, M.S. (1963). On the behavior and breeding biology of the African pipid frog *Hymenochirus boettgeri*. *Z. Tierpsychol.* 20: 215–241.

Ridewood, W.G. (1898). On the structure and development of the hyobranchial skeleton and larynx in *Xenopus* and *Pipa*; with remarks on the affinities of the Aglossa. *J. Linn. Soc. (Zool.)* **26**: 53–128.

Ridewood, W.G. (1899). On the hyobranchial skeleton and larynx of the new aglossal toad, *Hymenochirus boettgeri*. *J. Linn. Soc. (Zool.)* **27**: 454–460.

Russell, W.M.S. (1954). Experimental studies of the reproductive behaviour of *Xenopus laevis*. I. The control mechanisms for clasping and unclasping, and the specificity of hormone action. *Behaviour* **7**: 113–188.

Simmonds, M.P. (1985). Interactions between *Xenopus* species in the southwestern Cape Province, South Africa. *S. Afr. J. Sci.* **81**: 200.

Tinsley, R.C. (1973). Studies on the ecology and systematics of a new species of clawed toad, the genus *Xenopus*, from western Uganda. *J. Zool., Lond.* **169**: 1–27.

Tinsley, R.C. (1975). The morphology and distribution of *Xenopus vestitus* (Anura: Pipidae) in Central Africa. *J. Zool., Lond.* **175**: 473–492.

Tinsley, R.C. (1981). Interactions between *Xenopus* species (Anura Pipidae). *Monit. zool. ital. (N.S.) (Suppl.)* **15**: 133–150.

Trewavas, E. (1933). The hyoid and larynx of the Anura. *Phil. Trans. R. Soc. (B)* **222**: 401–527.

Vigny, C. (1977). *Étude comparée de 12 espèces et sous-espèces du genre* Xenopus. PhD thesis No. 1770: Université de Genève.

Vigny, C. (1979). The mating calls of 12 species and sub-species of the genus *Xenopus* (Amphibia: Anura). *J. Zool., Lond.* **188**: 103–122.

Wells, K.D. (1977). The social behavior of anuran amphibians. *Anim. Behav.* **25**: 666–693.

Weygoldt, P. (1976). Beobachtungen zur Biologie und Ethologie von *Pipa (Hemipipa) carvalboi* Mir. Rib. 1937 (Anura, Pipidae). *Z. Tierpsychol.* **40**: 80–99.

Yager, D.D. (1992a). Underwater acoustic communication in the African pipid frog *Xenopus borealis*. *Bioacoustics* **4**: 1–24.

Yager, D.D. (1992b). A unique sound production mechanism in the pipid anuran *Xenopus borealis*. *Zool. J. Linn. Soc.* **104**: 351–375.

9 Sexual differentiation in *Xenopus laevis*

DARCY B. KELLEY

Synopsis

During primary sexual differentiation in *Xenopus laevis*, the initially 'indifferent' gonads develop as testes or ovaries. Males can be functionally sex-reversed by the administration of the ovarian steroid hormone oestradiol during the early stages of gonadal differentiation (tadpole stages 54 to 66). Females can be functionally sex-reversed by the implantation of a testis during this time; administration of the testicular steroid testosterone does not result in gonadal masculinization. The sexes of progeny from crosses of sex-reversed frogs suggest that males are the homogametic (ZZ) and females the heterogametic sex (ZW); individuals whose inferred genotype is WW are female.

Secondary sexual differentiation is directed by secretion of steroid hormones. In females, differentiation of secondary sex characteristics is independent of gonadal steroids until late juvenile stages when oestrogen secretion promotes development of the oviduct and the synthesis of vitellogenin from the liver. However, sensitivity to gonadal steroids develops earlier (tadpole stage 56) in both sexes and its onset requires thyroxine secretion. Sexual receptivity in adult females is increased by gonadotropins. Oestrogen and progesterone stimulate female receptive behaviours while prostaglandins suppress unreceptive behaviours including the release call, ticking. In males, androgen is secreted during late tadpole stages and initiates a cellular programme for masculinization in androgen target tissues that include the vocal organ and the central nervous system. In adult males, gonadotropin-evoked release of spermatozoa is accompanied by increases in circulating androgens required for clasping behaviour and mate-calling. The activation of clasping and courtship song is mediated by groups of synaptically interconnected central neurons the majority of which express high levels of steroid hormone receptors. Calling behaviour is produced by the vocal organ or larynx. If the ovaries of post-metamorphic females are replaced with a testicular transplant, the vocal neuromuscular system, including the ability to mate-call, is masculinized. To examine the molecular control of sexual differentiation we have cloned a *X. laevis* androgen receptor and a laryngeal-specific, myosin heavy chain gene whose androgen-regulated expression may underlie the production of male song.

Introduction

Because of the ease with which the earliest developmental stages can be studied, the Anura have contributed significantly to our present understanding of mechanisms of vertebrate sexual differentiation. The African clawed frog, *Xenopus laevis*, in particular has been extensively investigated with respect to both gonadal differentiation and, more recently, secondary sexual development. This latter process has been the subject of experimental studies in our laboratory. We have focused our attention on masculinization, the developmental transformation of the initially female-like phenotype into a masculine one, and on the role played by secretion of androgenic steroids in this transformation. The nature of steroid hormone receptors and the way in which they affect gene transcription within target cells has recently been clarified by the use of molecular cloning techniques. The molecular approach has allowed us to identify androgen-sensitive genes and to begin dissection of the role played by androgen action in the differentiation of the masculine phenotype.

Gonadal development

The development of *Xenopus laevis* from fertilization through the end of metamorphosis has been divided into 66 stages (Nieuwkoop & Faber 1956). The developing gonads are first recognizable at stage 46 as thickenings on both sides of the dorsal root of the dorsal mesentery; these thickenings form the genital ridges (see Witschi 1971; Merchant-Larios & Villapanido 1981; Iwasawa & Yamaguchi 1984). The primordial germ cells (PGCs) are visible in this region even earlier (stage 40, Nieuwkoop & Faber 1956) and by stage 49 have migrated into the developing gonad and are enveloped by epithelial cells. During early gonadal development (stages 46 to 55: see Iwasawa & Yamaguchi 1984) the testes cannot be distinguished from the ovaries and this period is thus called 'indifferent'. During the indifferent phase, cells are added to the proximal portion of the gonad and the PGCs are located distally. The gonad then enlarges and an interior medulla can be clearly distinguished from the exterior cortex; the germ cells are in active mitosis. In females, a cavity appears in the centre of the gonadal medulla which is then transformed to a simple epithelium; PGCs are confined to the gonadal cortex. In males, the PGCs migrate into the gonadal medulla and no cavity is formed. In females, between tadpole stages 56 and 66 (the end of metamorphosis), the developing PGCs become surrounded by follicle cells and differentiate into oocytes; oocytes differentiate and enter prophase. In males, seminiferous tubules develop and PGCs differentiate into spermatogonia. The tadpole stages of gonadal differentiation are illustrated in Fig. 9.1.

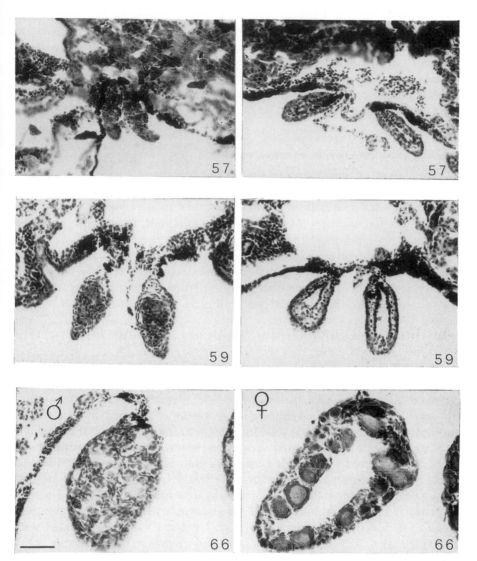

Fig. 9.1. Cross-sections of gonads from male and female tadpoles at stages 57, 59 and 66. Note the characteristic lumen that distinguishes the developing ovary from the testis. Scale bar 80 μm.

Genetic control of primary sexual differentiation

Chromosomes

The genus *Xenopus* is largely polyploid and *laevis* belongs to the tetraploid class (Graf & Kobel 1991): many genes are represented by two copies. The origin of tetraploidy is, however, sufficiently ancient that the species is functionally diploid; haploid 18-chromosome zygotes do not survive past stage 57 (Hamilton 1963). The mitotic karyotype consists of 18 pairs of chromosomes (Wickbom 1945), divided into seven groups based on medio- or acrocentricity (Tymowska & Kobel 1972). The sex chromosomes cannot be distinguished from the autosomes by heteromorphy (Mikamo & Witschi 1966) and their presence has been inferred from results of experiments using sex-reversed frogs.

Breeding experiments with sex-reversed frogs

Tadpoles grown in the presence of the steroid hormone oestradiol during gonadal differentiation will all develop as fertile females (Chang & Witschi 1956). Tadpoles exposed to a graft of testicular tissue during primary sexual differentiation develop as males (Mikamo & Witschi 1963). Because the expected sex ratio is 1:1, the most likely interpretation of these results is that half of the oestradiol-treated tadpoles are genotypically male but phenotypically female while half of the testis-implanted animals are genotypically female but phenotypically male. A series of breeding experiments with such sex-reversed individuals led Emil Witschi and his colleagues (Mikamo & Witschi 1963; Witschi 1971) to conclude that in X. laevis males are homogametic (ZZ) and females heterogametic (ZW). If oestrogen-treated tadpoles are reared and then bred with untreated males, they will give rise to male and female offspring in two patterns. Some individuals will produce entirely male offspring; this result is interpreted as the consequence of converting a genotypically (ZZ) male frog to female: ZZ male × ZZ female = 100% ZZ F1 males. Others will produce half male and half female offspring. This result is assumed to result from union of an oestrogen-treated ZW female (oestradiol does not affect female development) with an untreated ZZ male. If testis-implanted tadpoles are reared and then bred with untreated females, they will give rise to 75% female and 25% male offspring. One third of these females, when mated with normal (ZZ) males, give rise to all-female offspring, an expected result if this female sub-group were WW in genotype. The remaining two thirds of the females (presumed to be ZW in genotype) give rise to a 1:1 sex ratio when mated with genotypic males (ZZ). Results of such crosses are summarized in Table 9.1.

The sex-reversal experiments described above suggest that the ultimate fate of the primordial germ cells (PGCs)—ova or spermatozoa—is not determined by their genetic constitution (ZW = ova, ZZ = spermatozoa) but can instead be modified by the gonad into which they have migrated. This conclusion is supported by the experiments of Blackler (1965), who transplanted marked PGCs from genetic male X. laevis to genetic females and vice versa and observed that gametes adopted the sex of the host and not the donor. That genotype does not determine gamete phenotype is confirmed by the observation that WW frogs can produce either spermatozoa or eggs, depending upon whether the gamete developed in a testis or an ovary (Mikamo & Witschi 1964).

Table 9.1. Breeding combinations and composition of offspring

| Parents | Offspring (F_1) | |
	Proportions expected	Observed
♀ZZ × ♂ZZ	100% ZZ♂	2194♂
♀WW × ♂WW	100% WW♀	340♀
♀ZZ × ♂WW	100% ZW♀	318♀
♀WW × ♂ZZ	100% ZW♀	1377♀
♀ZZ × ♂ZW	50% ZW♀ + 50% ZZ♂	111♀ + 118♂
♀ZW × ♂ZZ	50% ZW♀ + 50% ZZ♂	1771♀ +1849♂
♀WW × ♂ZW	50% ZW♀ + 50% WW♀	621♀
♀ZW × ♂WW	Identical expectation	Not bred
♀ZW × ♂ZW	25% WW♀ + 50% ZW♀ + 25% ZZ♂	418♀ + 144♂

From Witschi (1971)

Determination of gonadal sex

Relation of chromosomal to gonadal sex

The mechanism whereby genetic sex is translated into gonadal sex is unclear. A testis-determining gene (TDF or SRY in humans, *Tdy* or *Sry* in mice) located on the Y chromosome has recently been cloned in mammals (Koopman *et al.* 1991). Expression of an *Sry* transgene (from the same species) in genetic females can result in a fertile masculine phenotype. While the *Sry* gene product is transiently expressed in somatic cells of the developing mouse genital ridge and again later in round spermatids, it is not known how *Sry* expression induces the formation of a testis (Lovell-Badge 1993). The *Sry* gene belongs to a family of high mobility group (HMG)-box containing, DNA-binding proteins (Harley *et al.* 1992); the SRY-like genes in this family have been named SOX (*SRY-box*; R. Lovell-Badge cited in Denny *et al.* 1992). Members of this family can bind to cruciform DNA with high affinity (10^{-9}) and induce a dramatic bending in DNA (Giese, Cox & Groschedl 1992; Harley *et al.* 1992). SRY may participate as a testis-specific architectural protein in the formation of a functional transcription unit for some as yet unidentified genes essential for initiating testis development. Members of the SOX family (XSox-5 and 11–13) have been identified in *X. laevis* by using the polymerase chain reaction and a cDNA library derived from oocytes (Denny *et al.* 1992). It is not yet clear whether these homologues function in sex determination in *X. laevis*. In gulls (*Larus fuscus*), in which—like *X. laevis*—females are the heterogametic sex, female-unique copies of the SOX gene family are apparently absent, which suggests that an SRY-homologue is not expressed on the W chromosome and is thus unlikely to function specifically in ovarian determination (Griffiths 1991).

H-W antigen and sex determination

Expression of a cell surface antigen, H-Y, is correlated with testis formation in some strains of mice and was proposed to play a fundamental role in gonadal sex determination in mammals (Goldberg, Boyse *et al.* 1971; Wachtel, Ohno *et al.* 1975; Ohno *et al.* 1979). However, because testis determination in mice can be genetically segregated from H-Y antigen expression (Goldberg, McLaren & Reilly 1991) and because a different testis-determining gene has been identified (see above), a primary role for H-Y antigen in gonadal differentiation in mammals is unlikely. In *X. laevis*, the homologue of the H-Y antigen (H-W) is associated with a female phenotype (Wachtel, Bresler & Koide 1980). For example, H-W is expressed in ovaries of ZW females but not in testes of ZZ males. If a ZZ male is converted to a female phenotype by exposure to oestradiol during the critical period for gonadal determination, the H-W antigen can be detected in 'his' ovary. In experimentally produced *Xenopus* hybrids (see also chapter by Kobel, this volume pp. 391–401), the expression of H-W antigen is inversely proportional to the number of copies of the Z chromosome, as is the probability of an individual animal exhibiting a female phenotype (Engel & Kobel 1980). It thus remains possible that H-W antigen expression will provide a useful marker for ovarian determination in *X. laevis* even if, as is the case in mice, the antigen itself does not direct gonadal sex.

Hormones and the determination of gonadal sex

In *X. laevis*, exogenous oestradiol, given between tadpole stages 52 and 55 (Chang & Witschi 1956) converts testicular to ovarian development. Whether *endogenous* oestrogen secretion plays a role in ovarian determination is not known. In the adult, the ovary is the major source of circulating oestrogen. However, the developing gonad appears incapable of synthesizing oestrogen (Witschi 1971; Iwasawa & Yamaguchi 1984); production of oestradiol by extragonadal tissues has not yet been studied.

The role of endogenous oestrogen has been investigated more extensively in reptiles, particularly turtles, in which gonadal sex is determined by the temperature of incubation of the eggs (Crews, Wibbels & Gutzke 1989; Dorizzi *et al.* 1991). Anti-oestrogen can block the feminizing action of a particular temperature range (Wibbels & Crews 1992). One interpretation of this result is that endogenous oestrogen is an intermediate in the sex-determination pathway controlled by the temperature of incubation. Possible sources of endogenous oestrogen include the gonads in some species (Dorizzi *et al.* 1991) and the adrenals or extragonadal tissues in others (Thomas *et al.* 1992). It is not yet clear where endogenous oestrogen acts since, in the only case where it has been investigated, oestrogen receptor is absent from the developing gonads (Gahr, Wibbels & Crews 1992). One possibility is that extragonadal tissues respond to oestrogen by producing a feminizing factor that acts on the developing gonad to induce ovarian differentiation. In *Xenopus* the primary sex ratio is not, however, affected by the temperature of incubation except in experimentally produced polyploid hybrids (Kobel this volume pp. 391–401).

In *X. laevis*, testicular grafts masculinize development of the indifferent gonad when present during primary sexual differentiation (Mikamo & Witschi

1963). The action of testicular grafts cannot be mimicked by administration of testosterone (Witschi 1950; Chang & Witschi 1956; Gallien 1962). Sex reversal can also be accomplished by parabiosis, which suggests that the testicular factor (or factors) that inhibits ovarian development is diffusible (reviewed in Witschi 1971). This masculinizing factor could be related to mammalian anti-Mullerian hormone. Anti-Mullerian hormone is a dimeric glycoprotein hormone member of the TGFβ family and has recently been cloned (Picard *et al.* 1986). Secretion of AMH by the developing testis is responsible for the regression of the Mullerian ducts. Anti-Mullerian hormone can induce cultured ovarian cells to differentiate into testicular tissue (Vigier, Watrin *et al.* 1987). Addition of AMH to the developing mammalian ovary can also masculinize the pattern of steroid secretion by interfering with aromatase, an enzyme complex responsible for converting androgen to oestrogen (Vigier, Forest *et al.* 1989).

A scheme for the determination of gonadal sex in *Xenopus laevis* is illustrated in Fig. 9.2. The available data suggest that in *X. laevis* the determination of sex (primary sexual differentiation) is genetically controlled (ZZ = male; ZW = female). Individuals with an inferred ZZ genotype develop testes which subsequently direct the development of the masculinized phenotype (secondary sexual differentiation). Individuals with an inferred ZW genotype develop ovaries responsible for the mature expression of the feminized phenotype. The process of secondary sexual differentiation is reviewed below. How genotypic sex controls gonadal sex is poorly understood; I have incorporated the available data on testicular and ovarian determination described above into a hypothetical, but testable, scheme (Fig. 9.2; boxed pathways). This scheme assumes that the developing gonad of both males and females is subject to the feminizing effects of endogenous oestrogen; ovarian development would thus constitute the default pathway for gonadal determination. In males, however, the pre-testis (i.e. the committed but morphologically indifferent gonad) secretes a masculinizing factor (e.g. AMH) which blocks ovarian differentiation by blocking the aromatization of androgen to oestrogen. If this scheme is correct, either AMH or aromatase inhibitors, administered to tadpoles during the critical stages of gonadal determination, should produce 100% phenotypic males.

Secondary sexual differentiation in females

In *Xenopus laevis*, as in other vertebrates (Wilson, George & Griffin 1981), secondary sexual differentiation is controlled by steroid hormones secreted by the gonads (Fig. 9.2). The direct targets of hormonal action can be identified because these cells express intracellular receptors for specific steroids. When the steroid diffuses through the membrane and enters the cell, it binds to a specific receptor protein which can interact with DNA hormone response elements and influence gene expression (Evans 1988). Thus we can characterize secondary sexual differentiation as a process in which gonadal steroid secretion

Darcy B. Kelley

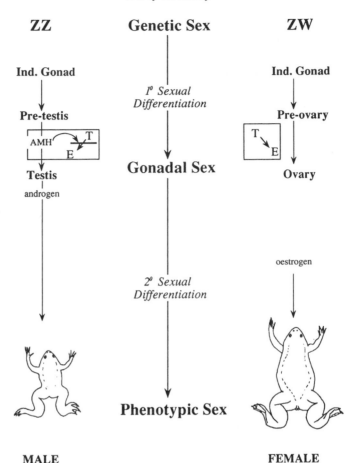

Fig. 9.2. A hypothetical scheme for sexual differentiation in *X. laevis* including a role for endogenous oestrogen in gonadal differentiation. During primary sexual differentiation the indifferent gonad is transformed to a testis or ovary. In males (genotype ZZ), anti-Mullerian hormone (AMH) secretion is postulated to block aromatization of testosterone (T) to oestradiol (E). In females (genotype ZW), oestrogen production is not blocked by AMH and the indifferent gonad is transformed to an ovary.

during development alters gene expression in target cells. Despite this general characterization, we know little about the cellular mechanisms that underlie hormonal control of the developmental programme in target cells.

For secondary sexual differentiation, several categories of hormonal responsiveness can be distinguished. One requires continual secretion of gonadal steroids; in their absence (e.g. immature animals, non-breeding season, gonadectomy) the secondary sex characteristics are not expressed. The characteristics exhibiting this pattern of differentiation include the oviducts of females, which grow in response to oestrogens and regress following ovariectomy, and the nuptial pads and clasping behaviour of males which are exhibited in response to androgens and regress following castration. Some structures or behaviours exhibit a sexually differentiated appearance in adults simply because the required hormones are absent. Good examples are the nuptial pads and clasping behaviours which can easily be observed in adult females given androgen (Kelley & Pfaff 1976). Other structures exhibit a sexually differentiated phenotype as the result of a hormonally controlled developmental process which differs in the sexes. In these cases, the characteristic cannot be exhibited in adults of one particular sex because the organ itself is no longer present. The oviduct, which is lost by seven months post-metamorphosis (PM) in males, is an extreme example. If males are castrated before this time, the oviducts are maintained. Oviduct regression in the male could be due to testicular secretion of anti-Mullerian hormone (see above) which is responsible for the loss of Mullerian ducts (oviducts) in male mammals.

Typically the development of phenotypic sex can proceed, up to a point, without gonadal participation. In mammals, the default programme (that executed in the absence of the gonads) is female and in birds the programme is male. This observation led to the suggestion that the default sex in vertebrates is the homogametic sex (female in XX mammals and male in ZZ birds; Adkins 1975). However, in *X. laevis* males are homogametic and females are the default sex with respect to phenotypic sex. The early phases of secondary sexual differentiation in females do not require the presence of the ovaries. For example, ovariectomy before seven months PM results in maintenance of the undeveloped oviduct (Witschi 1971). Thus, genetic determination of gonadal type does not mandate the subsequent development of secondary sexual characteristics; the homogametic sex is not necessarily the one that exhibits the default programme of secondary sexual differentiation.

Ovarian development after metamorphosis

The post-metamorphic development of the ovary includes follicular growth and oocyte maturation. The 14 or 15 gonomeres of the tadpole ovary enlarge and the oocytes contained within it mature, resulting in the multi-lobed ovary of the adult female (Fig. 9.3; Chang & Witschi 1956; Mikamo & Witschi 1963). In the laboratory, the adult ovary contains an asynchronous population of oocytes which have been divided into six stages of maturity based upon size (Dumont 1972) and production of yolk proteins (derived from vitellogenin synthesized in the liver; reviewed in Wallace 1978). In an adult female, progression through these stages requires a minimum of eight months (Dennis-Smith, Xu & Varnold 1991). Vitellogenesis occurs principally at stage IV. Depending on the conditions of rearing—most notably temperature, population density and food availability—females become fully sexually mature between 10 and 24 months after the completion of metamorphosis. Individual variability can be extreme; at 13 months PM some females have ovaries containing only immature oocytes (Mikamo & Witschi

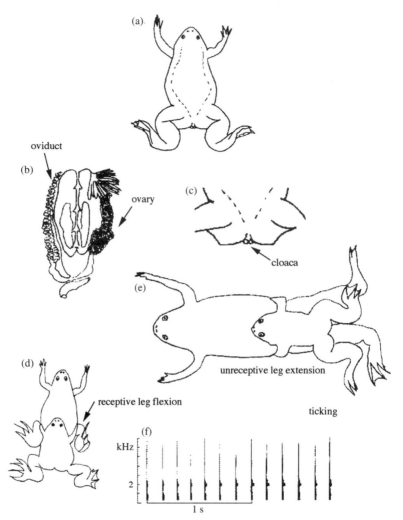

Fig. 9.3. Secondary sexual characteristics of female *X. laevis*. (a) The appearance of a sexually mature (PM6) female; dorsal view. (b) The reproductive organs include the ovary with an associated fat body (anterior to the ovary; right side); a cut-away view (left side) reveals the more dorsal oviduct which empties into the cloaca posteriorly. (c) An enlarged view of the female cloaca with its prominent labia. (d,e) A pair of frogs in amplexus, the male clasping the female in the inguinal position. Sexually receptive females (d) exhibit flexion of the hind limbs at the knee and abduction of the thighs, and are silent; sexually unreceptive females (e) extend the rear legs and vocalize. (f) Vocalizations characteristic of sexually unreceptive females: ticking. In this sound spectrogram, ticking is revealed to be a slow (6–7 Hz) trill made up of brief clicks with dominant sound energies between 1 and 2 kHz.

1963) while others contain an adult-like proportion of fully mature stage VI oocytes.

Ovarian steroid secretion

It is not clear when oestrogens are first secreted by the developing ovary. Witschi (1971) reported that when tadpole gonads were incubated with radioactive steroid precursors, oestradiol was not detected. The liver becomes competent to synthesize vitellogenin in response to oestrogen at tadpole stage 62 (May & Knowland 1980; Kawahara, Kohara *et al.* 1987) and yet endogenous yolk deposition does not occur until five or six months PM, a time when the oviducts also present evidence of oestrogen-dependent maturation (Witschi 1971). These observations suggest that adult-like secretion of oestrogens in females does not occur until six months PM although lower levels (see those for adult males below) may be present at earlier stages of development (Gallien & Chalumeau-le-Foulgoc 1960). In adults, circulating levels of oestradiol are 2.61 nmol in females and 0.16 nmol in males; values for oestrone are 2.39 and 0.13 nmol, respectively (Wright, Wright & Knowland 1983). These levels of circulating oestrogens in adult males are not sufficient to induce vitellogenin synthesis.

In the female ovary, another steroid, progesterone, is responsible for oocyte maturation including the final reduction divisions of meiosis (Baulieu *et al.* 1978; Smith 1989). Circulating progesterone concentrations associated with ovulation are 10 nmol and the hormone is believed to act via interaction with the cell membrane followed by an intracellular cascade with cyclic AMP and diacylglycerol as second messengers (Smith 1989).

Female secondary sexual differentiation: reproductive tissues

Secondary sexual characteristics of females include the organs of reproduction. In *X. laevis* these include the oviducts (Fig. 9.3b) and the cloacal labia which become swollen and red prior to oviposition (Fig. 9.3c). These secondary sexual characteristics become feminized during the first year after metamorphosis. The oviducts begin development during late tadpole stages (57 to 66) and continue growth for the next four to six months (Witschi 1971) as do the cloacal labia of females. At about seven months, the oviducts of females become convoluted and their walls start to thicken while those of males begin to regress.

Female reproductive behaviours

Mature, sexually unreceptive females exhibit a constellation of behaviours (including a vocalization called ticking, Fig. 9.3f, and extension of the rear legs, Fig. 9.3e) in response to clasp attempts by males; sexually receptive females are silent and respond to an inguinal clasp with knee flexion (Fig.

9.3d; Kelley 1982; Weintraub, Kelley & Bockman 1985; Hannigan & Kelley 1986). Ticking is a very slow (6 Hz) trill. It is not known when ticking or behavioural indications of sexual receptivity first appear during development. Under laboratory conditions females typically cannot be successfully bred until one year PM; two-year-PM females are usually fully sexually mature.

The production of reproductive behaviours in adults requires the presence of gonadal hormones. In females, injection of gonadotropins (including human chorionic gonadotropin) promotes sexual receptivity which is eliminated by ovariectomy (Kelley 1982). We have thus assumed that gonadotropin-induced sexual receptivity is due to the stimulation of steroid secretion from the ovary. In support, the production of sexually receptive postures (leg flexion) can be restored by the combined administration of oestradiol and progesterone; neither hormone alone is effective (Kelley 1982). Receptivity is enhanced in ovariectomized, steroid-treated females by concomitant administration of luteinizing hormone releasing hormone (Kelley 1982). In intact females, LHRH produced in the brain stimulates secretion of LH from the pituitary which promotes ovarian oestradiol synthesis. This endocrine route is not available in an ovariectomized female and exogenous LHRH must act elsewhere, perhaps via the central nervous system as it does in rats (Pfaff 1973; Moss & McCann 1973).

Females use a specific release call, ticking (Russell 1954), to terminate a male's clasp assault (Weintraub et al. 1985). Oestrogen and progesterone do not suppress ticking in females; suppression can be achieved by administration of androgen (Hannigan & Kelley 1986) or of prostaglandin E_2 which also promotes a receptive posture (Weintraub et al. 1985). Do either of these hormones play an endogenous role in control of female receptive behaviours? Circulating androgen can be detected in adult females and is increased by HCG administration (Lambdin & Kelley 1986) as it is in males (Lecouteux 1988). Since androgen is an oestrogen precursor within the ovary, the increase in circulating androgen that accompanies ovulation could contribute to the suppression of ticking in gravid females. Neurons in the central nervous system vocalization circuit contain receptor (see below) for androgen in both females and males (Kelley 1980, 1981) and androgen may act to suppress ticking by influencing these neurons (Hannigan & Kelley 1986). Prostaglandin synthase has been detected immunocytochemically in cells lining the lumen of the oviduct (Kelley, Weintraub & Bockman 1987) and we have speculated that prostaglandin release during the oviductal distension that accompanies oviposition acutely suppresses ticking. The target of prostaglandin's effects on receptive behaviours is not known but does not include either the ovary or the oviduct itself (Weintraub et al. 1985); the rapidity with which prostaglandin promotes receptive behaviours again suggests a CNS site of action (seen in Rana pipiens by Schmidt 1993).

Secondary sexual differentiation in males

Testicular development after metamorphosis

The gonomeres of the tadpole testes compact and enlarge during post-metamorphic development to form the kidney-bean-shaped testis characteristic of the adult male (Fig. 9.4; Mikamo & Witschi 1963). Earlier descriptions of spermatogenesis indicated that the first spermatocytes are present as early as tadpole stage 59 (Nieuwkoop & Faber 1956); this observation has not been confirmed by electron microscopic methods (Iwasawa & Yamaguchi 1984). Witschi (1971) reported that the first spermatocytes are found two to three months after the end of metamorphosis. Under exceptional conditions, viable spermatozoa can be produced as early as six months after the end of metamorphosis but are more typically not present in quantity until one year post-metamorphosis (Mikamo & Witschi 1963).

Testicular steroid secretion

The first production of C19 steroids (androstenedione) was detected in testes at stage 57 (Witschi 1971). We have observed axonal outgrowth from androgen-sensitive laryngeal motor neurons between tadpole stages 59 and 62 (Kelley & Dennison 1990). This outgrowth is blocked by anti-androgens in males and mimicked by exogenous androgens in both sexes; females do not exhibit this axonal outgrowth and are unaffected by anti-androgens (Kelley & Dennison 1990; Robertson, Watson & Kelley 1991). These observations, together with the detection of androgen in tadpole stage 62 liver (Kang, Marin & Kelley 1994), suggest that male tadpoles secrete androgen between stages 56 and 62. However, Iwasawa & Yamaguchi (1984) did not observe the characteristic ultrastructure of steroid-producing cells in the gonads of tadpoles, and tadpole androgens may derive from an extra-gonadal source.

We have also measured, using radioimmunoassay, circulating androgens at PM stages 1–5 (PM stages of Tobias, Marin & Kelley 1991a; see Table 9.2). Levels range from 0.9 to 5.2 nmol and are higher in males (M.L. Marin & D.B. Kelley unpubl. obs.). In adult males (PM6), radioimmunoassay reveals circulating androgens recognized by antisera to testosterone and dihydrotestosterone (DHT; Kelley 1980; Wetzel & Kelley 1983; M.L. Marin & D.B. Kelley unpublished). Levels of both hormones are higher during the summer months than in winter (testosterone: 4 versus 90 nmol; dihydrotestosterone: 5 versus 147 nmol). Androgen levels in adult females are ~5 nmol and do not exhibit a seasonal fluctuation.

Male secondary sexual differentiation: reproductive tissues

In males, secondary sexual characteristics include the nuptial pads used to clasp females and behaviour associated with amplexus. As was the case for females, male secondary sexual characteristics also emerge during the post-metamorphic period. The nuptial pads (black, thickened epidermal spines on the inner surfaces of the forearms; Fig. 9.4d and e) first appear between six and 10 months PM, when males first spontaneously exhibit clasping behaviours and the testes produce spermatozoa. One of the most dramatic secondary sex characteristics in *X. laevis* is the vocal organ, or larynx (Fig. 9.5), which is used by both sexes to produce ticking (Fig. 9.3f), the release call, and by males to produce courtship song, or mate-calling (Fig. 9.4c). The masculinization of the larynx and its control by gonadal androgen are described in detail below.

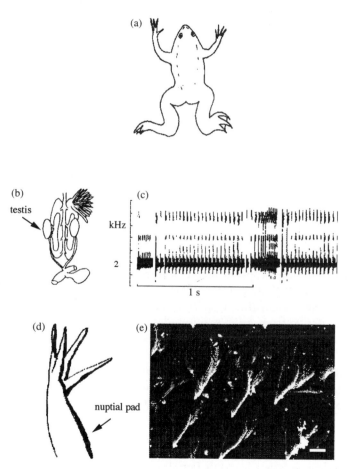

Fig. 9.4. Secondary sexual characteristics of male *X. laevis*. (a) The appearance of a sexually mature (PM6) male; dorsal view. (b) The reproductive organs include the testis with its associated fat body (anterior to the testis; right side); a cut-away view (left side) reveals the efferent ductules which empty into the cloaca posteriorly. (c) Vocalizations characteristic of sexually active males: mate calling. In this sound spectrogram, mate calling is revealed to include alternating fast (70 Hz) and slow (35 Hz) trills made up of brief clicks with dominant sound energies between 1 and 2 kHz. (d) An enlarged view of the forearm with the nuptial pads characteristic of sexually mature (PM6) males. (e) Scanning electron micrograph of the thickened epidermal spines which make up the nuptial pads. These spines facilitate the male's clasp during amplexus. Scale bar: 10 μm.

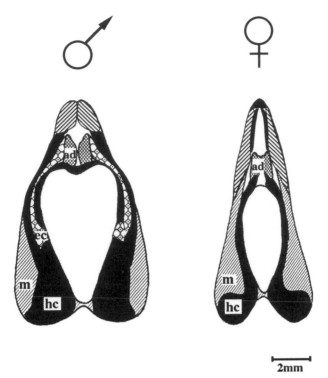

Fig. 9.5. The vocal organ of male and female *X. laevis*: the larynx. The larynx is a hollow box of hyaline cartilage (hc), flanked bilaterally by bipennate muscles (m). The sound-producing elements, the arytenoid discs (ad), located anteriorly, are pulled apart when the muscles contract, creating the clicks which comprise ticking and mate calling. The larynx is markedly dimorphic in cell number and type. For example, males have elastic cartilage (ec) while females do not. Muscle fibre number is greater in males than in females.

Male secondary sexual differentiation: reproductive behaviours

Male *X. laevis* can also tick when clasped by another male. While the endocrine control of ticking has not been studied in males, we suspect that androgen may suppress ticking as it does in females. Ticking is a very powerful release signal to clasping males. In the laboratory, sexually active males often form 'daisy chain' arrays of clasping animals and we have speculated that this behaviour reflects the suppression of ticking by high levels of circulating androgen (Hannigan & Kelley 1986).

Castration in adulthood abolishes clasping (Kelley & Pfaff 1976; Wetzel & Kelley 1983; Watson & Kelley 1992). Administration of exogenous testosterone or DHT restores reproductive behaviours although levels of activity do not completely attain those seen in intact males. Oestradiol is entirely without effect on clasping. Injection of thyrotropin releasing hormone results in longer clasp durations in males (Taylor & Boyd 1991). Male-typical clasping can be induced in adult female frogs that are ovariectomized and given testosterone or DHT; the quality and duration of clasping cannot be distinguished from similarly treated males (Kelley & Pfaff 1976). Dihydrotestosterone is more effective than testosterone in this behavioural assay as it is in other assays of masculinization (Segil, Silverman & Kelley 1987; Tobias, Marin & Kelley 1993).

Neural control of reproductive behaviours

The brain is a major target of gonadal steroids (Fig. 9.6) and believed to be responsible for steroid-induced activation of reproductive behaviours in adults. With the exception of vocal behaviour (see below), our knowledge of neural pathways specific for reproductive behaviours in X. laevis is primitive. Several features of the neural control of sexual activity are, however, well conserved in vertebrates and probably apply to frogs as well. These include participation of anterior hypothalamic (male) and more posterior hypothalamic (female) nuclei in copulatory behaviours as well as the control of gonadotropin release by the ventral infundibulum (Kelley & Pfaff 1978). The importance of acoustic signals for attraction of females and for male–male competition argues that CNS auditory nuclei (Wilczynski & Capranica 1984) are components of the CNS circuitry for vocal behaviours. In toads (Bufo americanus), dorsal telencephalic, diencephalic and mesencephalic auditory nuclei do not appear essential for responding to mate calls, whereas the ventral lateral midbrain tegmentum and posterior structures are required (Schmidt 1988). In X. laevis, clasping survives transections of the neuraxis in the anterior medulla (Hutchison & Poynton 1963); the motor neurons which innervate muscles involved in male clasping behaviour are located in the anterior spinal cord (Erulkar et al. 1981). For most reproductive behaviours in both sexes, the anterior preoptic area must be intact (e.g. Schmidt 1989).

Steroid hormone targets in the central nervous system

Provided that competing endogenous steroids are absent, cellular targets for gonadal steroids can be identified by providing radioactive hormones and observing the subsequent retention of the steroid by its receptor, a nuclear protein, autoradiographically or biochemically. We have used autoradiography to identify targets for ovarian steroids—oestradiol and progesterone—in the central nervous system of X. laevis. The locations of these cells are summarized schematically in Fig. 9.6.

Following the administration of radioactive oestrogen, labelled cells can be demonstrated in three telencephalic areas (ventral striatum, ventral-lateral septum and amygdala), the anterior preoptic area (APOA), the ventral thalamus, the ventral infundibular nucleus (VIN) and the laminar nucleus of the torus semicircularis (Morrell, Kelley & Pfaff 1975). The distribution of labelled cells is the same in males and females. A subset of oestrogen-target

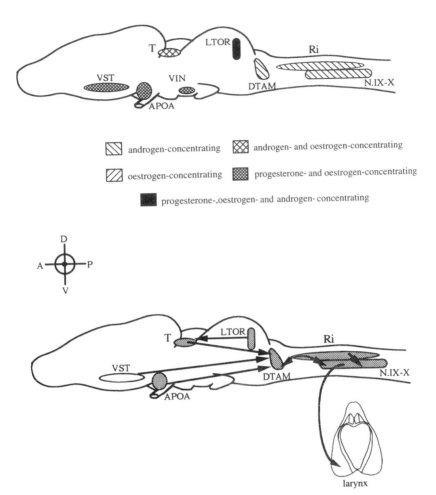

Fig. 9.6. The locations of steroid-concentrating neurons (upper panel) versus the neural circuit for vocal behaviour (lower panel) are depicted schematically in a side view of the frog brain. Upper panel: certain brain nuclei such as that containing laryngeal motor neurons (N. IX–X), the inferior reticular formation (Ri) or the pattern generator for mate calling (DTAM: pretrigeminal nucleus of the dorsal tegmental area of the medulla) contain labelled cells after radioactive androgen (testosterone or dihydrotestosterone) administration but not after administration of other steroids. Others, such as the ventral striatum (VST), anterior preoptic area (APOA) and ventral infundibular nucleus (VIN) contain both oestrogen- and progesterone-concentrating neurons. A third group of brain nuclei contain labelled cells after administration of oestrogens, progestins and androgens: these are the laminar nucleus of the torus semicircularis (LTOR) and thalamic (T) nuclei. Cells which concentrate radioactive androgen have also been shown to express androgen receptor mRNA; we thus presume that specific labelling patterns reflect specific patterns of steroid hormone receptor expression. Lower panel: the locations and interconnection of brain nuclei implicated in the generation of vocal behaviours in *X. laevis*. Many of these brain nuclei are labelled after administration of radioactive steroids, particularly androgens.

neurons contains progesterone-concentrating cells: the APOA and VIN (Roy, Wilson & Kelley 1986). The number and intensity of progesterone-labelled cells increases following oestrogen administration and the distribution of labelled cells expands to include telencephalic, thalamic and midbrain nuclei, which suggests that progesterone receptors are oestrogen-inducible in *X. laevis* as they are in mammals.

Many CNS steroid targets participate in the neural control of sex behaviours

The location of androgen target cells in the central nervous system has been surveyed autoradiographically following administration of testosterone (Kelley, Morrell & Pfaff 1975) or DHT (Kelley 1981) to gonadectomized frogs and confirmed by *in situ* hybridization of a cloned *X. laevis* androgen receptor probe to brain sections of intact animals (Cohen, Fischer & Kelley 1991). Brain nuclei which contain labelled cells after administration of radioactive androgen have been implicated in the neural control of male reproductive behaviours (Fig. 9.6). In the anterior spinal cord, many cells are labelled following testosterone and DHT administration (Kelley, Morrell *et al.* 1975; Erulkar *et al.* 1981). Some have been identified as motor neurons which project to arm muscles active in the clasp reflex: flexor carpi radialis and sternoradialis. The majority of androgen-labelled cells in the brain, however, are found in nuclei associated with the generation of vocal behaviours (Fig. 9.6; see description of the vocal neuromuscular system below). For example, the medullary nucleus, cranial nerve nucleus N.IX–X, which contains the motor neurons innervating the vocal organ, contains androgen-concentrating and androgen receptor mRNA-expressing cells (Kelley 1980; Cohen *et al.* 1991). Counts of laryngeal axons and numbers of labelled cells after tracer injection into the laryngeal muscles suggest that about half of the cells in N.IX–X are laryngeal motor neurons; at least some of the neurons in N.IX–X that contain androgen receptor are believed to be laryngeal motor neurons (Kelley 1980). The two major sources of innervation to laryngeal motor neurons, the DTAM nucleus of the superior reticular formation and the adjacent inferior reticular formation, are labelled following both testosterone and DHT adminstration (Kelley, Morrell *et al.* 1975; Kelley 1981). The reticular neurons have been implicated in generating mate-call patterns in anurans (Schmidt 1976). None of the above brain nuclei are labelled following administration of radioactive oestrogen or progesterone (see above).

Within the midbrain, cells in the laminar nucleus of the torus semicircularis, an auditory nucleus which may play a role in sound localization, concentrate DHT as they do oestradiol. Other sensory nuclei are also labelled following DHT administration: a presumed vestibular nucleus, several thalamic nuclei and an auditory nucleus. There are no DHT-labelled cells in the anterior preoptic area. Neurons of the telencephalon do not concentrate androgen nor do they express androgen receptor mRNA (Kelley 1981; Cohen *et al.* 1991). It is not yet clear what role CNS oestrogen and progesterone targets play in female reproductive behaviour. The VIN has been implicated in control of gonadotropin secretion by the pituitary; the torus is an auditory nucleus which could play a role in sound localization by gravid females (see discussion by Morrell *et al.* 1975; Kelley 1981). We do not yet know when during development oestrogen and progesterone receptors appear in the central nervous system.

One way in which a diversity of hormone effects can be generated from a limited variety of steroids is via control of enzymes which metabolize a less active hormone, such as testosterone, to a more active form, such as dihydrotestosterone. Cells with these specific enzymatic identities can be revealed by comparing the patterns of labelled cells following administration of different radioactive steroids (see Fig. 9.6) and by utilizing information gained, primarily in avian and mammalian systems, on the metabolic pathways of the steroids. For example, because DHT and oestradiol are not mutually interconvertible, a similar distribution of labelled cells (e.g. in the torus semicircularis) suggests the presence of oestrogen and androgen receptors rather than an enzymatic capability. Since labelled cells are found in APOA after testosterone administration and oestradiol administration but not after DHT, we infer that APOA neurons can aromatize testosterone to oestradiol but not reduce testosterone to DHT. In the anterior spinal cord, many cells are labelled following testosterone and DHT administration (Kelley, Morrell *et al.* 1975; Erulkar *et al.* 1981). Their location coincides with marked 5α reductase activity, suggesting that spinal cord neurons can metabolize testosterone to DHT (Jurman, Erulkar & Krieger 1982).

The onset of sensitivity to steroid hormones

Thyroxine secretion is required for the onset of sensitivity to oestrogen and androgen

How is the onset of sensitivity to gonadal steroid hormones controlled? Several studies suggest that competence to respond to gonadal steroids is established by thyroxine secretion. The ability of tadpoles to respond to oestrogen with induction of the vitellogenin gene begins at stage 62 and is regulated by thyroxine secretion (May & Knowland 1980). Thyroxine does not directly confer inducibility on the vitellogenin gene but instead produces a change in the population of competent hepatocytes (Kawahara, Kohara *et al.* 1987). The ability of DHT to induce laryngeal growth (see below) also appears to depend upon thyroxine. The larynges of tadpoles arrested at stage 54 by blockade of thyroxine synthesis do not respond at all to exogenous DHT (Robertson & Kelley 1992). Thyroxine probably does not act on the gonads to promote steroid secretions since the gonads do not appear to express thyroxine receptor (see Kawahara, Baker & Tata 1991) nor is their development affected, during tadpole treatment, by lack of thyroxine (Robertson & Kelley 1992). Thyroxine secretion is not required for continued sensitivity to oestrogen or androgen since levels are extremely low or undetectable after metamorphosis (Leloup & Buscaglia 1977).

The onset of receptor expression is independent of hormone secretion

The androgen and oestrogen receptors belong to a large gene superfamily that also includes the receptor for thyroid hormone (TH). In *X. laevis*, the oestrogen (ER), androgen (AR) and thyroid hormone (TR) receptors have been cloned (Barton & Shapiro 1988; He *et al.* 1990; Yaoita, Shi & Brown 1990), permitting detection of their mRNAs during early development.

Two forms of the thyroid hormone receptor have been cloned in *X. laevis*, α and β (Yaoita *et al.* 1990). The β form is of particular interest because it is thyroxine-regulated in *Xenopus* (Kanamori & Brown 1992) and is expressed in a tissue and development stage-dependent fashion (Yaoita & Brown 1990). In tadpoles, expression of the α form is very abundant in nervous system and epithelia, less abundant in connective tissue and muscle, and apparently absent from developing gonads (Kawahara, Baker *et al.* 1991). In this study, the extent and intensity of hybridization using a 3′ fragment of the TR α receptor increased between stages 44 and 58; a subsequent decline was noted in the central nervous system. Northern analyses suggest that oestrogen receptor mRNA is present beginning at tadpole stage 56 (Baker & Tata 1990; this analysis was performed with probes to the chicken ER and the tissue distribution of ER is not described). Expression of TR mRNA, measured by the primer extension method in whole tadpoles, was detected from tadpole stage 38 onwards (Yaoita & Brown 1990) and may be present earlier, perhaps even as a maternal transcript (Banker, Bigler & Eisenman 1991).

In situ hybridization of *Xenopus* AR riboprobes to developing androgen-sensitive tissues demonstrates the presence of AR mRNA in larynx as early as tadpole stage 50 and in brain as early as tadpole stage 54 (M. Cohen & D.B. Kelley unpublished; Cohen *et al.* 1991). Thyroxine secretion is initiated at tadpole stage 54 (Leloup & Buscaglia 1977) and androgen secretion by stage 48, seven days after fertilization (Kang *et al.* 1994). Thus for thyroxine (at least the α form) and steroid receptors initial expression of mRNAs precedes secretion of ligand. The earliest appearing receptor of the three is that for thyroid hormone. Thyroxine secretion regulates the onset of androgen-induced laryngeal growth (Robertson & Kelley 1992) and oestrogen-induced vitellogenin synthesis (Knowland 1978). A possible mechanism, thyroxine-induced expression of steroid receptors, appears to be ruled out by the observation that their mRNAs are expressed prior to thyroxine secretion. Thyroxine might, however, control the expression of an additional transcription factor whose presence is required for androgen or oestrogen regulation of target tissue development.

Androgen-induced masculinization of the vocal system

During secondary sexual differentiation in *Xenopus laevis*, the phenotype of males diverges from that of females owing to the secretion of gonadal androgen. Our studies have focused on masculinization of the vocal system (larynx and laryngeal motor neurons) because the differentiation of cellular targets can readily be manipulated by providing or withholding androgenic steroids at different developmental stages. The extreme androgen-sensitivity of the larynx and the neurons which control vocal behaviour provide a powerful experimental system for identification of the molecular basis of sexual differentiation.

How are mate-calls produced?

Sexually receptive males sing (mate-call) to attract and excite females (Wetzel & Kelley 1983; Picker 1983). Mate calls consist of rapid trills with alternating fast (70 Hz) and slow (35 Hz) phases (Fig. 9.4c). The fast portion of the mate call becomes progressively louder during the trill and we have recently demonstrated that this amplitude modulation is an essential attractive feature of the male's song (Tobias, Bivens *et al.* 1991). Trills are produced by contractions of the intrinsic muscles of the vocal organ, the larynx (Fig. 9.5; Ridewood 1898; Yager 1992), a boxlike structure of muscle and cartilage that contains two sound-producing discs of arytenoid cartilage that are normally tightly apposed. When the bipennate laryngeal muscles contract, the discs pop apart and a click is produced; repeated clicks constitute the trills of mate calling.

Laryngeal muscles and motor neurons make up a neuromuscular system dedicated exclusively to vocalization. Laryngeal muscles contract to produce calls in response to activity of the laryngeal motor neurons located in cranial nerve nucleus IX–X in the caudal medulla (Fig. 9.6; Kelley 1980). We have outlined an anatomically connected system of brain nuclei that provides input to the calling motor neurons (Wetzel, Haerter & Kelley 1985). On the basis of our studies in *X. laevis* and those of Schmidt in other anurans (cf. Schmidt 1976), we believe that the CNS motor pathway for song includes a vocal pattern generator in the superior reticular formation (DTAM or the pretrigeminal nucleus of the dorsal tegmental area of the medulla), interneurons within the motor nucleus and in inferior reticular formation, sensory nuclei in the thalamus (auditory, lateral line and somatosensory) and certain nuclei of ventral diencephalon (anterior preoptic area) and telencephalon (ventral striatum). Many of these brain nuclei are targets for steroid hormones (Fig. 9.6).

Ontogeny of vocal neuroeffectors

The vocal organ, or larynx, becomes masculinized during PM development (Sassoon & Kelley 1986; Fischer & Kelley 1991). The adult male larynx differs markedly from that of the female in size and cellular composition (Fig. 9.5).

Sexually dimorphic characteristics of the adult larynx include the type of cartilage present (see Yager this volume pp. 121–141), the number and twitch characteristics of laryngeal muscle fibres and the strength of synaptic connections between laryngeal motor neurons and muscle fibres.

We have divided the PM development of males into seven stages based on the relation between laryngeal weight and attainment of masculine properties (Tobias, Marin & Kelley 1991a; Table 9.2). At the end of metamorphosis (PM0), the number and type of laryngeal muscle fibres is the same in the sexes, as is the morphology of laryngeal cartilages. In males, the larynx then undergoes a complex transformation that involves changes in cartilage composition, the addition of laryngeal muscle fibres, their conversion to all fast twitch, and alterations in the synaptic efficacy of the laryngeal neuromuscular junction (Sassoon & Kelley 1986; Sassoon, Segil & Kelley 1986; Marin, Tobias & Kelley 1990; Tobias, Marin et al. 1991a,b; Fischer & Kelley 1991). In females, the larynx enlarges slowly and does not undergo the dramatic transformation seen in males.

While the most dramatic changes in the male larynx occur after metamorphosis is complete, they originate during tadpole stages. The number of axons innervating laryngeal muscle is greater in males than in females at PM0 and this sex difference can be traced to a period of axonal outgrowth between tadpole stages 59 and 62; such axonal outgrowth is absent in females (Kelley & Dennison 1990). Thus the masculine phenotype begins differentiation during late tadpole development and is realized during juvenile life.

Hormonal control of vocal system masculinization

Most characteristics of the male larynx are established by exposure to male-typical hormones during development; once the masculine phenotype is established, continued exposure to hormone is no longer necessary. We have examined the larynges of adult male X. *laevis* for several years after castration. Their weights do not differ from intact males (Segil et al. 1987), they maintain a completely fast-twitch complement of muscle fibres (Tobias & Kelley 1987; Tobias, Marin et al. 1991b) and muscle fibre number (Marin et al. 1990; Watson et al. 1993). However, although the larynges of castrated males are capable of producing male songs (Tobias & Kelley 1987), castrated males do not actually sing (Wetzel & Kelley 1983). Mate-calls are produced by a synaptically connected set of brain neurons many of which are, like the larynx, regulated by androgen (Fig. 9.6, upper panel; Kelley, Morrell et al. 1975; Kelley 1981; Kelley & Dennison 1990). Patterned activity in these neurons produces mate-calling and requires the secretion of testicular androgen in adulthood (Wetzel & Kelley 1983; Tobias & Kelley 1987).

The developing male larynx retains, even in the absence of the testes, the capacity to respond to exogenous androgen with laryngeal masculinization while the response of the female larynx decreases somewhat with age. The masculine contractile properties of laryngeal muscle fibres are normally established between PM2 and PM5 owing to testicular androgen secretion (Tobias, Marin *et al.* 1991a,b). If males are castrated at PM2 a female-like complement of mostly slow muscle fibres is maintained until exposure to androgen, when all fibres convert to fast twitch; the exogenous androgen can be provided at any stage of PM development in males and fibre conversion will follow (Tobias, Marin *et al.* 1991b). Females can also respond to androgen early in PM development by rapid and complete conversion of fibres to all fast twitch but this capacity is lost in adulthood (Sassoon, Gray & Kelley 1987) and months of androgen treatment are required for complete fibre type switching (Tobias, Marin *et al.* 1991b). The adult female also has a greatly diminished capacity to show male-like mate-calling in response to exogenous androgen (Hannigan & Kelley 1986; Watson & Kelley 1992). The later in development that females are given a testicular transplant, the less masculinized the sound frequencies that the larynx can produce (Watson & Kelley 1992). The loss of responsiveness to androgen experienced by females reflects changes in the cellular composition of the larynx, and perhaps also of the nervous system, resulting from lack of androgen exposure (see below).

Endocrine control of muscle fibre number

The muscular skeleton of the larynx is formed by 30 primary myofibres which appear at tadpole stage 40 (2.75 days post-fertilization, Pf). At tadpole stage 47 (5.5 days Pf), secondary myoblasts appear and surround the parent muscle fibre (J. Robertson & D.B. Kelley unpubl. obs.). Secondary myoblasts then proliferate and fuse to form new muscle fibres; about 7000 fibres are present at tadpole stage 66 (58 days Pf), the end of metamorphosis, in both males and females (Marin *et al.* 1990). Muscle fibre number is not controlled by exogenous androgen secretion during tadpole stages (Watson *et al.* 1993). Sexual differentiation of laryngeal muscle fibre number occurs after metamorphosis is complete. Sexually dimorphic numbers of fibres are

Table 9.2. Stages in the post-metamorphic masculinization of the larynx

Postmetamorphic stage	Larynx weight (mg)	Median body weight (g)	SIR[a] (g)	Age (months post-metamorphic)
0	3–10	1.5	0.98–1.8	0[b]
1	11–35	4.0	3.0–5.8	3
2	36–70	7.3	6.3–8.5	6
3	71–125	9.3	7.8–10.2	9
4	126–160	11.2	8.7–12.7	11
5	161–350	14.6	12.5–18.8	>12
6	>350	36.2	26.1–40.4	>24

[a]Semi-interquartile range.
[b]The end of metamorphosis, tadpole stage 66 of Nieuwkoop & Faber (1956).
From Tobias, Marin *et al.* (1991a).

present at PM1; by PM2, fibre number in the inner bipennate averages 31 353 in males and 10 175 in females (Marin *et al.* 1990) and essentially adult numbers have been established. Post-metamorphic fibre addition is due to myoblast proliferation and fusion to form new fibres (Sassoon, Segil *et al.* 1986; Sassoon & Kelley 1986).

Sex differences in the rate of post-metamorphic muscle fibre addition are attributable to androgen secretion in males. The antiandrogen, flutamide, blocks the dramatic muscle fibre addition characteristic of juvenile males (Sassoon & Kelley 1986). In juvenile frogs, androgen administration causes myoblasts to proliferate; newly generated myoblast nuclei are found in immature muscle fibres (Sassoon, Segil *et al.* 1986). More recently (Marin *et al.* 1990), we have shown that, within a critical developmental period, castration blocks muscle fibre addition in males, and androgen administration increases muscle fibre number in females; implantation of a testis can induce masculine muscle fibre numbers in juvenile females.

Ontogeny of laryngeal motor neurons

Laryngeal motor neurons are generated from tadpole stage 10½ until 54 (Gorlick & Kelley 1987). Their axons reach the larynx at stage 40. The gonads differentiate at tadpole stage 56, permitting sexual identification of tadpoles (Nieuwkoop & Faber 1956). At stage 56, the number of axons in each laryngeal nerve is high (~650) and the same for males and females (Kelley & Dennison 1990). In females, axon numbers *decrease* from tadpole stage 56 on, reaching lower (~225) adult values at PM2. In males, axon numbers *increase* between tadpole stages 59 and 62 (~750 axons) and then decline; adult values (~350 axons) are reached sometime after PM2. Adult male *X. laevis* have ~350 and adult female ~250 laryngeal motor neurons, as determined by retrograde labelling from laryngeal muscle (H.B. Simpson & D.B. Kelley unpublished); each neuron sends a single axon to the larynx (Simpson, Tobias & Kelley 1986). Thus, sex differences in axon numbers are present from stage 59 through adulthood and are attributable to two processes: more gain in males during early tadpole development followed by more loss in females. Sex differences in numbers of axons entering laryngeal muscle precede sex differences in muscle fibre numbers; males maintain elevated axon numbers through the period of post-metamorphic muscle fibre addition. These observations suggested that innervation could contribute to sex differences in laryngeal myogenesis. We have thus denervated larynges at PM2 and have induced muscle fibre loss in males but not in females; denervation blocks androgen-induced muscle fibre addition in females (Tobias, Marin *et al.* 1993). These findings suggest that innervation contributes to sexual differentiation of laryngeal muscle fibre number.

Endocrine control of axon number

The sex difference in axon outgrowth of laryngeal motor neurons during tadpole development is not due to sex differences in muscle fibre number

(Kelley & Dennison 1990; Marin *et al.* 1990). The elevated numbers of axons present in male tadpoles could, however, contribute to the survival or proliferation of myoblasts before metamorphosis and the greater number of muscle fibres in males could contribute to the maintenance of greater numbers of motor neurons in males during post-metamorphic development.

The increase in axon number observed in male tadpoles between stages 59 and 62 is controlled by androgen (Robertson *et al.* 1991). To investigate if androgen could be responsible for the early sex difference in axon number, we treated tadpoles with anti-androgen or androgen during pre-metamorphic development and counted laryngeal nerve axons at stage 62 using electron microscopy. Male tadpoles treated with the anti-androgen hydroxyflutamide (HF) have the same number of axons as control females. Anti-androgen treatment does not affect female axon number. However, treatment with androgen (DHT) causes an increase in female axon number above control male levels. It is not yet clear how the sexual differentiation of laryngeal axon number relates to the sexual differentiation of laryngeal motor neuron number. Neurogenesis in cranial nerve nucleus IX–X occurs until tadpole stage 54 and is not sexually dimorphic (Gorlick & Kelley 1986). Therefore, subsequent sex differences in axon numbers cannot result from sex differences in neurogenesis. The increase in number of axons in males that occurs between tadpole stages 59 and 62 must be due either to the formation of additional axons by neurons already projecting to the larynx or a late outgrowth of axons from previously non-projecting neurons.

The decrease in axon numbers that occurs in both sexes between tadpole stages 62 and PM2 probably results from ontogenetic cell death which occurs in *X. laevis* as in other vertebrates (Prestige & Wilson 1972). In one sexually dimorphic neuromuscular system which has been extensively investigated, the spinal nucleus of the bulbocavernosus in rats (SNB), sex differences in cell number are due to ontogenetic cell death (Nordeen *et al.* 1985). Motor neuron loss is attenuated in males by the secretion of testicular androgen which rescues motor neurons by acting on their synaptic targets, the LA/BC muscles (Forger *et al.* 1992). In the SNB system androgen exerts an indirect trophic effect on motor neuron survival via motor neuron/muscle interactions. This mechanism could also explain sex differences in the extent of axon loss in the laryngeal nerve of *X. laevis*.

Molecular analyses of androgen-directed development

The vocal organ of an adult male *X. laevis* has more and different cells than that of the female. For example, the male has many more muscle fibres, all of which are fast twitch; the fewer fibres of the female are predominantly slow twitch (Sassoon & Kelley 1986; Sassoon, Gray *et al.* 1987). In addition, the male cartilage is also more extensive than that of the female and contains a cell type, elastic cartilage, which the female lacks (see Yager this volume pp.

121–141). The presence of male-like cell numbers and types in adults depends upon testicular androgen secretion earlier in post-metamorphic development. We are investigating the cellular and molecular mechanisms which underlie androgen control of cell proliferation and cell differentiation in the larynx.

An androgen receptor isoform associated with hormone-induced laryngeal cell proliferation

Secondary sexual differentiation requires the expression of a functional androgen receptor (AR); in mammals, when the AR is absent or defective, genetic males develop phenotypically as females. The extreme sensitivity of the developing larynx to androgen together with very high levels of androgen binding (Kelley, Sassoon *et al.* 1989) suggested that this tissue might express high levels of receptor. Using the polymerase chain reaction and templates derived from the human AR, we cloned a fragment of the *Xenopus* AR and demonstrated that developing larynx expresses extraordinarily high levels of AR mRNA (He *et al.* 1990).

We have used the cloned *Xenopus* AR fragment as a probe in studies of the temporal and spatial expression of AR in developing larynx (Fischer, Catz & Kelley 1993). Two AR mRNAs are expressed in developing larynx: AR mRNAα (9.6 kb) and AR mRNAβ (8.0 kb). Southern analyses reveal the presence of two AR genes; it is not yet clear whether AR mRNAα and AR mRNAβ originate from these genes or are instead the alternatively spliced products of one gene. Steroid receptor proteins contain discrete functional domains: a hormone-binding domain located at the carboxy terminus, a centrally located DNA-binding domain, and a domain (the A/B or hypervariable domain) located at the amino terminus. We wished to determine if AR α and β mRNAs differ in regions encoding these domains. Using PCR-derived cDNA probes encoding different domains (hypervariable, DNA-binding, and ligand-binding) of the AR we have found that the 5′ A/B fragment recognizes only the 9.6 kb AR mRNAα while more 3′ fragments recognize both transcripts. Thus, the β transcript differs from the α transcript within the 5′ hypervariable domain.

Northern analyses reveal that the β transcript is transiently expressed while cells in muscle and cartilage are proliferating, whereas the α transcript is expressed throughout post-metamorphic development. Because its expression is temporally restricted to early post-metamorphic stages, we conclude that AR β is a developmentally regulated androgen receptor mRNA isoform. To localize AR mRNAα and AR mRNAβ, two AR probes were hybridized *in situ* to sections of a juvenile larynx: one that recognizes both mRNAs and the 5′ A/B fragment that recognizes only the α form. The highest levels of α/β expression occur in a proliferative zone that is induced to form elastic cartilage in males in response to androgen secretion and in developing muscle. Hybridization with the α-specific probe reveals markedly less expression in the proliferating elastic cartilage precursor cells and developing muscle. We conclude that expression of the developmentally regulated AR mRNA isoform, AR mRNAβ, is associated with a specific phase of laryngeal masculinization, the androgen-evoked cell proliferation that occurs between PM0 and PM2, while the later androgen-induced differentiation of specific muscle or cartilage types that occurs between PM2 and PM5 is associated with the constitutively expressed regulated AR mRNA isoform, AR mRNAα. Androgenic control of cell numbers may require the expression of the developmentally regulated AR subtype coded for by AR mRNAβ while androgenic control of cell types may require only the expression of the constitutively expressed protein derived from AR mRNAα.

A laryngeal-specific, androgen-regulated myosin heavy chain isoform expressed in fast twitch muscle fibres

A striking feature of the adult larynx is the sexual differentiation of fibre type (see above). Fibre type switching is regulated by androgen secretion during post-metamorphic development (Tobias, Marin *et al.* 1991b) and we have thus sought to identify the myosins expressed by male and female muscles in order to understand the molecular basis for androgen-directed sexual differentiation in this system. We have isolated cDNA clones encoding portions of a new *Xenopus* myosin heavy chain (MHC) gene and have detected expression of this gene only in laryngeal muscle and specifically in males (Catz *et al.* 1992). All adult male laryngeal muscle fibres express the laryngeal myosin (LM) and all adult male fibres are fast twitch. Adult female laryngeal muscle expresses LM only in some fibres; on the basis of histochemical and size criteria, LM-expressing fibres in females are fast twitch. Thus LM is a candidate laryngeal-specific fast twitch myosin in *Xenopus laevis*. Because LM could be co-expressed with other myosins, it is not yet clear whether expression of this particular myosin heavy chain is responsible for the extremely rapid (70 Hz) rate of contraction that can be achieved by male laryngeal muscle.

At the end of metamorphosis (PM0) both male and female laryngeal muscle contains a mixed population of fibres, mostly slow twitch (Sassoon, Gray *et al.* 1987). This fibre composition is maintained through PM2 (about six months later), after which slow fibres in males convert to fast twitch (Tobias, Marin *et al.* 1991a). If LM contributes to fibre type expression, we would expect a sexually differentiated pattern of expression that precedes changes in fibre type. Northern blots and *in situ* hybridization revealed that males attain adult levels of LM expression by PM3, while in females LM expression peaks transiently at PM2 (Catz *et al.* 1992). Treatment of juvenile female frogs with the androgen dihydrotestosterone masculinizes LM expression. Thus our current evidence suggests that LM encodes a laryngeal-specific, androgen-regulated myosin heavy chain isoform whose expression accounts for the sexual differentiation of muscle fibre types and the fast twitch characteristics of male muscle. The LM gene is a useful molecular marker for androgenic control of cell type in developing larynx.

Conclusions: future directions for research

The process of sexual differentiation in X. *laèvis* is one shared by most vertebrates: genetic determination of gonadal sex (primary sexual differentiation)
followed by gonadal hormone determination of phenotypic sex (secondary
sexual differentiation). Recent progress in molecular biology together with
the advantages of this amphibian system for developmental studies suggest
that X. *laevis* could provide a powerful experimental system in which to
examine vertebrate sexual differentiation. However, we can identify several
conceptual and experimental barriers which must be removed before the utility
of this system can be realized. Studies of primary sexual differentiation are
currently difficult because the sex chromosomes have not been identified.
In mammals, the androgen receptor gene is on the X chromosome; if this
location is evolutionarily conserved, the recent cloning of the X. *laevis*
androgen receptor may provide a tool for sex chromosome identification.
The relation between genetic control of gonadal sex (inferred from studies
with sex-reversed individuals) and the ability of exogenous oestradiol to
ensure ovarian development is mysterious and must be resolved. With regard
to secondary sexual differentiation, the masculinization of the vocal organ
provides a useful focus of investigation. However, while we are beginning to
identify a number of genes whose expression is regulated by androgen, the
mechanisms for this regulation are poorly understood. In particular, it will be
important to identify genes whose response to androgen is independent of *de
novo* protein synthesis; these direct-response genes will permit identification
of the molecular cascade activated by androgen binding to its receptor and
ultimately responsible for expression of the masculine phenotype.

References

Adkins, E. (1975). Hormonal basis of sexual differentiation in Japanese quail. *J. comp.
    physiol. Psychol.* **89**: 61–71.
Baker, B. & Tata, J. (1990). Accumulation of proto-oncogene c-erb-A related
    transcripts during *Xenopus* development: association with early acquisition of
    response to thyroid hormone and estrogen. *EMBO J.* **9**: 879–885.
Banker, D., Bigler, J. & Eisenman, R. (1991). The thyroid hormone receptor gene
    (c-*erb* Aα) is expressed in advance of thyroid hormone maturation during the early
    embryonic development of *Xenopus laevis*. *Molec. cell. Biol.* **11**: 5079–5089.
Barton, M. & Shapiro, D. (1988). Transient administration of estradiol-17β establishes
    an autoregulatory loop permanently inducing estrogen receptor mRNA. *Proc. natn.
    Acad. Sci. USA* **85**: 7119–7123.
Baulieu, E.-E., Godeau, F., Scoderet, M. & Schodert-Slatkine, S. (1978). Steroid-
    induced meiotic division in *Xenopus laevis* oocytes: surface and calcium. *Nature,
    Lond.* **275**: 593–598.
Blackler, A. (1965). Germ-cell transfer and sex ratio in *Xenopus laevis*. *J. Embryol.
    exp. Morph.* **13**: 51–61.

Catz, D., Fischer, L., Tobias, M., Moschella, T. & Kelley, D. (1992). Sexually dimorphic expression of a laryngeal-specific, androgen-regulated myosin heavy chain gene during *Xenopus laevis* development. *Devl Biol.* **154**: 366–376.

Chang, C.Y. & Witschi, E. (1956). Genic control and hormonal reversal of sex differentiation in *Xenopus*. *Proc. Soc. exp. Biol. Med.* **93**: 140–144.

Cohen, M., Fischer, L. & Kelley, D. (1991). Onset of androgen receptor mRNA expression in the CNS of *Xenopus laevis*: localization using *in situ* hybridization. *Abstr. Soc. Neurosci.* **17**: 1317.

Crews, D., Wibbels, T. & Gutzke, W.H.N. (1989). Action of steroid sex hormones on temperature-induced sex determination in the snapping turtle (*Chelydra serpentina*). *Gen. comp. Endocr.* **76**: 158–166.

Dennis-Smith, L., Xu, W. & Varnold, R. (1991). Oogenesis and oocyte isolation. In Xenopus laevis: *practical uses in cell and molecular biology*: 45–60. (Eds Kay, B. & Peng, H.B.). Academic Press, New York.

Denny, P., Swift, S., Brand, N., Dabhade, N., Barton, P. & Ashworth, A. (1992). A conserved family of genes related to the testis determining gene, SRY. *Nucleic Acids Res.* **20**: 2887.

Dorizzi, L., Mignot, T.-M., Guichard, A., Desvages, G. & Pieau, C. (1991). Involvement of oestrogens in sexual differentiation of gonads as a function of temperature in turtles. *Differentiation* **47**: 9–17.

Dumont, J.N. (1972). Oogenesis in *Xenopus laevis*. 1. Stages of oocyte development in laboratory maintained animals. *J. Morph.* **136**: 153–180.

Engel, W. & Kobel, H.R. (1980). The z-chromosome is involved in the regulation of H-W (H-Y) antigen gene expression in *Xenopus*. *Cytogenet. Cell Genet.* **35**: 28–33.

Erulkar, S., Kelley, D.B., Jurman, M., Zemlan, F., Schneider, G. & Krieger, N. (1981). The modulation of the neural control of the clasp reflex in male *Xenopus laevis* by androgens. *Proc. natn. Acad. Sci. USA* **78**: 5876–5880.

Evans, R.M. (1988). The steroid and thyroid hormone receptor superfamily. *Science* **240**: 889–895.

Fischer, L., Catz, D. & Kelley, D. (1993). An androgen receptor mRNA isoform associated with hormone-inducible cell proliferation. *Proc. natn. Acad. Sci. USA* **90**: 8254–8258.

Fischer, L. & Kelley, D. (1991). Androgen receptor expression and sexual differentiation of effectors for courtship song in *Xenopus laevis*. *Semin. Neurosci.* **3**: 469–480.

Forger, N., Hodges, L., Roberts, S. & Breedlove, S. (1992). Regulation of motoneuron death in the spinal nucleus of the bulbocavernosus. *J. Neurobiol.* **23**: 1192–1203.

Gahr, M., Wibbels, T. & Crews, D. (1992). Sites of estrogen uptake in embryonic *Trachemys scripta*, a turtle with temperature-dependent sex determination. *Biol. Reprod.* **46**: 458–463.

Gallien, L. (1962). Comparative activity of sexual steroids and genetic constitution in sexual differentiation of amphibian embryos. *Gen. comp. Endocr. Suppl.* **1**: 346–355.

Gallien, L. & Chalumeau-le-Foulgoc, M. (1960). Mise en évidence de stéroîdes oestrogènes dans l'ovaire juvénile de *Xenopus laevis* Daudin, et cycle des oestrogènes au cours de la ponte. *C.r. hebd. Seanc. Acad. Sci., Paris (Ser. D)* **251**: 460–462.

Giese, K., Cox, J. & Groschedl, R. (1992). The HMG domain of lymphoid enhancer factor 1 bends DNA and facilitates assembly of functional nucleoprotein structures. *Cell* **69**: 185–195.

Goldberg, E., Boyse, E., Bennett, D., Scheid, M. & Carswell, E. (1971). Serological demonstration of H-Y antigen on mouse sperm. *Nature, Lond.* **232**: 478–480.

Goldberg, E., McLaren, A. & Reilly, B. (1991). Male antigen defined serologically does not identify a factor responsible for testicular development. *J. reprod. Immunol.* **20**: 305–309.

Gorlick, D. & Kelley, D. (1986). The ontogeny of androgen receptors in the CNS of *Xenopus laevis. Devl Brain Res.* **26**: 193–201.

Gorlick, D. & Kelley, D. (1987). Neurogenesis in the vocalization pathway of *Xenopus laevis. J. comp. Neurol.* **254**: 614–627.

Graf, J.-D. & Kobel, H. (1991). Genetics of *Xenopus laevis.* In Xenopus laevis: *practical uses in cell and molecular biology*: 19–34. (Eds Kay, B. & Peng, H.B.). Academic Press, New York.

Griffiths, R. (1991). The isolation of conserved DNA sequences related to the human sex-determining region Y gene from the lesser black-backed gull (*Larus fuscus*). *Proc. R. Soc. (B)* **244**: 123–128.

Hamilton, L. (1963). An experimental analysis of the development of the haploid syndrome in embryos of *Xenopus laevis. J. Embryol. exp. Morph.* **11**: 267–278.

Hannigan, P. & Kelley, D. (1986). Androgen-induced alterations in vocalizations of female *Xenopus laevis*: modifiability and constraints. *J. comp. Physiol. (A)* **158**: 517–527.

Harley, V.R., Jackson, D.I., Hextall, P.J., Hawkins, J.R., Berkovitz, G.D., Sockanathan, S., Lovell-Badge, R. & Goodfellow, P.N. (1992). DNA binding activity of recombinant SRY from normal males and XY females. *Science* **255**: 453–456.

He, W.-W., Fischer, L., Sun, S., Bilhartz, D., Zhu, X., Young, C., Kelley, D. & Tindall, D. (1990). Molecular cloning of androgen receptor from divergent species with the PCR technique: complete cDNA sequence of the mouse androgen receptor and isolation of cDNA probes from dog, guinea pig and frog. *Biochem. biophys. Res. Communs* **171**: 697–704.

Hutchison, J.B. & Poynton, J.C. (1963). A neurological study of the clasp reflex in *Xenopus laevis* (Daudin). *Behaviour* **22**: 41–63.

Iwasawa, H. & Yamaguchi, K. (1984). Ultrastructural study of gonadal development in *Xenopus laevis. Zool. Sci.* **1**: 591–600.

Jurman, M., Erulkar, S. & Krieger, N. (1982). Testosterone 5 alpha-reductase in spinal cord of *Xenopus laevis. J. Neurochem.* **38**: 657–661.

Kanamori, A. & Brown, D. (1992). The regulation of thyroid hormone receptor β genes by thyroid hormone in *Xenopus laevis. J. biol. Chem.* **267**: 739–745.

Kang, L., Marin, M. & Kelley, D. (1994). Androgen secretion and biosynthesis in developing *Xenopus laevis. Abstr. Soc. Neurosci.* **20**: 460.

Kawahara, A., Baker, B. & Tata, J. (1991). Developmental and regional expression of thyroid hormone receptor genes during *Xenopus* metamorphosis. *Development* **112**: 933–943.

Kawahara, A., Kohara, S., Sugimoto, Y. & Amano, M. (1987). A change of the hepatocyte population is responsible for the progressive increase of vitellogenin synthetic capacity at and after metamorphosis of *Xenopus laevis. Devl Biol.* **122**: 139–145.

Kelley, D. (1980). Auditory and vocal nuclei of frog brain concentrate sex hormones. *Science* **207**: 553–555.

Kelley, D. (1981). Locations of androgen-concentrating cells in the brain of *Xenopus laevis*: autoradiography with ³H-dihydrotestosterone. *J. comp. Neurol.* **199**: 221–231.

Kelley, D. (1982). Hormone control of female sex behavior in South African clawed frogs, *Xenopus laevis*. *Horm. Behav.* **1**: 158–174.

Kelley, D.B. & Dennison, J. (1990). The vocal motor neurons of *Xenopus laevis*: development of sex differences in axon number. *J. Neurobiol.* **21**: 869–882.

Kelley, D.B., Morrell, J.I. & Pfaff, D.W. (1975). Autoradiographic localization of hormone-concentrating cells in the brain of an amphibian, *Xenopus laevis*. I. Testosterone. *J. comp. Neurol.* **164**: 63–78.

Kelley, D.B. & Pfaff, D.W. (1976). Hormone effects on male sex behavior in adult South African clawed frogs, *Xenopus laevis*. *Horm. Behav.* **7**: 159–182.

Kelley, D.B. & Pfaff, D.W. (1978). Generalizations from comparative studies on neuroanatomical and endocrine mechanisms of sexual behavior. In *Biological determinants of sexual behaviour*: 225–254. (Ed. Hutchison, J.). Wiley and Sons, Chichester.

Kelley, D., Sassoon, D., Segil, N. & Scudder, M. (1989). Development and hormone regulation of androgen receptor levels in the sexually dimorphic larynx of *Xenopus laevis*. *Devl Biol.* **131**: 111–118.

Kelley, D.B., Weintraub, A.S. & Bockman, R.S. (1987). Oviductal prostaglandin synthesis and female sexual receptivity in *Xenopus laevis*. *Adv. Prostaglandin, Thromboxane, Leukotriene Res.* **17B**: 1133–1135.

Knowland, J. (1978). Induction of vitellogenin synthesis in *Xenopus laevis* tadpoles. *Differentiation* **12**: 47–51.

Koopman, P., Gubbay, J., Vivian, N., Goodfellow, P. & Lovell-Badge, R. (1991). Male development of chromosomally female mice transgenic for *Sry*. *Nature, Lond.* **351**: 117–121.

Lambdin, L. & Kelley, D. (1986). Organization and activation of sexually dimorphic vocalizations: androgen levels in developing and adult *Xenopus laevis*. *Abstr. Soc. Neurosci.* **12**: 1213.

Lecouteux, A. (1988). Gonadotropin action on androgen synthesis in short-term incubation of explants and dispersed cells in the testis of *Xenopus laevis*. *J. exp. Zool.* **245**: 187–193.

Leloup, J. & Buscaglia, M. (1977). La triiodothyronine, hormone de la métamorphose des amphibiens. *C.r. hebd. Seanc. Acad. Sci. Paris (Sér. D.)* **284**: 2261–2263.

Lovell-Badge, R. (1993). Sex determining gene expression during embryogenesis. *Phil. Trans. R. Soc. (B)* **339**: 159–164.

Marin, M.L., Tobias, M.L. & Kelley, D.B. (1990). Hormone-sensitive stages in the sexual differentiation of laryngeal muscle fiber number in *Xenopus laevis*. *Development* **110**: 703–712.

May, F. & Knowland, J. (1980). The role of thyroxine in the transition of vitellogenin synthesis from noninducibility to inducibility during metamorphosis in *Xenopus laevis*. *Devl Biol.* **77**: 419–430.

Merchant-Larios, H. & Villapanido, I. (1981). Ultrastructural events during early gonadal development in *Rana pipiens* and *Xenopus laevis*. *Anat. Rec.* **199**: 349–360.

Mikamo, K. & Witschi, E. (1963). Functional sex-reversal in genetic females of *Xenopus laevis*, induced by implanted testes. *Genetics* **48**: 1411–1421.

Mikamo, K. & Witschi, E. (1964). Masculinization and breeding of the WW *Xenopus*. *Experientia* **20**: 622–623.

Mikamo, K. & Witschi, E. (1966). The mitotic chromosomes in *Xenopus laevis* (Daudin): normal, sex reversed and female WW. *Cytogenetics* **5**: 1–19.

Morrell, J.I., Kelley, D.B. & Pfaff, D.W. (1975). Autoradiographic localization of hormone-concentrating cells in the brain of the amphibian, *Xenopus laevis*. II. Estradiol. *J. comp. Neurol.* **164**: 63–78.

Moss, R. & McCann, S. (1973). Induction of mating behavior in rats by luteinizing hormone-releasing factor. *Science* **181**: 177–179.

Nieuwkoop, P.D. & Faber, J. (1956). *Normal table of* Xenopus laevis *(Daudin)*. North-Holland, Amsterdam.

Nordeen, E., Nordeen, K., Sengelaub, D. & Arnold, A. (1985). Androgens prevent normally occurring cell death in a sexually dimorphic spinal nucleus. *Science* **229**: 671–673.

Ohno, S., Nagai, Y., Ciccarese, S. & Iwata, H. (1979). Testis-organizing H-Y antigen and the primary sex-determining mechanism of mammals. *Recent Prog. Horm. Res.* **35**: 449–476.

Pfaff, D. (1973). Luteinizing hormone-releasing hormone factor potentiates lordosis behavior in hypophysectomized ovariectomized female rats. *Science* **182**: 1148–1149.

Picard, J-Y., Benarous, R., Guerrier, D., Josso, N. & Kahn, A. (1986). Cloning and expression of a cDNA for anti-Mullerian hormone. *Proc. natn. Acad. Sci. USA* **83**: 5464–5468.

Picker, M.D. (1983). Hormonal induction of the aquatic phonotactic response of *Xenopus*. *Behaviour* **84**: 74–90.

Prestige, M. & Wilson, M. (1972). Loss of axons from ventral roots during development. *Brain Res.* **41**: 467–470.

Ridewood, W. (1898). On the structure and development of the hyobranchial skeleton and larynx in *Xenopus* and *Pipa*; with remarks on the affinities of the Aglossa. *J. Linn. Soc. (Zool.)* **26**: 53–128.

Robertson, J. & Kelley, D.B. (1992). Gonadal and laryngeal development in hypothyroid tadpoles. *Am. Zool.* **32**: 86A.

Robertson, J., Watson, J. & Kelley, D.B. (1991). Laryngeal nerve axon number in premetamorphic *Xenopus laevis* is androgen dependent. *Abstr. Soc. Neurosci.* **17**: 1320.

Roy, E.J., Wilson, M.A. & Kelley, D.B. (1986). Estrogen-induced progestin receptors in the brain and pituitary of the South African clawed frog, *Xenopus laevis*. *Neuroendocrinology* **42**: 51–56.

Russell, W.M.S. (1954). Experimental studies of the reproductive behaviour of *Xenopus laevis*. I. The control mechanisms for clasping and unclasping, and the specificity of hormone action. *Behaviour* **7**: 113–188.

Sassoon, D., Gray, G. & Kelley, D. (1987). Androgen regulation of muscle fiber type in the sexually dimorphic larynx of *Xenopus laevis*. *J. Neurosci.* **7**: 3198–3206.

Sassoon, D. & Kelley, D. (1986). The sexually dimorphic larynx of *Xenopus laevis*: development and androgen regulation. *Am. J. Anat.* **177**: 457–472.

Sassoon, D., Segil, N. & Kelley, D. (1986). Androgen-induced myogenesis and chondrogenesis in the larynx of *Xenopus laevis*. *Devl Biol.* **113**: 135–140.

Schmidt, R. (1976). Neural correlates of frog calling. Isolated brainstem. *J. comp. Physiol.* **108**: 99–113.

Schmidt, R. (1988). Mating call phonotaxis in female American toads: lesions of central auditory system. *Brain Behav. Evol.* **32**: 119–128.

Schmidt, R. (1989). Mating call phonotaxis in female American toad: lesions of anterior preoptic nucleus. *Horm. Behav.* **23**: 1–9.

Schmidt, R. (1993). Anuran calling circuits: inhibition of pretrigeminal nucleus by prostaglandin. *Horm. Behav.* **27**: 82–91.

Segil, N., Silverman, L. & Kelley, D.B. (1987). Androgen binding levels in a sexually dimorphic muscle of *Xenopus laevis. Gen. comp. Endocrinol.* **66**: 95–101.

Simpson, H.B., Tobias, M.L. & Kelley, D.B. (1986). Origin and identification of fibers in the cranial nerve IX-X complex of *Xenopus laevis*: Lucifer Yellow backfills *in vitro. J. comp. Neurol.* **244**: 430–444.

Smith, L. (1989). The induction of oocyte maturation: transmembrane signaling events and regulation of the cell cycle. *Development* **107**: 685–699.

Taylor, J.A. & Boyd, S.K. (1991). Thyrotropin-releasing hormone facilitates display of reproductive behavior and locomotor behavior in an amphibian. *Horm. Behav.* **25**: 128–136.

Thomas, E.O., Licht, P., Wibbels, T. & Crews, D. (1992). Hydroxysteroid dehydrogenase activity associated with sexual differentiation in embryos of the turtle *Trachemys scripta. Biol. Reprod.* **46**: 140–145.

Tobias, M., Bivens, R., Nowicke, S. & Kelley, D.B. (1991). Amplitude modulation is an attractive feature of *X. laevis* song. *Abstr. Soc. Neurosci.* **17**: 1403.

Tobias, M.L. & Kelley, D.B. (1987). Vocalizations by a sexually dimorphic isolated larynx: peripheral constraints on behavioral expression. *J. Neurosci.* **7**: 3191–3197.

Tobias, M.L., Marin, M.L. & Kelley, D.B. (1991a). Development of functional sex differences in the larynx of *Xenopus laevis. Devl Biol.* **147**: 251–259.

Tobias, M.L., Marin, M.L. & Kelley, D.B. (1991b). Temporal constraints on androgen directed laryngeal masculinization in *Xenopus laevis. Devl Biol.* **147**: 260–270.

Tobias, M., Marin, M. & Kelley, D.B. (1993). The roles of sex, innervation and androgen in laryngeal muscle of *Xenopus laevis. J. Neurosci.* **13**: 324–333.

Tymowska, J. & Kobel, H.R. (1972). Karyotype analysis of *Xenopus muelleri* (Peters) and *Xenopus laevis* (Daudin), Pipidae. *Cytogenetics* **11**: 270–278.

Vigier, B., Forest, M.G., Eychenne, B., Bezard, J., Garrigou, O., Robel, P. & Josso, N. (1989). Anti-Muellerian hormone produces endocrine sex reversal of fetal ovaries. *Proc. natn. Acad. Sci. USA* **86**: 3684–3688.

Vigier, B., Watrin, F., Magre, S., Tran, D. & Josso, N. (1987). Purified bovine AMH induces a characteristic freemartin effect in fetal rat prospective ovaries exposed to it *in vitro. Development* **100**: 43–55.

Wachtel, S., Bresler, P. & Koide, S. (1980). Does H-Y antigen induce the heterogametic ovary? *Cell* **20**: 859–864.

Wachtel, S., Ohno, S., Koo, G. & Boyse, E. (1975). Possible role for H-Y antigen in the primary determination of sex. *Nature, Lond.* **257**: 235–236.

Wallace, R. (1978). Oocyte growth in nonmammalian vertebrates. In *The vertebrate ovary: comparative biology and evolution*: 469–502. (Ed. Jones, R.E.). Plenum Press, New York.

Watson, J. & Kelley, D. (1992). Testicular masculinization of vocal behavior in juvenile female *Xenopus laevis*: prolonged sensitive period reveals component features of behavioral development. *J. comp. Physiol. (A)* **171**: 343–350.

Watson, J., Robertson, J., Sachdev, U. & Kelley, D. (1993). Laryngeal muscle and motor neuron plasticity in *Xenopus laevis*: testicular masculinization of a developing neuromuscular system. *J. Neurobiol.* **24**: 1615–1625.

Weintraub, A., Kelley, D. & Bockman, R. (1985). Prostaglandin induces sexual receptivity in female *Xenopus laevis. Horm. Behav.* **19**: 386–399.

Wetzel, D., Haerter, U. & Kelley, D. (1985). A proposed efferent pathway for mate calling in South African clawed frogs, *Xenopus laevis*: tracing afferents to laryngeal motor neurons with HRP-WGA. *J. comp. Physiol. (A)* **157**: 749–761.

Wetzel, D. & Kelley, D. (1983). Androgen and gonadotropin control of the mate calls of male South African clawed frogs, *Xenopus laevis*. *Horm. Behav.* **17**: 388–404.

Wibbels, T. & Crews, D. (1992). Specificity of steroid hormone-induced sex determination in a turtle. *J. Endocr.* **133**: 121–129.

Wickbom, T. (1945). Cytological studies on Dipnoi, Urodela, Anura, and *Emys*. *Hereditas* **31**: 241–346.

Wilczynski, W. & Capranica, R.R. (1984). The auditory system of anuran amphibians. *Prog. Neurobiol.* **22**: 1–38.

Wilson, J., George, F. & Griffin, J. (1981). The hormonal control of sexual development. *Science* **211**: 1278–1284.

Witschi, E. (1950). Génétique et physiologie de la différenciation du sexe. *Archs Anat. microsc. Morph. exp.* **39**: 215–246.

Witschi, E. (1971). Mechanisms of sexual differentiation. In *Hormones in development*: 601–618. (Eds Hamburg, M. & Barrington, E.). Appleton Century Crofts, New York.

Wright, C., Wright, S. & Knowland, J. (1983). Partial purification of estradiol receptor from *Xenopus laevis* liver and levels of receptor expression in relation to estradiol concentration. *EMBO J.* **2**: 973–977.

Yager, D. (1992). A unique sound production mechanism in the pipid anuran *Xenopus borealis*. *Zool. J. Linn. Soc.* **104**: 351–375.

Yaoita, Y. & Brown, D.D. (1990). A correlation of thyroid hormone receptor gene expression with amphibian metamorphosis. *Genes Dev.* **4**: 1917–1924.

Yaoita, Y., Shi, Y.-B. & Brown, D.D. (1990). *Xenopus laevis* α and β thyroid hormone receptors. *Proc. natn. Acad. Sci. USA* **87**: 7090–7094.

10 Underwater acoustics and hearing in the clawed frog, *Xenopus*

ANDREAS ELEPFANDT

Synopsis

Xenopus communicates acoustically in the same way as other frogs but completely under water. This requires that the communication and auditory system matches the properties of underwater sound in shallow waters. The ear shows the basic characteristics of a frog ear, but is modified. The tympanum is a cartilaginous disc beneath the skin, the stapes leaves the tympanic cavity caudally, and the two Eustachian tubes form a joint air-filled canal that connects to the mouth via one medial opening. In *Xenopus laevis*, males prefer to call from deeper parts of a water body. Their advertisement calls consist of clicks of which the dominant frequency is at 1600–2000 Hz. The hearing range is 200–3900 Hz, with optima at 600 Hz and 1400–1800 Hz. Sound is localized throughout the whole anterior semicircle. Frequency discrimination is found at 400–800 Hz and, with a limen attaining 2%, at 1400–2500 Hz. The latter range encompasses the dominant frequency of the advertisement call. The dominant frequency is constant for any male but differs between males so that the level of frequency discrimination allows for distinction between calling males. Since the calls of larger males tend to have lower dominant frequencies, choice of larger males might thus be possible. Biophysical studies show that the ear responds to sound pressure rather than particle velocity. The most probable mechanism is an air bubble resonance of the air in the Eustachian tubes and possibly the lungs. Tympanic vibration amplitude depends on sound direction, thus forming a basis for sound localization by the animal.

Introduction

For acoustic communication, three components are essential: a sender which produces the signal, a transmission channel through which the signal is broadcast, and a receiver which detects and analyses the signal. The capacities and constraints of each of these components as well as their interplay establish the range of potential communication. Acoustic communication under water in a frog such as *Xenopus*, therefore, requires not only a sound-producing apparatus and an ear constructed in such a way that sound can be radiated into and detected from the water, but also signals and communicative behaviour that correspond with the transmission properties of the aquatic environment. The chapter by Yager (this volume, pp. 121–141) deals with aspects of sound production. This account will first summarize the characteristics of underwater sound in the shallow waters where *Xenopus* lives and outline some potential adaptations of acoustic communication in *Xenopus*. Then the ear of *Xenopus* and its hearing abilities will be considered. Finally, initial investigations on the biophysics of hearing in *Xenopus* will be described.

Sound in shallow waters

Water is much denser and less compressible than air. As a consequence, a sound of a given energy yields a sound pressure that is 60 times, or 36 dB, higher in water than in air and, correspondingly, particle velocity that is 36 dB lower. To illustrate this difference, sound pressure in water is indicated not according to the SPL sound pressure scale of airborne sound (dB re 20 µPa;) but as dB re dyn/cm^2 or, more recently, as dB re µPa. Recalculation gives

$$0 \text{ dB re } \mu Pa = -100 \text{ dB re } dyn/cm^2 = -26 \text{ dB SPL,}$$

but when recalculating it is necessary to keep the differences in the two media in mind.

Sound propagation in water is nearly five times faster than in air (1480 m/s at 20 °C in freshwater). Correspondingly, wavelengths are nearly five times longer. Thus, the wavelength of a 1.5 kHz tone in water is approximately 1 m. The higher wavelength means that the acoustic near-field extends further, too. The acoustic near-field is an area in the vicinity of a sender where the normal far-field proportionality between sound pressure and particle velocity does not hold. Rather, particle velocity is higher relative to sound pressure than in the far-field, the more so the closer to the sender. In an infinite homogeneous medium, sound pressure decreases with distance from the sender as $1/r$ whereas particle velocity initially decays as $1/r^2$ in the near-field until it falls off as $1/r$, or proportionally to the pressure, in the far-field. The transition between near-field and far-field is smooth. Commonly, the far-field is considered to begin at 1–2 wavelengths, depending on the criterion. The larger extent of the acoustic near-field in water implies that a considerable part of acoustic communication in *Xenopus* occurs in the near-field.

Physically, water is a good medium for sound transmission. Energy loss by absorption is minimal so that acoustic signals can be transmitted over great distances. This applies, however, only to deep water. In shallow waters with depths in the range of the sound wavelength, where *Xenopus* lives, the situation is different and very complex, owing to factors that have little effect in deep water. First, sound is nearly completely reflected at the water–air interface, that is, at the water surface. Sound is also reflected from the bottom of the water and the reflection depends on the type of ground, for instance whether it is clay, sand or rock (Rogers & Cox 1988). Sound may, thus, reach a point through various pathways, that is, either directly or through single or multiple reflections at the surface and bottom. As a result, shallow water acts as a highpass filter, so that waves below a certain cut-off frequency are strongly attenuated and do not propagate far. The cut-off frequency is inversely proportional to water depth and strongly dependent on the type of ground. In water of 1 m depth, for instance, it has been calculated to be 2400 Hz on clayey ground and 400 Hz on rocky ground (Rogers & Cox 1988). In natural waters of irregular shape, varying depth and slope of the bottom, and maybe patchy ground structure, the change of sound with distance from the source is, therefore, very variable. With regard to the vertical axis, sound pressure is small near the water surface because reflection at the interface with the air inverts the phase of the reflected wave, so that the original and reflected waves largely cancel each other. For further references on underwater sound see Hawkins & Myrberg (1983), Rogers & Cox (1988) and Coates (1990).

Acoustic communication in *Xenopus* has to cope with this complexity and variety of physical sound propagation properties in its environment. In addition, sound from other animals may interfere: in standing or slow-running African waters, there may be considerable background noise from aquatic insects that produce underwater sounds for communication. Some ponds in which we measured sound during a field study in South Africa were as noisy as tropical forests at night (unpubl. obs.). The erroneous impression that ponds are silent is due to the nearly total reflection of underwater sound at the water surface.

Adaptations to sound transmission characteristics in the water

Because of the multipath propagation of sound mentioned above, the audibility of tonal signals with specific frequencies may be unreliable at some distance. More dependable are signals that contain a band of frequencies and show strong amplitude modulations, such as clicks. Indeed, the advertisement calls of all *Xenopus* species are composed of clicks with differing temporal sequences (Vigny 1979). The clicks contain a band of frequencies centred around a species-specific dominant frequency (Vigny 1979; Yager 1992). The dominant frequency is generally high, that is, in the hearing range of the basilar papilla, and spectral components below 1000 Hz are mostly absent or minor (Vigny 1979; Kobel, Loumont & Tinsley, this volume pp. 9–33). This absence of low frequencies matches the highpass characteristics of shallow waters. Low-frequency signals that do not propagate might be advantageous for close-range communication because they are not heard from a distance. In *Xenopus laevis*, whose advertisement call has a dominant frequency between 1600 and 2000 Hz, soft clicks with dominant frequencies below 1000 Hz are produced in some close encounters (Jansen 1992).

Adaptations may also exist in the calling behaviour of the frogs. *Xenopus* lives in standing or slowly flowing water of any depth. One might wonder, however, whether there is a minimal depth for *calling*, in accordance with the highpass properties of shallow water. Our field studies on *Xenopus laevis* showed no such minimal depth, and occasionally individuals were seen calling in water only 15 cm deep. However, in any given water body we found calling males mainly in the deeper areas, where their calls propagate better.

In a pond with clear water where we had marked all the approximately 40 individuals and could follow them individually night after night with a torch, we also obtained evidence for adaptation among females. They divided roughly into two groups. Some females were found in the deeper part of the pond, where the males were calling. Others went at late dusk to the edges of the pond where they stayed, sometimes for hours, in very shallow water, often with their heads protruding slightly so that the ears were at surface level. These edges are so shallow that the advertisement call hardly propagates into them. It is conceivable, therefore, that the females at the edges are unreceptive females that go to the edges to avoid the attention of males and perhaps to evade hearing the advertisement calls. This suggestion is corroborated by two observations. First, when a male met a female in the pond, attempts to clasp were observed only in deeper areas of the pond but not at its edges (the only exception was when a male met a couple in amplexus that had come into this shallow water for laying eggs; in this case the male immediately started fighting in order to replace the other male). Second, in our laboratory experiments on sound localization in *Xenopus laevis* (see Schanz & Elepfandt 1988), the female tested remained on the bottom of the basin as long as it showed phonotactic reactions. Occasionally, however, it surfaced after some time of experimentation and swam to the edge. Such a female would not show any further phonotaxis during that day. Even when pushed down by hand it rose again and returned to the edge. That is, the end of the readiness for phonotactic responses to male calls was indicated by coming up and swimming to the edge.

Morphology of the ear

The most obvious difference of the ear in *Xenopus* from that in terrestrial frogs is the absence of an externally visible tympanum. The unmodified skin in the tympanic area does not give any indication of the existence of a tympanum. However, after removal of the skin a cartilaginous tympanic disc becomes apparent at the usual place. The skin overlies the tympanum loosely without any attachment to it. Being soft and consisting essentially of water, the skin is transparent to water-borne sound and does not interfere with it.

The structure of the ear in *Xenopus* was described as early as 1876 by Parker (Fig. 10.1a); more recent accounts are given by de Villiers (1932) and Wever

(1985). Middle ear development was examined by Spannhof (1954), Sedra & Michael (1957) and Roy & Elepfandt (1990). Paterson (1949, 1960) studied the inner ear and its development; data on the auditory papillae and their innervation are scattered in Schucker (1972), Baird (1974a,b), Lewis (1978, 1984), Lewis, Leverenz & Bialek (1985), and Will & Fritzsch (1988). Basically, the *Xenopus* ear possesses the characteristics of an anuran ear. The inner ear is equipped with the usual sensory structures of anurans, including the amphibian papilla and basilar papilla for sound detection, and has an oval and a round window for sound input and release. There is also the middle ear complex with Eustachian tube, tympanic cavity, tympanum and stapes. The existence of these features provides evidence that *Xenopus* had terrestrial ancestors. The middle ear complex is an evolutionary invention for the detection of airborne sound. Without ancestors whose ears were developed for the detection of aerial sound no reason would exist for the anuran-type middle ear in *Xenopus*. The ear of recent *Xenopus* is, therefore, an adaptation to underwater sound of an ear designed originally to detect airborne sound.

Within the setting of the anuran ear, the ear of *Xenopus* shows some remarkable modifications (Fig. 10.1). First, as mentioned above, the tympanum is not membraneous and not part of the skin. It is an oval cartilaginous disc covering the outer side of the funnel-shaped tympanic cavity. It is connected around its circumference to the tympanic annulus by thin connective tissue, so that it can vibrate as a disc. The stapes is attached internally on the centre of the tympanum, from where it runs posteriorly and, after leaving the tympanic cavity, bends along the caudal edge of the skull towards the oval window. The operculum, the other ossicle covering the oval window in amphibians, probably involved in the detection of ground vibrations (Jaslow, Hetherington & Lombard 1988), has united with the wall of the inner ear and thus lost its sensory function. Ground vibrations would, via the water, vibrate the whole animal so that no specific pathway is needed in aquatic amphibians.

The Eustachian tubes are also modified (Fig. 10.1b). In terrestrial frogs the tubes are short and open into the lateral buccal chamber, one on either side. The close connection between mouth and tympanic cavity has important functional implications: sound impinges on the tympanum not only from the outside but also, via the mouth and the Eustachian tube, from the inside, and it is the difference between external and internal sound pressures that makes the tympanic membrane of terrestrial frogs vibrate (Eggermont 1988). In *Xenopus*, however, the Eustachian tubes extend medially forming one joint tube between the left and right ears that shows only one medial opening to the pharynx. The Eustachian tubes are always filled with air, even when the frog is submerged (Wever 1985; pers. obs.). This construction suggests that the pathway through the mouth is reduced or abolished and that there might be an interaction between the left and right ears.

The mechanisms through which these morphological changes affect hearing have still to be explored (see section on biophysical studies). However, some

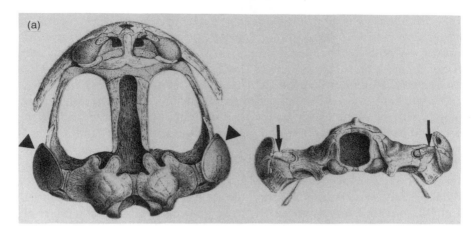

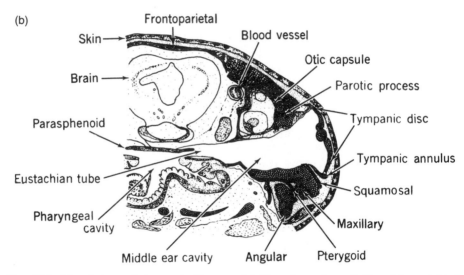

Fig. 10.1. Morphology of the ear in *Xenopus*. (a) View from above (left) and from behind (right), showing the cartilaginous tympanic disc (arrowhead) and the caudal location of the stapes (arrow). (After Parker 1876.) (b) A transverse section through the head showing the tympanic region (Wever 1985). The tympanic disc is covered by the skin. At its centre the stapes is attached. The Eustachian tube extends to the midline.

features, such as solidification of the tympanic membrane into a cartilaginous disc and the burying of the ear beneath soft tissue, are also seen in aquatic reptiles and mammals (Witschi, Bruner & van Bergeijk 1953; Henson 1974) and appear to be general adaptations to the properties of underwater sound and conditions of living under water.

The adaptation of the ear in *Xenopus* to underwater sound is also shown by the ear's relative sensitivities to airborne or underwater sound. In *Rana catesbeiana*, a terrestrial frog, the hearing threshold for sound pressure is

30–40 dB higher in water than in air (Lombard, Fay & Werner 1981). In *Xenopus*, we found nearly equal hearing thresholds for sound pressure in air and in water (Bibikov & Elepfandt in prep.). With regard to sound energy, the ear of *Xenopus* is thus much more sensitive to underwater than to airborne sound, in contrast to the ear of *Rana catesbeiana*.

The size of the tympanic disc in *Xenopus* differs between the sexes. In relation to skull size, the area of the male tympanic disc is about twice as large as the female disc. The difference is compensated only a little by the larger size of the female's skull. The occurrence of larger tympana in males is not unfamiliar in anurans, but its functional implications are unknown (Capranica 1976).

In the inner ear, the amphibian papilla is of the shorter, U-shaped type, suggesting an upper limit of its frequency range at 800–900 Hz (Lewis 1978, 1984; Lewis & Lombard 1988). The number of sensory hair cells in the amphibian papilla is 400–500 (e.g. *Xenopus laevis* 487; *Xenopus borealis* 390; *Xenopus tropicalis* 486; E.R. Lewis pers. comm.). The basilar papilla possesses some 50–70 sensory hair cells (Baird 1974b). An unusual feature is a relatively high number of afferent fibres from the basilar papilla. In *Xenopus laevis*, there are 310, which is approximately 65% of the fibres projecting from the amphibian papilla (Will & Fritzsch 1988). In terrestrial frogs, the afferent neurons from the basilar papilla amount to only 15–25% of those from the amphibian papilla (Will & Fritzsch 1988). Since the basilar papilla is the structure through which the higher frequencies are detected (Capranica 1976), the shift in the ratio of afferent neurons from the amphibian and basilar papillae may suggest some emphasis in *Xenopus* on processing signals with higher-frequency components.

Hearing abilities

In *Xenopus laevis*, we have examined hearing abilities by behavioural testing. Hearing range and frequency discrimination were investigated by conditioning, and sound localization was studied by analysis of phonotaxis response.

Hearing range and threshold

The most direct method to determine hearing abilities of animals is conditioning. However, attempts at acoustic conditioning of frogs have so far failed. Thus, only indirect evidence is available on hearing abilities in terrestrial frogs, based either on spontaneous reactions of the animals or on determination of response thresholds in electrophysiological recordings from auditory neurons.

We succeeded in conditioning *Xenopus* to tones. The procedure makes use of the turn response of *Xenopus* towards the origin of an impinging water wave (Elepfandt 1985). The frog was trained in a go/no-go procedure to turn towards a wave presented simultaneously with a tone but not towards a wave presented without a tone. Hearing threshold could subsequently be determined by reducing sound intensity.

Experiments were carried out in a circular tank of 60 cm diameter filled with water 60 cm deep. On the inside, the bottom and walls of the tank were lined completely with air bubble foil that isolated the water from all structures that could resonate. Sound was presented from an underwater loudspeaker (University Sound UW-30) placed on the bottom. The frogs were tested while sitting on a net 8 cm below the water surface. At this depth, it was possible to obtain pure tones for testing. Tests were done with tones 3 s long and having 100 ms rise and fall times; interstimulus intervals were 30 s.

The hearing range of *Xenopus laevis* was 200–3900 Hz (Elepfandt & Günther 1986). This is similar to that of other frogs of comparable size (Capranica 1976). Hearing optima were found at 600 Hz, 1400–1800 Hz and 3400 Hz. The threshold at these optima was about 90 dB re μPa (Elepfandt & Günther 1986). Under far-field conditions which, however, do not apply here, this pressure would correspond in energy to airborne sound of approximately 30 dB SPL, a value not much different from hearing thresholds of terrestrial frogs (Fay 1988).

For the detection of sounds consisting of a frequency band, such as the click in *Xenopus*, the existence of critical bands in auditory processing is important: sub-threshold tones presented simultaneously that fall into the same critical band (which means, typically, that they differ by less than 10–20% in frequency) produce an audible sound if their total energy equals the energy necessary to detect a pure tone in that band (Gässler 1954). Bandpass noise may thus still be detected even if its frequency components are individually sub-threshold. In amphibians, critical bands have been shown so far only in *Hyla cinerea* (Ehret & Gerhardt 1980). In *Xenopus laevis*, we determined the threshold for 600 Hz and for a 1/3-octave band centred around 600 Hz. The threshold for the band was 25 dB below that for the pure tone, so that the total energy at threshold was roughly equal in both cases (Hepperle 1990). This demonstrates the existence of auditory processing with critical band characteristics in *Xenopus*, an advantage for hearing clicks.

The clicks in the advertisement call of *Xenopus laevis* can attain 165 dB re μPa, measured 2 cm in front of the mouth (Jansen 1992). This is sufficiently far above the hearing threshold for the call to be heard over some distance in the pond even under non-optimal conditions. The call is confined to the caller's pond anyway, and usually *Xenopus* lives in dense populations within a pond. Thus, there is no need to attract females over long distances.

Sound localization

When a sound is detected it is important to establish its direction in order to locate, or escape, the sender. The importance is increased by the lack of visual cues in turbid water and at night. Localization of underwater sound, however, poses serious problems. It has been postulated that, on physical grounds, localization of underwater sound is impossible with ears which have a small interaural distance (van Bergeijk 1964), as in *Xenopus*. Time and intensity differences between the ears would be too small to be detected. With an average interaural distance of 1.5 cm in adult *Xenopus laevis*, the time of arrival of underwater sound at the two ears will differ by at most 10 µs, too short to be resolved by a frog nervous system. Also, the head is too small to produce sufficient sound shadow (which would require a head diameter of at least a fifth of the wavelength) and, having nearly the same sound impedance as water, it is hardly an obstacle at all for underwater sound. Rather, the sound will run undiminished through the animal's head.

On the other hand, a central function of advertisement calls is to let a female find the calling male. Indeed, Picker (1983) demonstrated that *Xenopus laevis* does find an underwater loudspeaker from which the call is broadcast. Since in his experiment the detection of the loudspeaker could be based either on sound localization or on mere lateralization (i.e., distinction between left and right side), we made specific tests of the frog's localization abilities. For the experiments, a circular basin of 4 m diameter was constructed and filled with water 35 cm deep (Schanz & Elepfandt 1988). In South Africa, I had seen *Xenopus* calling in ponds of this size. The slope of the basin's walls was made very gentle so that sound from the basin's centre would be reflected up and outwards rather than back to the centre. By this means, directionality of sounds was achieved. At four sides of the basin, underwater loudspeakers were placed from each of which the advertisement call could be broadcast. Females, injected with gonadotropin for sexual stimulation, were tested in the central area with diameter 1.20 m. Their turns towards the broadcasting loudspeaker were examined under open loop conditions, that is, no sound was broadcast while the animal was turning so that its turn angle reflects the previously determined sound direction. Turn angles were recorded by an overhead camera and subsequently analysed as a function of stimulus angle.

The tests showed that *Xenopus laevis* locates sound from anterior directions, up to 90° laterally (Fig. 10.2) (Schanz & Elepfandt 1988). This sector of localization is wider than in *Hyla gratiosa*, the only terrestrial frog tested specifically for localization ability. In *Hyla gratiosa*, localization has only been found in a sector extending up to 45° laterally (Klump & Gerhardt 1989). The mean angular deviation in the anterior sector was 19° in *Xenopus laevis*, roughly the same as in *Hyla gratiosa*. Turn response angles of *Xenopus laevis* to posterior stimuli also depended on sound direction but showed some ambiguity: turns tended to be either roughly correct or to fall short of 90°. The latter might be a behavioural trait: after midbrain lesions some individuals located sound up to 180°, that is, a physical basis exists for localizing sound from all directions.

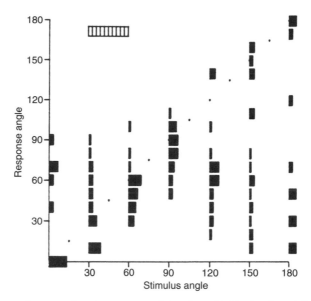

Fig. 10.2. Localization of underwater sound (mating calls) in the clawed frog, *Xenopus laevis*, under open loop conditions. Angles indicate right or left deviations from the frontal axis, which is 0°. The lengths of the bars indicate the number of turns, the hatched inset bar indicates 10 responses. The dots on the diagonal show where exact turns would be expected. Sound is localized up to 90°. (Courtesy of B. Schanz.)

Frequency discrimination

Beyond the detection and localization of sound, sound identification is crucial for communication as well as other purposes. The only method to test the acuity of stimulus discrimination in an animal and, from this, the possible differentiation of its sensory detection, is conditioning. Because attempts to condition amphibians to sound have so far failed, it is not known to what extent terrestrial frogs can distinguish between sounds. Only indirect evidence is available. All frogs identify their conspecific advertisement call which, however, implies only an acoustic distinction between the two categories 'own' and 'not own'. In choice tests, when a female is presented with two conspecific advertisement calls that differ in acoustic parameters, considerable differences are required to induce a preference (Gerhardt 1988). Even these preferences can often be reversed by making the less preferred signal louder (Gerhardt 1988) so that the acoustical basis for the preference remains uncertain. Electrophysiological recordings in frogs have demonstrated the existence of afferent auditory neurons with differing best frequencies in the range of the amphibian papillae, that is, below approximately 1000 Hz. At higher frequencies, however, in the range of the basilar papilla, afferent neurons exhibit identical, or nearly identical, best frequencies in each individual. From these results it has been concluded that some kind of frequency discrimination may be possible in the range of the amphibian papilla, but that frogs are tone-deaf for higher frequencies (Zakon & Wilczynski 1988).

We succeeded in determining the acuity of frequency discrimination in *Xenopus laevis* by conditioning. The tank and the test procedure were the same as in the determination of hearing threshold. Only the task was changed in that the frog was trained to distinguish a water wave presented in association with tone A from a wave presented with tone B. Sound pressure was 40 dB above threshold. When the discrimination had been learned the frequencies were brought closer to each other until discrimination failed. In the range of the amphibian papilla, that is, from 400–800 Hz, we found frequency discrimination with a limen of approximately 5% (Fig. 10.3) (Elepfandt & Hainich 1988). Such a limen corresponds to the acuity of frequency discrimination shown by fish (Fay 1988). At 1000 Hz, frequency discrimination was very poor (Traub & Elepfandt 1987). Between 1400 and 2500 Hz, however, frequency discrimination improved, attaining a 2% limen at 2000 Hz (Elepfandt, Fleig, Hainich & Traub 1989). That is, 1960 Hz was distinguished from 2000 Hz. To my knowledge, this is the best frequency discrimination found in an ectotherm vertebrate. Towards higher frequencies discrimination deteriorated again.

The range of 1400–2500 Hz, in which the good frequency discrimination was found, encompasses the dominant-frequency range in the species' advertisement call. Therefore, we tested whether the discrimination could be used to distinguish between individual males on the basis of their dominant frequency in the advertisement call. For this, advertisement calls of 31 males were tested and analysed spectrally (Jansen & Elepfandt 1992). Dominant frequencies ranged between 1600 and 2000 Hz. For any individual, however,

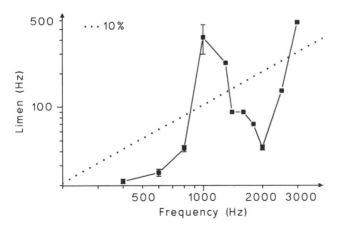

Fig. 10.3. Tone frequency discrimination in the clawed frog, *Xenopus laevis*. The dotted line indicates a relative discrimination acuity of 10%. Discrimination is found at 400–800 Hz and 1400–2500 Hz. The optimum at 2000 Hz is 2% discrimination limen.

the dominant frequency was constant. Neither injection of various dosages of sexual hormone in order to affect the frog's motivational state nor change of water temperature by 5 °C, which is beyond the diurnal fluctuation on the bottom of ponds where *Xenopus* lives, shifted the frequency by more than 20–25 Hz, which is below the frog's frequency resolution. The dominant frequency of the advertisement call is, therefore, constant for any individual male but differs between males. Similar results were obtained recently in calls of *Xenopus borealis* (Yager 1992). The dominant frequency can therefore be used by these frogs, or at least *Xenopus laevis*, to distinguish reliably between individuals. Other parameters of the advertisement call such as call duration, call sequence duration, or interclick intervals are more modifiable by the sender (Jansen 1992). Comparison with animal size showed that, on average, larger animals tend to exhibit lower dominant frequencies and that the dominant frequency decreases with the growth of the animal (Jansen & Elepfandt 1992). The high frequency resolution in the range of the advertisement call's dominant frequency thus would enable choice of larger males. The existence of mating preferences in *Xenopus* is, however, as yet unexplored.

Studies on hearing biophysics

As mentioned above, the adaptation of the frog to underwater sound required changes in the biophysics of sound detection by the ear. We have started to test these changes by measuring tympanic vibrations by means of laser vibrometry. For this, a small piece of skin overlying the tympanum was removed in the anaesthetized frog and a miniature piece of reflector foil was glued on the tympanic disc. Since the skin is transparent to underwater sound and the mass of the foil is minimal in comparison to the water pressing on the tympanum, these manipulations do not affect hearing sensitivity. Tympanic vibration under various conditions was then measured with a low-energy laser beam projected on the tympanum from a laser vibrometer (DISA X55).

Many aquatic organisms including fish hear by detecting sound particle velocity, despite the fact that in water sound pressure is much higher than in air and particle velocity smaller. In *Xenopus*, however, pressure detection appeared more probable for several reasons. First, the middle ear is, by its construction, a pressure-to-motion transducer. Second, the upper frequency limit for sound detection by particle velocity is, in all organisms studied so far, below or just above 1000 Hz (Fay 1988). *Xenopus laevis*, however, hears up to 3900 Hz, and sound localization with undiminished accuracy was found when advertisement calls were presented whose frequencies below 1200 Hz had been cut out. Finally, Hetherington & Lombard (1982) have shown that at 220 Hz the pressure of the sound was received in an isolated head preparation of *Xenopus*.

We tested the mechanism of sound detection in *Xenopus laevis* by measuring tympanic vibration in response to sound and comparing it to water particle vibration in the vicinity of the ear (Christensen-Dalsgaard, Breithaupt & Elepfandt 1990). If the ear were a particle velocity receiver its vibration amplitude should equal or be less than the vibration amplitude of the water particles. This was not found. Instead, throughout the frequency range 500–4000 Hz, tympanic vibrations exceeded the amplitude of water particle oscillations by approximately 30 dB (Fig. 10.4). Thus, the middle ear acts as a pressure receiver.

If *Xenopus* detects sound pressure, what is the physical basis for it? In terrestrial frogs, airborne sound impinges on the tympanum from the outside and, via the mouth and the Eustachian tube, from the inside (Eggermont 1988). This cannot be the case in *Xenopus*. Its middle ear cavity and Eustachian tubes are filled with air but not its mouth. Water-borne sound impinges on the tympanum from the outside, but on the inside it will end at the water–air interface at the pharyngeal opening of the Eustachian tubes. Therefore, the most plausible mechanism is that the air in the joint Eustachian tubes reacts as an air bubble whose volume oscillates in correlation with the sound pressure oscillation, similar to the mechanism in the swimbladder of fish.

The mechanism of sound localization in *Xenopus* is puzzling. Localization by comparing the intensities or times of arrival of the sound at the two ears is excluded (van Bergeijk 1964). Perception of a pressure gradient between sound impinging on the tympanum from the outside and sound impinging via mouth

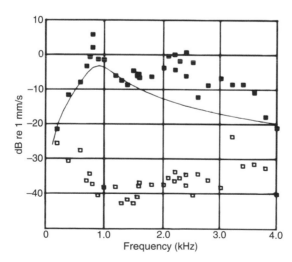

Fig. 10.4. Vibration velocity of the tympanic disc in *Xenopus laevis* (filled squares) and of water particles (open squares) in response to pure tones with sound pressure 10 Pa. Tympanic vibrations exceed the corresponding water vibrations by about 30 dB. The thin continuous line indicates tympanic velocities that would be expected from an air bubble model. (Christensen-Dalsgaard *et al.* 1990.)

and Eustachian tube from the inside, as in terrestrial frogs, is also not possible because the sound that could impinge from the inside will be blocked at the water–air interface between the mouth and the Eustachian tubes. Further, *Xenopus* localizes calls whose frequency components below 1200 Hz have been cut off. Since auditory afferent discharges of frogs are not phase-coupled to frequencies above approximately 1000 Hz (Zakon & Wilczynski 1988), mechanisms that would transform sound differences into phase differences of tympanic oscillations cannot be used by the animal. The basis of localization, therefore, must be bilateral differences of tympanic response *amplitudes* depending on sound direction. Air bubble pulsation, however, the mechanism suggested to produce the tympanic response amplitude in the *Xenopus* ear, is intrinsically non-directional because it results from sound pressure, which is a scalar quantity.

As a first step towards resolving the riddle, we measured, by laser vibrometry, tympanic vibrations in response to sound presented from various directions. Alternatively, we measured sound pressure oscillation in the Eustachian tube by means of a miniature microphone (Bruel & Kjaer 4182) that was projected through a small hole drilled in the frog's skull, so that the microphone's frontal end was flush with the wall of the Eustachian tube[1]. The measurements were made in the same pond as that in which we had made the behavioural tests of sound localization. The results of both types of measurements show a dependence of tympanic vibration amplitude and phase on the direction of the sound, and the kind and degree of directionality depend on stimulus frequency (Rangol, Golden, Bischof, Brändle, Elepfandt & Plassmann 1992). Since the amplitude of tympanic vibrations is conveyed via the stapes to the inner ear, directionality of tympanic response amplitudes provides a basis for sound direction determination in the frog by comparing the vibration amplitudes of left and right tympana.

In summary, the biophysical studies show that the tympanum responds to sound pressure, a scalar quantity. The most probable mechanism is an air bubble resonance. Tympanic response amplitudes, on the other hand, depend on the direction of the sound. This should not be possible with the mechanism just outlined. Further studies are necessary to resolve this contradiction.

Acknowledgements

The author's field study was supported by Deutsche Forschungsgemeinschaft grant El 75/7. I thank A.D. Hawkins for helpful comments on the manuscript and for checking the English text.

[1]The experiments described in this chapter were conducted with permission from the German Governmental Committee for Animal Experimentation.

References

Baird, I.L. (1974a). Anatomical features of the inner ear in submammalian vertebrates. In *Handbook of sensory physiology 5. 1. Auditory system: anatomy, physiology (ear)*: 159–212. (Eds Keidel, W.D. & Neff, W.D.). Springer-Verlag, Berlin.

Baird, I.L. (1974b). Some aspects of the comparative anatomy and evolution of the inner ear in submammalian vertebrates. *Brain Behav. Evol.* 10: 11–36.

Bibikov, N.G. & Elepfandt, A. (In preparation). *Auditory evoked potentials in the clawed frog*, Xenopus laevis.

Capranica, R.R. (1976). Morphology and physiology of the auditory system. In *Frog neurobiology: a handbook*: 551–575. (Eds Llinás, R. & Precht, W.). Springer-Verlag, Berlin.

Christensen-Dalsgaard, J., Breithaupt, T. & Elepfandt, A. (1990). Underwater hearing in the clawed frog, *Xenopus laevis*: tympanic motion studied with laser vibrometry. *Naturwissenschaften* 77: 135–137.

Coates, R.F.W. (1990). *Underwater acoustic systems*. Macmillan, London.

de Villiers, C.G.S. (1932). Über das Gehörskelett der aglossen Anuren. *Anat. Anz.* 74: 33–55.

Eggermont, J.J. (1988). Mechanisms of sound localization in anurans. In *The evolution of the amphibian auditory system*: 307–336. (Eds Fritzsch, B., Ryan, M.J., Wilczynski, W., Hetherington, T.E. & Walkowiak, W.). Wiley, New York.

Ehret, G. & Gerhardt, H.C. (1980). Auditory masking and effects of noise on responses of the green treefrog (*Hyla cinerea*) to synthetic mating calls. *J. comp. Physiol.* 141: 13–18.

Elepfandt, A. (1985). Naturalistic conditioning reveals good learning in a frog (*Xenopus laevis*). *Naturwissenschaften* 72: 492–493.

Elepfandt, A., Fleig, A., Hainich, M. & Traub, B. (1989). Good tone-frequency discrimination in a frog (*Xenopus laevis*, Pipidae). In *Dynamics and plasticity in neuronal systems. Proceedings of the 17th Göttingen neurobiology conference*: 275. (Eds Elsner, N. & Singer, W.). Georg Thieme Verlag, Stuttgart.

Elepfandt, A. & Günther, E. (1986). Behavioral determination of the auditory threshold in the clawed frog, *Xenopus laevis* Daudin. *Neurosci. Lett. Suppl.* 26: 380.

Elepfandt, A. & Hainich, M. (1988). Determination of low-frequency tone discrimination in the clawed frog, *Xenopus laevis*, by conditioning. *Eur. J. Neurosci. Suppl.* 1: 163.

Fay, R.R. (1988). *Hearing in vertebrates: a psychophysics databook*. Hill-Fay Associates, Winnetka, Illinois.

Gässler, G. (1954). Über die Hörschwelle für Schallereignisse mit verschieden breitem Frequenzspektrum. *Acustica* 4: 408–414.

Gerhardt, H.C. (1988). Acoustic properties used in call recognition by frogs and toads. In *The evolution of the amphibian auditory system*: 455–483. (Eds Fritzsch, B., Ryan, M.J., Wilczynski, W., Hetherington, T.E. & Walkowiak, W.). Wiley, New York.

Hawkins, A.D. & Myrberg, A.A. Jr. (1983). Hearing and sound communication under water. In *Bioacoustics: a comparative approach*: 347–405. (Ed. Lewis, B.). Academic Press, London.

Henson, O.W. (1974). Comparative anatomy of the middle ear. In *Handbook of sensory physiology 5. 1. Auditory system: anatomy, physiology (ear)*: 40–110. (Eds Keidel, W.S. & Neff, W.D.) Springer-Verlag, Berlin.

Hepperle, S. (1990). *Untersuchungen zum Einfluss von Reizparametern auf die Hörschwelle beim Krallenfrosch* Xenopus laevis. Diploma thesis: Universität Konstanz.

Hetherington, T.E. & Lombard, R.E. (1982). Biophysics of underwater hearing in anuran amphibians. *J. exp. Biol.* 98: 49–66.

Jansen, S. (1992). *Hydro-akustische Untersuchungen an Rufen des Krallenfrosches,* Xenopus laevis. Diploma thesis: Universität Konstanz.

Jansen, S. & Elepfandt, A. (1992). Dominant frequency of mating call: a possible cue for interindividual discrimination in the clawed frog, *Xenopus laevis.* In *Rhythmogenesis in neurones and networks. Proceedings of the 20th Göttingen neurobiology conference*: 200. (Eds Elsner, N. & Richter, D.). Georg Thieme Verlag, Stuttgart.

Jaslow, A.P., Hetherington, T.E. & Lombard, R.E. (1988). Structure and function of the amphibian middle ear. In *The evolution of the amphibian auditory system*: 69–91. (Eds Fritzsch, B., Ryan, M.J., Wilczynski, W., Hetherington, T.E. & Walkowiak, W.). Wiley, New York.

Klump, G.M. & Gerhardt, H.C. (1989). Sound localization in the barking treefrog. *Naturwissenschaften* 76: 35–37.

Lewis, E.R. (1978). Comparative studies of the anuran auditory papillae. *Scann. Electron Microsc.* 1978: 633–642.

Lewis, E.R. (1984). On the frog amphibian papilla. *Scann. Electron Microsc.* 1984: 1899–1913.

Lewis, E.R., Leverenz, E.L. & Bialek, W.S. (1985). *The vertebrate inner ear.* CRC Press, Boca Raton, Florida.

Lewis, E.R. & Lombard, R.E. (1988). The amphibian inner ear. In *The evolution of the amphibian auditory system*: 93–123. (Eds Fritzsch, B., Ryan, M.J., Wilczynski, W., Hetherington, T.E. & Walkowiak, W.). Wiley, New York.

Lombard, R.E., Fay, R.R. & Werner, Y.L. (1981). Underwater hearing in the frog, *Rana catesbeiana. J. exp. Biol.* 91: 57–71.

Parker, W.K. (1876). On the structure and development of the skull in the Batrachia. II. *Phil. Trans. R. Soc.* 166: 601–669.

Paterson, N.F. (1949). The development of the inner ear of *Xenopus laevis. Proc. zool. Soc. Lond.* 119: 269–291.

Paterson, N.F. (1960). The inner ear of some members of the Pipidae (Amphibia). *Proc. zool. Soc. Lond.* 134: 509–546.

Picker, M.D. (1983). Hormonal induction of the aquatic phonotactic response of *Xenopus.* Behaviour 84: 74–90.

Rangol, H.-P., Golden, J., Bischof, H.-J., Brändle, K., Elepfandt, A. & Plassmann, W. (1992). Middle ear transferfunction and directional characteristic in clawed frog and zebra finch. *Soc. Neurosci. Abstr.* 18: 1192.

Rogers, P.H. & Cox, M. (1988). Underwater sound as a biological stimulus. In *Sensory biology of aquatic animals*: 131–149. (Eds Atema, J., Fay, R.R., Popper, A.N. & Tavolga, W.N.). Springer-Verlag, New York.

Roy, D. & Elepfandt, A. (1990). Untersuchung der Entwicklung des Mittelohres beim Krallenfrosch (*Xenopus laevis*). *Verh. dt. zool. Ges.* 83: 469–470.

Schanz, B. & Elepfandt, A. (1988). Untersuchungen zur Ortung von Unterwasserschall durch den Krallenfrosch (*Xenopus laevis*). *Verh. dt. zool. Ges.* 81: 211–212.

Schucker, F.A. (1972). A preliminary report on light and electron microscopic studies of the basilar papilla in the toad, *Xenopus laevis. Anat. Rec.* 172: 400.

Sedra, S.N. & Michael, M.I. (1957). The development of the skull, visceral arches, larynx and visceral muscles of the South African clawed toad, *Xenopus laevis* (Daudin) during the process of metamorphosis (from stage 55 to stage 66). *Verh. K. ned. Akad. Wet. Afd. Natuurk. (Sect.* 2) **51**: 1–80.

Spannhof, L. (1954). Die Entwicklung des Mittelohres und des schalleitenden Apparates bei *Xenopus laevis* Daudin. *Z. wiss. Zool.* **158**: 1–30.

Traub, B. & Elepfandt, A. (1987). Conditioning of tone-discrimination in the clawed frog, *Xenopus laevis. Neurosci. Suppl.* **22**: 170.

van Bergeijk, W.A. (1964). Directional and nondirectional hearing in fish. In *Marine bio-acoustics*: 281–299. (Ed. Tavolga, W.N.). Pergamon Press, Oxford.

Vigny, C. (1979). The mating calls of 12 species and sub-species of the genus *Xenopus* (Amphibia: Anura). *J. Zool., Lond.* **188**: 103–122.

Wever, E.G. (1985). *The amphibian ear.* Princeton University Press, Princeton.

Will, U. & Fritzsch, B. (1988). The eighth nerve of amphibians: peripheral and central distribution. In *The evolution of the amphibian auditory system*: 125–155. (Eds Fritzsch, B., Ryan, M.J., Wilczynski, W., Hetherington, T.E. & Walkowiak, W.). Wiley, New York.

Witschi, E., Bruner, J.A. & van Bergeijk, W.A. (1953). The ear of the adult *Xenopus. Anat. Rec.* **117**: 602–603.

Yager, D.D. (1992). Underwater acoustic communication in the African pipid frog *Xenopus borealis. Bioacoustics* **4**: 1–24.

Zakon, H.H. & Wilczynski, W. (1988). The physiology of the anuran eighth nerve. In *The evolution of the amphibian auditory system*: 125–155. (Eds Fritzsch, B., Ryan, M.J., Wilczynski, W., Hetherington, T.E. & Walkowiak, W.). Wiley, New York.

11 The biology of *Xenopus* tadpoles

RICHARD WASSERSUG

Synopsis

Virtually all that is known about the biology of *Xenopus* larvae comes from laboratory studies of one species, *Xenopus laevis*. This account reviews several aspects of larval biology, especially feeding, respiration, schooling behaviour, swimming mechanics and certain visceral functions.

Compared to tadpoles in other genera, *Xenopus* larvae are highly specialized, obligatory microphagous tadpoles. Many of their unique morphological features testify to their efficiency as suspension feeders. *Xenopus* larvae are also obligatorily air-breathers and must intermittently surface for air in order to have normal growth rates. *Xenopus* tadpoles are social and use both visual and lateral line information to maintain school geometry.

Xenopus larvae are positively buoyant and swim in a characteristic, head-downward posture off the bottom. They regulate their swimming speed in a fundamentally different manner from more typical, benthic (e.g., *Rana*) tadpoles. The caudal myology and spinal cord of *Xenopus* differ radically from those of *Rana* tadpoles, although the spinal cord differences appear to reflect an evolutionary rather than a functional difference.

All visceral behaviours that involve forceful ejection of materials from the body cavity, such as coughing and vomiting, are absent in tadpoles, including those of *Xenopus*. Metamorphic changes in the musculoskeletal system, associated with development of the adult locomotor system, are essential for the massive elevation in abdominal pressure required for these visceral behaviours.

The metamorphic process in anurans is discussed in light of (1) an evolutionary trend in the absolute amount of morphological change and abruptness of the metamorphic transformation and (2) the ecological/endocrinological regulation of the process. A hypothesis is reviewed about how tadpoles might assess environmental quality and incorporate that information into the endocrinological 'decision' to metamorphose. It is suggested that regulatory peptides, such as epidermal growth factor, found in the oral mucus of *Xenopus* tadpoles, may play a role in preventing premature metamorphosis.

Tadpoles are complex, highly integrated vertebrates and not simply a transitional embryonic stage from egg to frog. Field studies on the biology of *Xenopus* tadpoles, including *X. laevis*, lag woefully behind laboratory investigations.

Introduction

This essay focuses on the *Xenopus* larva as an independent, free-living organism, rather than on either the transition of the egg into a tadpole (primary embryogenesis) or the transformation of the tadpole into a frog (metamorphosis). In organizing this essay, I have elected to follow the history of my own work on *Xenopus* functional and evolutionary morphology and behaviour. This research programme has held to the creed of organismal biology, namely that organisms function as integrated wholes. Feeding adaptations and locomotor patterns, for example, must function together for any animal to meet its energy requirements and not itself end up as prey. I try to demonstrate this integrated nature of *Xenopus* larval design in a series of selected topics. Among the many important areas which I do not address are the massive fields of *Xenopus* biochemistry and molecular biology.

For those completely unfamiliar with anuran larvae, Duellman & Trueb's (1986) textbook introduces the topic. The bibliography in Nieuwkoop & Faber (1967) still constitutes the best single introduction to *Xenopus* larval biology, while Deuchar (1975) and Fox (1984) also provide general information. Regrettably, two newer books dedicated to *Xenopus* biology say little (Kay & Peng 1991) or nothing (Hausen & Riebsell 1991) about tadpoles.

Taxonomic boundaries and descriptive history

About 20 species of *Xenopus* are currently recognized but descriptions of larvae are available for only 10 (Nieuwkoop & Faber 1967; Arnoult & Lamotte 1968; Rau 1978; Vigny 1979). Of these the *X. laevis* tadpole is the best known.

Although adults of most *Xenopus* species can be distinguished on gross anatomical features and call characteristics (Kobel, Loumont & Tinsley this volume pp. 9–33), the larvae are more of a problem. Altig & Johnston (1986), who compiled and codified descriptions of known tadpoles, provided no qualitative features to separate the larvae of one *Xenopus* species from another. Arnoult & Lamotte (1968) distinguished *X.* (now *Silurana*) *tropicalis* and *X. fraseri* larvae from *X. laevis* and *X. muelleri* larvae according to size. Rau (1978) distinguished *X. gilli* from *X. laevis* primarily on differences in the density of melanophores, but also suggested that subtle differences in eye diameter and the spacing of the nasal capsules may differentiate the two tadpoles. Eye diameter may further distinguish *laevis* from *muelleri* tadpoles (Arnoult & Lamotte 1968). In the most extensive effort to separate the tadpoles of *Xenopus* species, Vigny (1979) relied almost entirely on pigment pattern and melanophore density, which are characters that vary greatly in tadpoles raised in captivity.

There are no studies of geographical variation in the few features used to differentiate *Xenopus* tadpoles to give confidence in the characters that have been used to date. Environmental features that relate to the zoogeographical distribution of *Xenopus* species, such as humidity and rainfall (Loumont 1984; Tinsley, Loumont & Kobel, this volume pp. 35–59) may manifest themselves in tadpole adaptations over evolutionary time. However, there is no information on differences in internal features of even the better known *Xenopus* larvae that may correlate with the broader climatic patterns. In the absence of any in-depth investigation of variation within and among *Xenopus* larvae, we may tentatively assume that *Xenopus* larvae, regardless of species group, are all very similar. Thus, this review deals only with *X. laevis*, on the presumption that features which distinguish its larvae from those of other anuran genera are true for *Xenopus* as a whole.

There is abundant information on the larval biology of a few species in the genus *Rana* (e.g., *catesbeiana, pipiens, temporaria*), but otherwise research on the biology of tadpoles is sparse. Our knowledge about the tadpole of one species of *Xenopus—laevis—*is a clear exception. Information on *X. laevis* is increasing rapidly and may soon surpass the data on all *Rana* combined.

Feeding structures: functional and evolutionary implications

Most tadpoles have keratinized external mouthparts for holding onto surfaces and shearing food matter, but these structures are absent in *X. laevis* and other pipoid larvae. Most non-pipoid tadpoles also have a single, usually sinistral, opening in their body wall, the spiracle, through which water that has passed across the branchial chamber ultimately exits. In pipids (and *Rhinophrynus*) there are instead two spiracles, symmetrically arranged, one on each side at the caudal end of the branchial baskets. The Microhylidae share with pipids the absence of keratinized mouthparts, but have a single spiracle on the mid-ventral line. These features led Orton (1953, 1957) to sort anurans into four suprafamilial groups based solely on larval spiracle position and external mouthparts. The pipoids, including all known *Xenopus* larvae, make up Orton's first group; the microhylids her second. Her third group is formed of the archaeobatrachian genera *Ascaphus* (plus *Leiopelma*), *Discoglossus*, *Bombina* and *Alytes*, which have keratinized mouthparts but a single, mid-ventral spiracle. Orton's last group includes the remaining tadpoles, i.e., the 90% or so of anuran genera with keratinized mouthparts and a single, sinistral spiracle.

Orton's classification prompted much debate in the 1960s and 70s (reviewed in Sokol 1975, 1977). Some workers felt that it revealed true phylogenetic information on relationships among anuran families whereas others doubted whether Orton's two characters deserved much weight in suprageneric systematics. In the 1970s I began to examine anuran larvae and discovered that the buccopharyngeal design of *Xenopus* tadpoles is radically different from that of other larvae. The current view of Orton's characters is that they may have a role in anuran higher systematics but that the absence of keratinized mouthparts in microhylid and pipoid frogs is strictly convergent (Wassersug 1989a).

The mechanism for drawing water into the mouth in all tadpoles is a buccal pump (Kenny 1969a; Gradwell 1971; Wassersug & Hoff 1979). The piston of the pump is the ceratohyal cartilage embedded on each side of the buccal floor. When the orbitohyoideus muscles, which attach to the lateral side of the palatoquadrate and the lateral arm of the ceratohyal, contract, the medial portion of the buccal floor drops. As the floor is depressed, the mouth is opened. One-way flow through the mouth is achieved by a transverse flap on the floor of the mouth called the ventral velum. This flap separates the oral cavity from the pharynx. As the floor of the mouth is depressed the free edge of the ventral velum rotates upward (de Jongh 1968) to seal off the buccal cavity from the pharynx, preventing backflow. After the oral cavity fills, the mouth is closed and the floor elevated. This last action causes the ventral velum's free edge to drop down and water to be propelled over it and into the pharynx. Pressure fluctuates in the atrial cavity, just inside the spiracle, but always remains positive. Thus, as the floor of the buccal cavity goes up and down, water is continually ejected from the spiracle.

Xenopus and other pipids (Sokol 1977) do not have a ventral velum, which means that they have instead a single, common buccopharyngeal cavity (cf. figures in Sokol 1977; Seale, Hoff & Wassersug 1982; Wassersug & Heyer 1988). How, then, do they maintain one-way flow through their mouths? The answer is that the margins of their spiracles act as valves. As the floor of the mouth oscillates in a *Xenopus* tadpole, the spiracles rhythmically open and close (Gradwell 1975). In *Xenopus*, when the floor of the mouth is lowered the pressure in the atrial chamber becomes negative and the spiracle walls collapse inward. Thus *Xenopus* and its pipid relatives maintain one-way flow, like all other suspension-feeding tadpoles, but in a radically different fashion.

The internal oral region of *Xenopus* larvae differs from that of all other tadpoles in many other ways. Secretory ridges on the buccopharyngeal floor in *Xenopus* appear to be homologous with structures called 'branchial food traps' (Kenny 1969a, b) in the mouths of other tadpoles, but are confined to the pharyngeal (i.e., ventral) surface of the ventral velum in the other forms (Wassersug & Rosenberg 1979; Viertel 1987). In *Xenopus*, they cover the whole common buccopharyngeal floor. The secretory ridges themselves are narrower, straighter, and more extensive in *Xenopus* than in any other tadpole (Kenny 1969b; Wassersug & Rosenberg 1979; Viertel 1987). Mucus extruded from these ridges is used to capture small food particles from the water (Wassersug 1972; Gradwell 1975; Viertel 1987), although how material is transported away from these surfaces to gill filters, and ultimately to the oesophagus, has yet to be determined (Sanderson & Wassersug 1993).

Surprisingly little is known about the feeding ecology of *Xenopus* tadpoles in the wild, but their oral morphology clearly suggests, and laboratory data confirm, that they are superior suspension-feeding vertebrates (Wassersug 1972; Seale 1982; Seale *et al.* 1982; Viertel 1992). *X. laevis* can extract virus-size particles from water and filtration rates are far superior to those of other tadpoles. For example, in a single pump stroke a *X. laevis* tadpole can clear a volume of water approximately five times greater than that cleared by a *Rana sphenocephala* larva of the same size (Seale 1982).

If the particulate concentration in the water is too high, suspension-feeding tadpoles must reduce the volume of water that they process per unit time to avoid clogging their gill filters. In theory, tadpoles could do this in two ways: they could modulate either the amplitude (AM) or the frequency (FM) of their buccal pump stroke. *Rana* are primarily AM tadpoles. *Xenopus* tadpoles regulate their ingestion rate primarily by varying pumping frequency (i.e. FM) (Seale *et al.* 1982). The relatively small size of the muscles that operate the buccal pump in *X. laevis* (see Satel & Wassersug 1981) and the proportions of the ceratohyals—with a short, lateral, lever arm—favour this type of regulation in *Xenopus* (see discussion in Wassersug & Hoff 1979).

At very low concentrations of food particles, tadpoles reduce the amount of energy they expend on feeding by reducing pumping frequency (Seale *et al.* 1982). There is some evidence that at low particle concentrations they may also reduce the amount of mucus that they extrude from their branchial food traps (Viertel 1992).

Respiratory patterns

Xenopus tadpoles lack true, internal gill filaments, and this has led to speculation on the relative role of lungs versus gill filters in their respiration (reviewed in Gradwell 1971). Only in the past decade has the topic been rigorously explored. There is a complex interaction between the regulation of respiration, feeding and swimming in *Xenopus* tadpoles. *Xenopus* larvae surface to fill their lungs with air within the first day or two after they become free-swimming (pers. obs.). From then until metamorphosis they are positively buoyant and swim in their characteristic, head-downward posture against their own buoyancy. If maintained without access to air, even in normoxic water, *Xenopus* tadpoles become negatively buoyant and pay a price in reduced growth rates (Wassersug & Murphy 1987). In normoxic water ($PO_2 > 100$ Torr) 17% of the total oxygen consumed by *X. laevis* larvae is obtained from the air (Feder & Wassersug 1984). This can be achieved by surfacing to breathe less than twice an hour. However, as the amount of dissolved oxygen decreases *Xenopus* tadpoles rely more and more on aerial respiration. When PO_2 falls below approximately 50 Torr they rely entirely on aerial respiration and must then surface on average 10 times an hour. Some larger, older tadpoles in extremely hypoxic water have been observed surfacing more than 30 times an hour.

At low PO_2, tadpoles risk losing O_2 to the water if they increase the flow of water through their buccopharyngeal cavity. Thus buccal pumping rates stay relatively low and constant across the hypoxic to hyperoxic spectrum for aquatic PO_2. Essentially, *X. laevis* larvae modulate feeding by varying aquatic pumping frequency, but modulate respiration by varying aerial surfacing rates. If exposed to the double challenge of high concentrations of particulate matter in the water while being simultaneously denied access to the surface, *X. laevis* tadpoles increase buccal pumping rates dramatically to facilitate aquatic gas exchange (Feder *et al.* 1984). They also spit food trapped in mucus out of their mouths, presumably in an attempt to clear buccopharyngeal surfaces for respiration. If the concentration of particulate matter is high enough, tadpoles without access to the surface will die, no matter how well oxygenated the water. (See Orlando & Pinder (1995) for an update on this topic.)

In conclusion, *Xenopus* tadpoles are obligatorily air-breathers, even though their aerial respiratory rates seem low in normoxic water. It is possible to raise a *X. laevis* tadpole through metamorphosis that has never inflated its lungs, but the growth is abnormally slow and the resulting froglet is afflicted with pulmonary and cardiac abnormalities (Pronych & Wassersug 1994).

Paradoxically, *Xenopus* tadpoles with deflated lungs have improved stamina in normoxic water when forced to swim against a current (Wassersug & Feder 1983). This relates to the advantages of being negatively buoyant and near the bottom of a stream, where current is reduced because of surface drag. Most tadpoles that live in currents have reduced lung volumes (Wassersug & Heyer 1988) and do not inflate their lungs until metamorphosis (Nodzenski, Wassersug & Inger 1989).

Xenopus tadpoles are not designed for sustained locomotion under any circumstances (Hoff & Wassersug 1986). When they sprint they cease buccal pumping. Given their large, anteriorly-directed mouths and dense gill filters, any buccal pumping during sprinting would probably increase pressure drag substantially in the Reynolds numbers range that they normally swim (Wassersug 1989b).

Many other topics pertaining to respiration in *Xenopus* tadpoles have yet to be explored. For example, cardio-respiratory coupling has been reported in restrained *X. laevis* larvae (Wassersug, Paul & Feder 1981), but whether this is common in free-swimming animals is not known (cf. Burggren & Pinder 1991). Future studies of respiration and metabolic activity in *Xenopus* should consider the diurnal periodicity in their respiratory rates (Abel, Seale & Boraas 1992), and the sensitivity of *Xenopus* tadpoles to disturbances. A simple tap on the side of an aquarium can cause these tadpoles to reduce their surfacing rate for the next hour, even when their swimming and other behaviours appear normal (pers. obs.).

Schooling behaviour

Undisturbed *X. laevis* larvae in the laboratory orient parallel to each other, prompting questions about the mechanisms they use to maintain school structure and also its adaptive significance. *Xenopus* tadpoles are attracted toward conspecifics, although they avoid direct contact. They use both their eyes and their lateral line neuromasts to assess their orientation to, and distance from, their nearest neighbours (Katz, Potel & Wassersug 1981). For *X. laevis* tadpoles, the interactive distance—the distance over which significantly non-random orientation is maintained—is about two body lengths in the light but half that in the dark. The tendency for *X. laevis* larvae to orient parallel to each other increases with tadpole size and density. The fact that the tadpoles still orient parallel in the dark suggests that the lateral line system is involved. This is confirmed by the observation that *Xenopus* tadpoles that have had the hair cells of their lateral line system pharmacologically blocked position themselves closer to each other than do control individuals (Lum *et al.* 1982).

The adaptive significance of schooling behaviour in *Xenopus* larvae is still unknown but they can discriminate between sibs and non-sibs, which supports a kin selection model for tadpole schooling (Blaustein & Waldman 1992). Katz *et al.* (1981) suggested that *X. laevis* larvae metamorphose at a larger size when raised in groups than when raised as isolates.

This implies that schooling in this species either facilitates energy capture or reduces energy expenditure. Channing (1976) suggested that *Phrynomerus* tadpoles, which are morphologically and behaviourally rather similar to *Xenopus*, swim slowly in a common direction because it offers 'advantages in utilizing the available water for feeding'. This is an attractive hypothesis but one that is difficult to test. Many highly specialized invertebrate suspension-feeders are social organisms that collectively generate a common feeding current. Using fluorescein dye as a tracer, I have tried to establish whether larger volumes of water are transported past *Xenopus* larvae when they are in parallel schools than when they are either isolated or not parallel. The results were not conclusive, leaving open the question of whether schooling facilitates the transport of water to and 'through' *Xenopus* larvae. It should be emphasized that all of the work on the schooling behaviour of *Xenopus* to date has been undertaken in the laboratory.

Swimming mechanics and the regulation of locomotion

Studies on the mechanism of schooling in tadpoles lead to questions about how tadpoles regulate their position in the water column and the kinematics of locomotion. Tadpoles are simple organisms compared to fishes in terms of their locomotor structures. Until they are near metamorphosis, they rely solely on axial bending without limb assistance for propulsion. There are two different ways that tadpoles can regulate their locomotion. Pipids, and a few tadpoles in other families (e.g., some hylids and microhylids) which have converged on *Xenopus*'s mid-water way of life, hang in the water column by rapidly oscillating their tail tip. Most other tadpoles, which are more benthic, swim with a fuller tail stroke when they are not resting on the bottom or other surfaces. For convenience, we can call the former mode of swimming the '*Xenopus* method' and the latter, the '*Rana* method'.

The neuroanatomy and neurophysiology of swimming in *Xenopus* embryos have been studied extensively in recent years (e.g. van Mier, Armstrong & Roberts 1989; Sillar, Wedderburn & Simmers 1991). Within a day or two of hatching, *X. laevis* change from exhibiting an embryonic pattern of locomotion, which resembles a sporadic series of startle responses, to continuously swimming within the water column. The tail-beat frequency in very young, free-swimming tadpoles is close to 14 Hz (Sillar *et al*. 1991), but declines to about 10 Hz as the tadpoles get larger. For virtually all aquatic vertebrates, including common tadpoles like *Rana*, the higher their swimming velocity, the higher their tail-beat frequency. For *Xenopus* this pattern is only partially true. What *Xenopus* does, as it increases velocity, is to incorporate more of its tail in the propulsive stroke, i.e., the length of the wave travelling down the tail increases. The wavelength increases from tail tip forward, and at swimming speeds above six body lengths per second the whole tail oscillates. To swim any faster, *Xenopus* must then do what other tadpoles do and increase its overall tail-beat frequency. At maximum sprint speeds, the tail-beat frequency in *X. laevis* exceeds 20 Hz.

Other differences in swimming between *Xenopus* and *Rana* tadpoles can be seen in their fast-starts and startle responses (Hoff 1987; Will 1991). For example, *Xenopus* differ from *Rana* in the distribution of muscle mass along the tail and consequently the flexibility of various sections of their tails. This affects how abruptly they can turn. All tadpoles lack an osseous or cartilaginous caudal skeleton except at the very base of their tail. Thus they rely on muscle activity to change tail stiffness in an asymmetric fashion in order to execute turns in the horizontal plane (Hoff 1987; Wassersug 1989b). *Xenopus* tadpole tails are thicker and thus stiffer at the base than those of *Rana* larvae of comparable size. On the other hand, the terminal portion of the tail is thinner and thus more flexible in *Xenopus* than in *Rana*. One consequence is that *Xenopus* tadpoles can bend the caudal end of their tails up or down to execute sharp turns in the vertical plane (Wassersug 1992).

The differences in swimming between *Xenopus* and *Rana* correspond to a wealth of other underlying anatomical differences. In *Xenopus*, for example, but less so in *Rana* (or *Bufo*), there is a continual addition of myotomes to the tail tip as the tadpole grows (Kordylewski 1986; Hoff 1987). There are also histological differences in musculature that reflect the differences in swimming (Muntz, Hornby & Dalooi 1989). Even the gross anatomy of the postsacral spinal cords differs radically between *Xenopus* and *Rana* tadpoles (Nishikawa & Wassersug 1988). In *Xenopus*, the spinal cord in the tail is large with paired spinal nerves exiting in a regular array to innervate caudal myotomes. The cell bodies of the motoneurons that serve those myotomes are found along the length of the tail. In contrast, in *Rana* the caudal spinal cord is a simple, thin filament—a filum terminale. The cell bodies for caudal motoneurons are restricted to the trunk region and axons only reach their caudal targets via a cauda equina.

Nishikawa & Wassersug (1988) speculated that differences in spinal cord construction between *Rana* and *Xenopus* may somehow account for their different kinematics. However, it is possible that those differences reflect phylogeny more than function. The caudal spinal cord in larval *Xenopus* is, in fact, more similar to that of salamanders than to that of frogs, whereas the spinal cord in *Rana* tadpoles is more like that of adult anurans (and mammals). Indirect support for the idea that the spinal cord of *Xenopus* is more primitive than in neobatrachian frogs is the fact that at least one specimen of *Xenopus* has accomplished the atavistic feat of transforming into a frog without losing its functional tail (Kinoshita & Watanabe 1987).

Nishikawa & Wassersug (1989) confirmed that there is a phylogenetic progression in the number of paired spinal nerves in tadpoles. 'Mature' *X. laevis* tadpoles have 26 spinal nerves per side, a rather high and thus 'primitive' number for anurans. *Rana* tadpoles have 21, a reduced and more derived pattern. Notably, some non-pipid tadpoles that swim like *Xenopus* also have a reduced number of spinal nerves (Nishikawa & Wassersug 1989), so the number of spinal nerves alone does not predict the swimming kinematics in tadpoles.

Visceral functions

Functional morphologists give more attention to the front and back ends of aquatic organisms than anything in between: it is easier to watch a fish or tadpole feed and swim than to watch it digest its food. Nevertheless, those less conspicuous, visceral functions are clearly as important for survival as ingestion and locomotion. What is most remarkable—almost definitional—about tadpoles is how few visceral behaviours they actually have.

The most conspicuous visceral behaviours involve ejection of substances from the body; e.g., egg laying, ejaculation, defecation, vomiting and coughing. By definition, tadpoles are non-reproductive and thus neither lay nor fertilize eggs. However, neither do they retain faecal material in their gut and then episodically excrete that material in large boluses. Instead, in *Xenopus* at least, faeces are excreted as a continuous stream, which depends almost entirely on the rate at which food enters the alimentary tract (Wassersug 1975). Transportation down the oesophagus of *X. laevis* larvae is under ciliary control (Dodd 1950; Gradwell 1975) and peristalsis is absent in the foregut (Naitoh, Wassersug & Leslie 1989). Except in the colon and rectum, *Xenopus* tadpoles exhibit little evidence of neural control of gut function (Naitoh, Miura *et al.* 1990). Furthermore, no tadpoles can vomit (Naitoh, Wassersug *et al.* 1989), although frogs can. Finally, although oral ejection of material from the pharynx has been called 'coughing' in tadpoles, it involves pharyngeal compression and buccal floor elevation, not lung compression. No tadpoles have been observed forcefully expelling air from their lungs.

All the visceral behaviours listed above require the ability to elevate pressure abruptly in the abdominal cavity, and tadpoles do not have the musculoskeletal hardware to do this. Elevation of abdominal pressure in postmetamorphic anurans, including *Xenopus*, requires movement at the sacroiliac joints (Naitoh, Wassersug *et al.* 1989)—the same joints which are so important for saltatory locomotion in frogs (Emerson 1982; Videler & Jorna 1985)—but those joints do not develop until late in metamorphic climax. Both the 'frog kick', which marks the transition from tadpole to frog, and all of the visceral behaviours mentioned above depend on sacral movements. Thus it is not surprising that premetamorphic anurans have so few conspicuous visceral behaviours.

It is surprising how little is known about the neural regulation of visceral behaviours in anurans. Do the neural centres that control the multitude of postmetamorphic visceral behaviours (reproduction, gastric digestion, defecation patterns, etc.) develop only at metamorphosis? The whole alimentary tract changes dramatically with metamorphosis in *Xenopus* (see Geigy & Engelmann 1954) and most other anurans (Hourdry & Beaumont 1985). Are there metamorphic changes in both the central and peripheral nervous systems associated with the emergence of postmetamorphic visceral functions, or does neural development precede visceral metamorphosis? Despite a classic review paper on metamorphic changes in the anuran nervous system (Kollros 1981) and much recent attention to neural metamorphosis in *Xenopus* (see articles in *J. Neurobiol.* 1990, Vol. 21, no. 7), the developmental biology of visceral regulation in anurans remains largely unstudied.

Metamorphosis

The previous section introduced the broad topic of metamorphosis. I address here two subtopics: (1) the evolution of the process within anurans, and (2) the relationship between the environmental factors that affect anuran metamorphosis and the internal, endocrinological control of the process.

Evolutionary considerations

Entomologists have long used the process of metamorphosis in their higher classification of insects. Holometabolous insects are distinguished from hemimetabolous insects by how much larvae and adults differ, and by the abruptness of the morphological change at metamorphosis (Brown 1977).

Among living and fossil amphibian orders a similar pattern is generally accepted. Palaeozoic amphibians did not have a marked metamorphosis and neither do urodeles compared to frogs. Can we detect evolution of the metamorphic process within the Anura proper? If so, do neobatrachian frogs change more than archaeobatrachian frogs at metamorphosis, and is their metamorphosis more abrupt? A graphic account of this hypothesis for amphibians is presented by Alberch (1987), and a rough test of it is given by Wassersug & Hoff (1982), focusing on metamorphosis of orientation of anuran jaw suspension.

Tadpoles can be distinguished from virtually all other vertebrates by their very long and horizontally rather than vertically oriented palatoquadrate bars. This suspensorium slings the jaws to the braincase and results in the typical small, anterior mouth found in tadpoles. With metamorphosis, the palatoquadrate both shortens and rotates to a more vertical position, while its articulation with the lower jaw moves backward as the jaws elongate. The result is the classic transformation of the 'small-mouthed' tadpole into a 'big-mouthed' frog.

On the basis of suspensorial reorientation, the archaeobatrachian frogs have a more generalized metamorphosis than neobatrachian frogs (Wassersug & Hoff 1982), i.e., the reorientation of the jaw suspension at metamorphosis is less in the former than in the latter taxa. Furthermore, that change is, on average, more gradual in the Archaeobatrachia than in the Neobatrachia.

Xenopus, however, does not fit cleanly into either group. The suspensorium in *Xenopus* tadpoles differs substantially from that of other genera, including other pipids. Its orientation changes greatly with metamorphosis, i.e., the 'advanced' or neobatrachian pattern, although the palatoquadrate does not shorten as much as in most neobatrachian frogs (Wassersug & Hoff 1982, and unpubl. data). On the other hand, in terms of abruptness, metamorphosis in *Xenopus* is clearly gradual compared to most Neobatrachia; i.e., more like the 'generalized' or archaeobatrachian pattern. These mixed results could be interpreted as support for Cannatella & Trueb (1988; see also Trueb this volume pp. 349–377), who remove the Pipidae from the Archaeobatrachia (see Hillis 1991 for a recent discussion of neobatrachian/archaeobatrachian relationships), but that would be based on incomplete evidence. With the criteria outlined by Wassersug & Hoff (1982), both *Hymenochirus* and *Pipa*, whose tadpoles differ radically from each other and from *Xenopus* (see Sokol 1977), have an archaic metamorphosis with a transformation that is both comparatively slight and gradual.

It may be concluded that the pipid genera are highly diverse in larval morphology and consequently in their metamorphic patterns. They retain primarily a generalized or archaic metamorphosis compared to neobatrachian frogs, but that may reflect the fact that the adults, as well as the larvae, are aquatic. Thus pipids require a less extensive and less abrupt metamorphosis than tadpoles in other families that face greater habitat shifts at metamorphosis.

Ecological and endocrinological regulation

Ecologists and endocrinologists have contributed greatly to understanding the factors that affect metamorphosis. Ecologists have shown that environmental degradation, such as the decline in water levels in a pond or changes in food concentration, alter the size at and time to metamorphosis for amphibians (e.g., Semlitsch 1987; Alford & Harris 1988). Endocrinologists have established the influence of hormones on metamorphosis (see Just & Kraus-Just this volume pp. 213–229). What has received far less attention is the interplay between the ecological and the endocrinological regulation of this process. How does a tadpole know that the environment is changing and to adjust its endocrinology accordingly?

Tadpoles may be able to assess food availability if the mucus from their branchial food traps, which they use to capture food particles, contains a metamorphic inhibitor (Wassersug 1986). This theory supposes that when food is abundant a tadpole would swallow a lot of food and concomitantly a lot of mucus. If the mucus is rich in an inhibitor of gut metamorphosis, then as long as food is abundant the tadpole gut would be exposed to the inhibitor, and a tadpole would stay a tadpole. If food becomes scarce, the tadpole would necessarily ingest less food and less of its own mucus, and consequently less of the hypothetical inhibitor. The gut and subsequently the whole tadpole would proceed to metamorphose. Because the surface area of the branchial food traps does not grow as quickly as the tadpole overall (Wassersug 1986), at some species-specific upper size limit the amount of inhibitor released would not be sufficient to block metamorphosis, no matter how much particulate matter was in the water. At that point metamorphosis would be unstoppable.

The idea that a secreted (*vide* excreted) compound, which could affect metamorphosis, might use the lumen of the alimentary tract as a transport route from source to target is not new. The thyroid gland primitively opened into the pharynx in vertebrates. The lumen of the gut is the one place where the external and internal environment of a tadpole meet in a somewhat controlled fashion. Within the gut lumen vertebrates can essentially titrate environmental factors, or at least food resources.

Recently, students in my laboratory have identified an epidermal growth factor (EGF)-like peptide as a secretory product from the branchial food traps of *X. laevis* larvae and have proposed that this growth factor, or one similar to it, may play an inhibitor role in metamorphosis (Lee *et al.* 1993). However, the crucial experiments have yet to be performed to confirm whether any peptide growth factors play a role in anuran metamorphosis.

Conclusions

In this review, I have tried to illustrate some of the special features of *Xenopus* larvae that distinguish them from other tadpoles. In so doing

I have tried to promote the view that tadpoles are whole organisms and not simply embryonic parts and pieces waiting to become frogs. *Xenopus* larvae are complex integrated organisms (albeit non-reproductive), like any other free-living vertebrates. For example, the feeding activities of X. *laevis* tadpoles interact with their respiratory activities; those respiratory activites can affect their swimming behaviour, swimming behaviour can affect school geometry; and—to come full circle—*Xenopus* schooling activities may affect their feeding rates. All of this may alter their size at and time to metamorphosis. Clearly, the integration of the *Xenopus* tadpole into a whole organism is far greater than our knowledge of that integration.

In reviewing my own and other work on *Xenopus* larval biology, I have been struck by how little we know about *Xenopus* larvae in the wild. Meanwhile the popularity of X. *laevis* as a model organism for laboratory investigations continues to rise. During this symposium, experimental work with X. *laevis* has reached astronomic heights, literally, as X. *laevis* eggs were fertilized on NASA's Space Shuttle. The first free-living vertebrates conceived in space and brought to earth were *Xenopus* tadpoles that developed from those eggs (Sousa, Black & Wassersug 1995). It seems ironic then that we know so little about the natural history of premetamorphic *Xenopus*, including X. *laevis*, given the world-wide interest in clawed frog biology.

Acknowledgements

I thank the many students and colleagues who have collaborated with me on studies of anuran larvae over the years. Ronn Altig, Luc Bourque, David Cannatella, Monika Fejtek, Edward Hitchcock, Kiisa Nishikawa, Scott Pronych and Dianne Seale all offered critical comment on drafts of this manuscript. Ronn Altig graciously provided access to his database on tadpole biology. I especially thank Monika Fejtek and Michelle Kelly for bibliographic and clerical help in the production of the manuscript.

Most of my research with tadpoles has been supported by the National Science Foundation in the USA and Natural Sciences and Engineering Council of Canada. Other support has come from Dalhousie University and the Canadian Space Agency.

References

Abel, D., Seale, D.B. & Boraas, M.E. (1992). Periodicities and transient shifts in anuran (*Xenopus laevis, Rana clamitans*) oxygen consumption revealed with flow-through respirometry. *Comp. Biochem. Physiol. (A)* 101: 425–432.
Alberch, P. (1987). Evolution of a developmental process: irreversibility and redundancy in amphibian metamorphosis. *MBL Lect. Biol.* 8: 23–46.

Alford, R.A. & Harris, R.N. (1988). Effects of larval growth history on anuran metamorphosis. *Am. Nat.* **131**: 91–106.

Altig, R. & Johnston, G.F. (1986). Major characteristics of free-living anuran tadpoles. *Smithson. herpet. Inf. Serv.* No. 67: 1–75.

Arnoult, J. & Lamotte, M. (1968). Les Pipidae de l'Ouest africain et du Cameroun. *Bull. Inst. fond. Afr. noire. (A)* **30**: 270–306.

Blaustein, A.R. & Waldman, B. (1992). Kin recognition in anuran amphibians. *Anim. Behav.* **44**: 207–221.

Brown, V.K. (1977). Metamorphosis: a morphometric description. *Int. J. Insect Morph. Embryol.* **6**: 221–223.

Burggren, W.W. & Pinder, A.W. (1991). Ontogeny of cardiovascular and respiratory physiology in lower vertebrates. *A. Rev. Physiol.* **53**: 107–135.

Cannatella, D.C. & Trueb, L. (1988). Evolution of pipoid frogs: intergeneric relationships of the aquatic frog family Pipidae (Anura). *Zool. J. Linn. Soc.* **94**: 1–38.

Channing, A. (1976). Life histories of frogs in the Namib Desert. *Zoologica afr.* **11**: 299–312.

de Jongh, H.J. (1968). Functional morphology of the jaw apparatus of larval and metamorphosing *Rana temporaria* L. *Neth. J. Zool.* **18**: 1–103.

Deuchar, E.M. (1975). Xenopus: *the African clawed frog.* John Wiley & Sons Ltd., London.

Dodd, J.M. (1950). Ciliary feeding mechanisms in anuran larvae. *Nature, Lond.* **165**: 283.

Duellman, W.E. & Trueb, L. (1986). *Biology of amphibians.* McGraw-Hill Book Co., New York.

Emerson, S.B. (1982). Frog postcranial morphology: identification of a functional complex. *Copeia* **1982**: 603–613.

Feder, M.E., Seale, D.B., Boraas, M.E., Wassersug, R.J. & Gibbs, A.G. (1984). Functional conflicts between feeding and gas exchange in suspension-feeding tadpoles, *Xenopus laevis. J. exp. Biol.* **110**: 91–98.

Feder, M.E. & Wassersug, R.J. (1984). Aerial versus aquatic oxygen consumption in larvae of the clawed frog, *Xenopus laevis. J. exp. Biol.* **108**: 231–245.

Fox, H. (1984). *Amphibian morphogenesis.* Humana Press Inc., Clifton.

Geigy, R. & Engelmann, F. (1954). Beitrag zur Entwicklung und Metamorphose des Darmes bei *Xenopus laevis* Daud. *Rev. suisse Zool.* **61**: 335–347.

Gradwell, N. (1971). *Xenopus* tadpole: on the water pumping mechanism. *Herpetologica* **27**: 107–123.

Gradwell, N. (1975). The bearing of filter feeding on the water pumping mechanism of *Xenopus* tadpoles (Anura: Pipidae). *Acta zool., Stockh.* **56**: 119–128.

Hausen, P. & Riebsell, M. (1991). *The early development of* Xenopus laevis: *An atlas of the histology.* Springer-Verlag, New York.

Hillis, D.M. (1991). The phylogeny of amphibians: current knowledge and the role of cytogenetics. In *Amphibian cytogenetics and evolution*: 7–31. (Eds Green, D.M. & Sessions, S.K.). Academic Press Inc., San Diego.

Hoff, K. (1987). *Morphological determinants of fast-start performance in anuran tadpoles.* PhD thesis: Dalhousie University.

Hoff, K. & Wassersug, R.J. (1986). The kinematics of swimming in larvae of the clawed frog, *Xenopus laevis. J. exp. Biol.* **122**: 1–12.

Hourdry, J. & Beaumont, A. (1985). *Les métamorphoses des amphibiens.* Masson, Paris.

Katz, L.C., Potel, M.J. & Wassersug, R.J. (1981). Structure and mechanisms of schooling in tadpoles of the clawed frog, *Xenopus laevis*. *Anim. Behav.* **29**: 20–33.

Kay, B.K. & Peng, H.B. (Eds) (1991). *Xenopus laevis*: practical uses in cell and molecular biology. *Methods Cell Biol.* **36**: 1–695.

Kenny, J.S. (1969a). Feeding mechanisms in anuran larvae. *J. Zool., Lond.* **157**: 225–246.

Kenny, J.S. (1969b). Pharyngeal mucous secreting epithelia of anuran larvae. *Acta zool., Stockh.* **50**: 143–153.

Kinoshita, T. & Watanabe, K. (1987). [Collecting and raising.] In [*Embryogenesis and metamorphosis of amphibians: electron microscopic observation in tadpoles*]: 185–200. (Ed. Watanabe, K.). Nishimura Shoten, Niigata. [In Japanese.]

Kollros, J.J. (1981). Transitions in the nervous system during amphibian metamorphosis. In *Metamorphosis: a problem in developmental biology*: 445–459. (Eds Gilbert, L.I. & Frieden, E.). Plenum Publishing Corp., New York.

Kordylewski, L. (1986). Differentiation of tail and trunk musculature in the tadpole of *Xenopus laevis*. *Z. mikrosk.-anat. Forsch.* **100**: 767–789.

Lee, C.W., Bitter-Suermann, B., Bourque, L.A. & Wassersug, R.J. (1993). Epidermal growth factor-like immunoreactivity in the buccopharyngeal mucous glands of *Xenopus laevis* tadpoles. *Gen. comp. Endocr.* **89**: 82–90.

Loumont, C. (1984). Current distribution of the genus *Xenopus* in Africa and future prospects. *Rev. suisse Zool.* **91**: 725–746.

Lum, A., Wassersug, R.J., Potel, M. & Lerner, S. (1982). Schooling behavior of tadpoles: a potential indicator of ototoxicity. *Pharm. Biochem. Behav.* **17**: 363–366.

Muntz, L., Hornby, J.E. & Dalooi, M.R.K. (1989). A comparison of the distribution of muscle type in the tadpole tails of *Xenopus laevis* and *Rana temporaria*: an histological and ultrastructural study. *Tissue Cell* **21**: 773–781.

Naitoh, T., Miura, A., Akiyoshi, H. & Wassersug, R.J. (1990). Movements of the large intestine in the anuran larvae, *Xenopus laevis*. *Comp. Biochem. Physiol. (C)* **97**: 201–207.

Naitoh, T., Wassersug, R.J. & Leslie, R.A. (1989). The physiology, morphology, and ontogeny of emetic behavior in anuran amphibians. *Physiol. Zool.* **62**: 819–843.

Nieuwkoop, P.D. & Faber, J. (1967). *Normal table of* Xenopus laevis (*Daudin*). (2nd edn). North-Holland Publishing Co., Amsterdam.

Nishikawa, K. & Wassersug, R.J. (1988). Morphology of the caudal spinal cord in *Rana* (Ranidae) and *Xenopus* (Pipidae) tadpoles. *J. comp. Neurol.* **269**: 193–202.

Nishikawa, K. & Wassersug, R.J. (1989). Evolution of spinal nerve number in anuran larvae. *Brain Behav. Evol.* **33**: 15–24.

Nodzenski, E., Wassersug, R.J. & Inger, R.F. (1989). Developmental differences in visceral morphology of megophrynine pelobatid tadpoles in relation to their body form and mode of life. *Biol. J. Linn. Soc.* **38**: 369–388.

Orlando, K. & Pinder, A.W. (1995). Larval cardiorespiratory ontogeny and allometry in *Xenopus laevis*. *Physiol. Zool.* **68**: 63–75.

Orton, G.L. (1953). The systematics of vertebrate larvae. *Syst. Zool.* **2**: 63–75.

Orton, G.L. (1957). The bearing of larval evolution on some problems in frog classification. *Syst. Zool.* **6**: 79–86.

Pronych, S. & Wassersug (1994). Lung use and development in *Xenopus laevis* tadpoles. *Can. J. Zool.* **72**: 738–743.

Rau, R.E. (1978). The development of *Xenopus gilli* Rose & Hewitt (Anura, Pipidae). *Ann. S. Afr. Mus.* **76**: 247–263.

Sanderson, S.L. & Wassersug, R.J. (1993). Convergent and alternative designs for suspension feeding in vertebrates. In *The vertebrate skull: function and evolutionary mechanisms* 3: 37–112. (Eds Hanken, J. & Hall, B.K.). University of Chicago Press, Chicago.

Satel, S.L. & Wassersug, R.J. (1981). On the relative sizes of buccal floor depressor and elevator musculature in tadpoles. *Copeia* 1981: 129–137.

Seale, D.B. (1982). Obligate and facultative suspension feeding in anuran larvae: feeding regulation in *Xenopus* and *Rana. Biol. Bull. mar. biol. Lab. Woods Hole* 162: 214–231.

Seale, D.B., Hoff, K. & Wassersug, R.J. (1982). *Xenopus laevis* larvae (Amphibia, Anura) as model suspension feeders. *Hydrobiologia* 87: 161–169.

Semlitsch, R.D. (1987). Paedomorphosis in *Ambystoma talpoideum*: effects of density, food, and pond drying. *Ecology* 68: 994–1002.

Sillar, K.T., Wedderburn, J.F.S. & Simmers, A.J. (1991). The development of swimming rhythmicity in post-embryonic *Xenopus laevis. Proc. R. Soc. (B)* 246: 147–153.

Sokol, O.M. (1975). The phylogeny of anuran larvae: a new look. *Copeia* 1975: 1–23.

Sokol, O.M. (1977). The free swimming *Pipa* larvae, with a review of pipid larvae and pipid phylogeny (Anura: Pipidae). *J. Morph.* 154: 357–426.

Souza, K.A., Black, S. & Wassersug, R.J. (1995). Amphibian development in the virtual absence of gravity. *Proc. natn. Acad. Sci. USA* 92: 1975–1978.

van Mier, P., Armstrong, J. & Roberts, A. (1989). Development of early swimming in *Xenopus laevis* embryos: myotomal musculature, its innervation and activation. *Neuroscience* 32: 113–126.

Videler, J.J. & Jorna, J.T. (1985). Functions of the sliding pelvis in *Xenopus laevis. Copeia* 1985: 251–254.

Viertel, B. (1987). The filter apparatus of *Xenopus laevis, Bombina variegata*, and *Bufo calamita* (Amphibia, Anura): a comparison of different larval types. *Zool. Jb. (Anat.)* 115: 425–452.

Viertel, B. (1992). Functional response of suspension feeding anuran larvae to different particle sizes at low concentrations (Amphibia). *Hydrobiologia* 234: 151–173.

Vigny, C. (1979). Morphologie larvaire de 12 espèces et sous-espèces du genre *Xenopus. Rev. suisse Zool.* 86: 877–891.

Wassersug, R.J. (1972). The mechanism of ultraplanktonic entrapment in anuran larvae. *J. Morph.* 137: 279–288.

Wassersug, R.J. (1975). The adaptive significance of the tadpole stage with comments on the maintenance of complex life cycles in anurans. *Am. Zool.* 15: 405–417.

Wassersug, R.J. (1986). How does a tadpole know when to metamorphose? A theory linking environmental and hormonal cues. *J. theor. Biol.* 118: 171–181.

Wassersug, R.J. (1989a). What, if anything, is a microhylid (Orton type II) tadpole? *Fortschr. Zool.* 35: 534–538.

Wassersug, R.J. (1989b). Locomotion in amphibian larvae (or 'Why aren't tadpoles built like fishes?'). *Am. Zool.* 29: 65–84.

Wassersug, R.J. (1992). The basic mechanics of ascent and descent by anuran larvae (*Xenopus laevis*). *Copeia* 1992: 890–894.

Wassersug, R.J. & Feder, M.E. (1983). The effects of aquatic oxygen concentration, body size and respiratory behaviour on the stamina of obligate aquatic (*Bufo americanus*) and facultative air-breathing (*Xenopus laevis* and *Rana berlandieri*) anuran larvae. *J. exp. Biol.* 105: 173–190.

Wassersug, R.J. & Heyer, W.R. (1988). A survey of internal oral features of leptodactyloid larvae (Amphibia: Anura). *Smithson. Contr. Zool.* No. 457: 1–99.

Wassersug, R.J. & Hoff, K. (1979). A comparative study of the buccal pumping mechanism of tadpoles. *Biol. J. Linn. Soc.* 12: 225–259.

Wassersug, R.J. & Hoff, K. (1982). Developmental changes in the orientation of the anuran jaw suspension. *Evol. Biol.* 15: 223–246.

Wassersug, R.J. & Murphy, A.M. (1987). Aerial respiration facilitates growth in suspension-feeding anuran larvae (*Xenopus laevis*). *Expl Biol.* 46: 141–147.

Wassersug, R.J., Paul, R.D. & Feder, M.E. (1981). Cardio-respiratory synchrony in anuran larvae (*Xenopus laevis, Pachymedusa dacnicolor,* and *Rana berlandieri*). *Comp. Biochem. Physiol. (A)* 70: 329–334.

Wassersug, R.J. & Rosenberg, K. (1979). Surface anatomy of branchial food traps of tadpoles: a comparative study. *J. Morph.* 159: 393–426.

Will, U. (1991). Amphibian Mauthner cells. *Brain Behav. Evol.* 37: 317–332.

12 Control of thyroid hormones and their involvement in haemoglobin transition during *Xenopus* and *Rana* metamorphosis

JOHN J. JUST and JEANNE KRAUS-JUST

Synopsis

It has been known for 75 years that thyroid hormones (THs) are involved in amphibian metamorphosis. In the last 25 years several investigators have shown that at or near the time of tail loss there is a sudden rise in circulating TH concentrations that is dependent on the pituitary and hypothalamus. The required pituitary hormone is thyroid stimulating hormone (TSH); however, circulating TSH concentrations have yet to be determined during metamorphosis. During the last decade much effort focused on measurements of thyrotropin releasing hormone (TRH) concentrations in the tissues of several anuran tadpoles, but there is no credible evidence showing that TRH can directly initiate metamorphic climax in any species. Recent data suggest that other hypothalamic hormones are involved in the control of TSH concentrations and therefore TH concentrations before climax. During metamorphosis THs are known to control cell death, stem cell stimulation and differentiation of existing cell populations. In order for larval tissues to respond to THs, they must have TH receptors. Several different types of TH receptors exist and all are members of a larger group of nuclear protein receptors that include receptors for THs, steroid hormones and retinoic acid. More recent evidence suggests that larval tissue TH receptor concentrations change during metamorphosis and that their concentrations are highest about the time of tail loss. The transition from larval to adult haemoglobin is used to document both hormonal and environmental factors involved in the control of this metamorphic process.

Introduction

Larval life in the vast majority of aquatic chordates begins with the release of enzymes from the frontal (hatching) glands. The release of these 'hatching enzymes' is triggered when the embryo experiences anoxia (Petranka, Just & Crawford 1982; J.J. Just, in prep.). In *Xenopus laevis* this event normally

occurs at N.F. Stage 36 (Nieuwkoop & Faber 1967). Many of the morphological changes that occur after hatching are dependent on three endocrine
glands: the thyroid, pituitary and hypothalamus. The larval morphological
stages that proceed in the absence of the thyroid gland, therefore presumably
with little or none of the thyroid hormones (THs), are called premetamorphic
stages. Prometamorphic stages are those that require the presence of THs and
the term metamorphic climax should be limited to larval stages that require the
hypothalamic-dependent surge of TH levels (Just, Kraus-Just & Check 1981).
This review demonstrates that the endocrine control and hormonal responses
of cells during these three periods of larval life are very similar in two anuran
genera, namely *Xenopus* and *Rana*.

Endocrinology

When the thyroid is surgically removed or chemically inhibited, development
proceeds for a few stages but then morphological differentiation stops (Fig.
12.1). In thyroidectomized *X. laevis* development proceeds to about N.F.
stage 50–54, and in *Rana catesbeiana* morphological changes proceed to T.K.
stage III–VI (Taylor & Kollros 1946). These morphological changes, which
are independent of TH, are called the premetamorphic period and account
for about 30% of the larval life-span in *X. laevis* and about 25% in *R.
catesbeiana*. When the pituitary anlage is removed, larvae develop to about the
same premetamorphic stages as do the thyroidectomized animals (Fig. 12.4).
  Several ecological models suggest that amphibian metamorphosis is affected
by the rate of larval growth and by the absolute size of larvae (Wilbur &
Collins 1973; see Wassersug this volume pp. 195–211). One principle of these
models is that the potential for growth or growth itself determines the time
of transformation. As can be seen in Fig. 12.1, thyroidectomized larvae of *X.
laevis* and *R. catesbeiana* have essentially the same growth rate as normal animals but never complete metamorphosis. Chemical 'thyroidectomy' probably
does not completely inhibit thyroid hormone production yet greatly decreases
the rate of external morphological differentiation of *X. laevis* larvae (N.F.
stage 53) while allowing continued normal animal growth rate (Fig. 12.1).
After surgical thyroidectomy of *R. catesbeiana* larvae, external morphological
differentiation stops completely (T.K. stage IV) even though growth continues
for several years. The growth rate of hypophysectomized tadpoles is lower
than that of thyroidectomized animals (Fig. 12.4) but they proceed to the
same premetamorphic stages. Therefore, the effects of thyroidectomy and
hypophysectomy on animals prove that growth rates *per se* do not determine
metamorphic progress. Another principle hypothesized by these models is that
certain minimum and maximum sizes exist for morphological metamorphosis.
Thyroid hormones can and do induce morphological changes in both the
smallest larvae and in thyroidless larvae 500% larger than normal animals
(Hsu, Yu & Liang 1971). Although much information is available on induced

metamorphosis in normal-sized prometamorphic tadpoles (for reference see Burggren & Just 1992), few or no physiological, biochemical or behavioural studies have been carried out on these very small or very large premetamorphic tadpoles. Wassersug (see this volume pp. 195–211) has proposed a theoretical model to help to explain how the food availability and/or growth rate could affect TH levels and thus control the timing of metamorphosis.

Increasing concentrations of THs were required in the external medium to match the morphological rate of metamorphosis of thyroidectomized and/or of hypophysectomized tadpoles with that of normal animals. These results led to the hypothesis that circulating TH concentrations rise during metamorphosis. This hypothesis was first confirmed in *R. pipiens* by Just (1972) and was later demonstrated in seven other anuran species (for background see Burggren & Just 1992). The relative percentages of circulating THs in both *Xenopus*

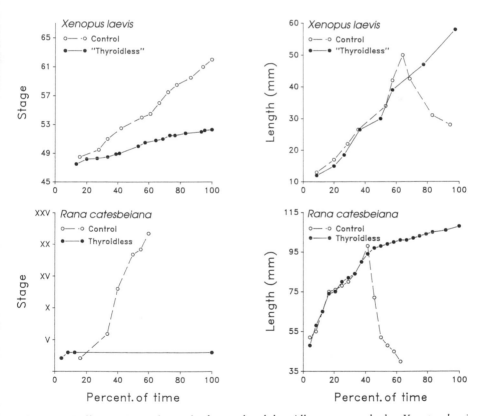

Fig. 12.1. Differentiation and growth of control and thyroidless anuran tadpoles. *Xenopus laevis* were 'thyroidectomized' in potassium perchlorate (KClO$_4$) solution (Streb 1967), while embryonic *Rana catesbeiana* were surgically thyroidectomized (Hsu & Yu 1967). The percentage of time represents a part of the total experimental period (90 days, *X. laevis*; 2 years, *R. catesbeiana*). The stages refer to N.F. stages for *X. laevis* (Nieuwkoop & Faber 1967) and to T.K. stages for *R. catesbeiana* (Taylor & Kollros 1946). The data on changes in stages of control *R. catesbeiana* come from unpublished studies (J.J. Just).

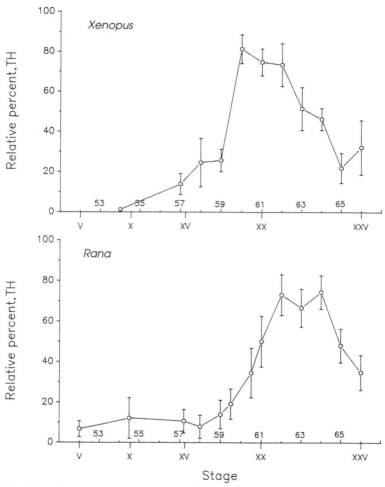

Fig. 12.2. The relative percentage of thyroid hormones (TH) in the plasma of several species of *Xenopus* and *Rana* during normal metamorphosis. Once the hormone concentration per stage in each species from each of eight publications was determined, the relative concentration per stage was calculated (hormone concentration for any stage/maximum hormone concentration at a specific stage × 100 = relative percentage of circulating hormones at a stage). After the relative concentration per stage for each species was calculated, averages for *Xenopus* (*borealis, clivii, laevis, muelleri*) and for *Rana* (*catesbeiana, clamitans, pipiens*) were determined (for details see Burggren & Just 1992). The abscissa represents the N.F. stages (Arabic numerals) and T.K. stages (Roman numerals) which were equated from morphological descriptions in the original staging papers by Just, Kraus-Just *et al.* (1981).

and *Rana* at different stages of development are shown in Fig. 12.2. In premetamorphic stages TH concentrations are about 5% of that found at their maximum. The circulating TH concentration rises gradually during prometamorphosis to about 30% of the maximum. Maximum circulating THs occur after forelimb emergence, and decline again before complete tail loss. Most of the original papers give no statistical analysis of the data; therefore, one cannot be certain at what specific stage the actual maximum concentration of THs occurs. Sequential blood samples from individual larvae throughout the climax period are needed to establish the exact stage at which the maximum circulating THs occur.

What causes the increase (Fig. 12.2) in circulating THs? In all species the rise in THs is dependent on thyroid stimulating hormone (TSH) from the pituitary gland. Remarkably, no data are available on circulating concentrations of TSH; however, attempts have been made to quantify TSH amounts in the pituitary gland. In all species studied, including *X. laevis* (see Moriceau-Hay, Doerr-Schott & Dubois 1982), cells containing TSH make their appearance in late embryonic stages or in the youngest larval stages. Both the number of cells containing TSH and the amount of TSH increase during the course of pre- and prometamorphosis (Hemme 1972; Dodd & Dodd 1976; Moriceau-Hay *et al.* 1982; Garcia-Navarro, Malagon & Garcia-Navarro 1988; Tanaka *et al.* 1991). The quantification of pituitary TSH levels as measured by TSH granule size (Hemme 1972), or the amount of the α-subunit of TSH, luteinizing hormone (LH) and follicle stimulating hormone (FSH) (Tanaka *et al.* 1991), shows that the amount of TSH increases throughout pre- and prometamorphosis and then suddenly declines at the beginning of climax (Fig. 12.3). The quantification of the α-subunit of TSH, FSH and LH indicates TSH amounts because the TSH cells are by far the most prevalent (55–85%) of these three pituitary cell populations. The quantity of β-subunit of FSH and LH is never more than 25% of the quantity of the total amount of the α-subunit (Fig. 12.3). Since the complete molecule of these hormones is composed of one α-subunit and one β-subunit, these data indicate that at least 75% of the α-subunits (Fig. 12.3) is either associated with TSH or is free inside the cells.

The precipitous release of TSH from the pituitary is dependent on the hypothalamus. Growth of hypothalectomized *R. pipiens* proceeds at the same rate as that of control animals, but the former do not enter metamorphic climax even if they are maintained until they are nearly twice the size of control animals (Fig. 12.4). *X. laevis* also do not enter climax stages when the hypothalamus is removed (Dodd & Dodd 1976). Much effort has been expended on identifying the hypothalamic hormone responsible for TSH release. It has been assumed that this hormone is thyrotropin-releasing hormone (TRH). Various authors have measured TRH levels in the hypothalamus, other brain regions and organs. In all *Xenopus* and *Rana* species, the organ TRH concentrations increase during prometamorphosis until climax stages and then either decline or continue to increase (for review see Burggren & Just 1992). Most published and personal unpublished results fail to show induced precocious metamorphosis by TRH administration. Recently it has been shown that corticotrophin-releasing hormone (CRH) can cause the release of TSH from stage XVII–XIX *R. catesbeiana* tadpoles (Denver & Licht 1989). We therefore propose that at the beginning of climax in all anuran larvae including *Xenopus* and *Rana*, the hypothalamus releases CRH and that the TSH and

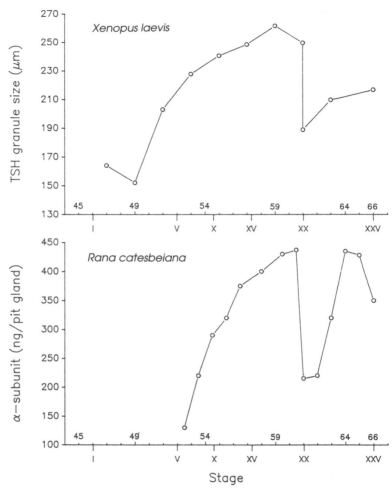

Fig. 12.3. Changes in thyroid stimulating hormone (TSH) content in the pituitary (pit) during anuran metamorphosis. Data for *Xenopus laevis* were obtained by measuring TSH granule size in pituitaries as seen in TEM (Hemme 1972). A radioimmunoassay was used to determine the amount of α-subunit of the glycoprotein hormones in the pituitary of *Rana catesbeiana* (see Tanaka *et al.* 1991). The values are the mean ± SE. Abscissa as in Fig. 12.2.

ACTH cells in the pituitary gland have CRH receptors. We hypothesize that CRH causes the release of both TSH and ACTH which in turn cause the surge in THs (Fig. 12.2) and in adrenal steroid hormones (see Burggren & Just 1992) at early climax stages of both *Xenopus* and *Rana* larvae. One or both of these groups of hormones (THs or steroid hormones) probably induce TRH receptors on pituitary TSH cells during late climax stages. This would explain why the pituitary of the froglet becomes responsive to TRH (Denver & Licht 1989) and why steroid hormones accelerate the metamorphic process (for background see Burggren & Just 1992).

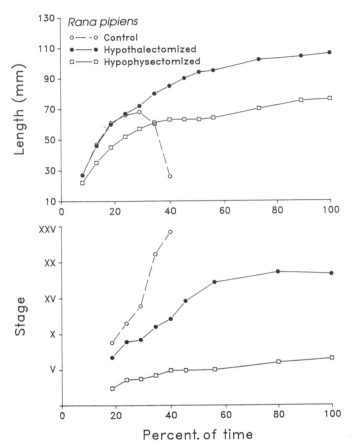

Fig. 12.4. Differentiation and growth of control, hypothalectomized and hypophysectomized *Rana pipiens* larvae. Embryos were either hypothalectomized or hypophysectomized and their growth and differentiation were monitored for a year (Hanaoka 1967). The changes in the duration of normal climax stages are from Taylor & Kollros (1946). Abscissa as in Fig. 12.1.

Haemoglobin transition

Most organs and organ systems respond in a specific fashion to the rising concentrations of THs during prometamorphosis and climax. The most obvious of these responses is the degeneration of the tail and gills and the growth of legs and lungs. Numerous cell populations change completely during this process, including red blood cells (RBCs) and intestinal and skin epithelial cells. In other organs or organ systems, like the brain and liver, changes induced by THs cause only partial changes in cell populations. This review will only address the changes that occur in the RBCs.

Over half a century ago it was documented that adult and larval haemoglobin have different oxygen saturation curves and that larval haemoglobins have a greater affinity for oxygen. It was later demonstrated that such physiological differences are caused by differences in the globin composition of adult and larval haemoglobin. In all amphibians there are more than one adult haemoglobin and several different types of larval haemoglobins. When these differences were documented, questions arose as to when the transition occurs and what causes it.

The timing of haemoglobin transition was relatively simple to document. Electrophoretic and immunological methods demonstrate that in both *X. laevis* and *R. catesbeiana* larval haemoglobin is replaced by adult haemoglobin at climax stages (Fig. 12.5). Since mature amphibian RBCs, unlike mammalian RBCs, are nucleated, many investigators presumed that this transition could occur in a single population of cells. Increases of thymidine incorporation into DNA of RBCs at climax stages suggested that a new population of cells arises during climax stages (Just & Atkinson 1972). In the 1980s investigators using either Renografin 60 (Fig. 12.5) or Percoll (Dorn & Broyles 1982; Flajnik & Du Pasquier 1988) density gradients successfully separated two populations of circulating RBCs in the late climax stages of both *Xenopus* and *Rana* tadpoles. One population contained only larval haemoglobin and the other contained only adult haemoglobin (*R. catesbeiana*: Forman & Just 1981; Dorn & Broyles 1982; *X. laevis*: J.J. Just unpubl. data not shown). When radioactive thymidine or amino acids are injected into climax tadpoles incorporation of radioactive compounds was observed only in the adult RBCs of *R. catesbeiana* (Forman & Just 1981; Dorn & Broyles 1982) and of *X. laevis* (amino acid data: Just, Schwager & Weber 1977; thymidine: J.J. Just unpubl. data not shown). Data from immunofluorescence studies of RBCs (Weber, Geiser *et al.* 1989) and *in vitro* incorporation studies of thymidine and amino acids into the RBCs of climax *X. laevis* and *R. catesbeiana* larvae are consistent with the hypothesis that a new cell population appears during late climax and that the biosynthetic activity is limited to the adult RBCs in late climax stages.

Since haemoglobin transition normally occurs during climax, it is assumed that this change might be caused by THs. When a pre- or prometamorphic larva is immersed in a solution containing THs, legs start to grow and differentiate and the tail starts to atrophy. As these external changes occur a new population of RBCs appears in circulation (Table 12.1). These new RBCs contain haemoglobin identified as adult haemoglobin on the basis of its immunological properties (*X. laevis*) or its electrophoretic mobility (*R. catesbeiana*). The old population of cells contains only larval haemoglobin in both species throughout the transition period. In induced metamorphosis, the complete replacement of larval RBCs with adult RBCs occurs within 28 days for *X. laevis* but takes 71 days in *R. catesbeiana* (Table 12.1). Measurement of amino acid (Table 12.1) and thymidine (data not shown) incorporation into these two RBC populations shows that in induced metamorphosis, as in natural metamorphosis, incorporation is almost exclusively into the adult RBC population. Significant increases in protein synthesis early in the induction period can be measured before a significant number of circulating adult RBCs can be separated physically (Table 12.1, *X. laevis*, day 7; *R. catesbeiana*, day

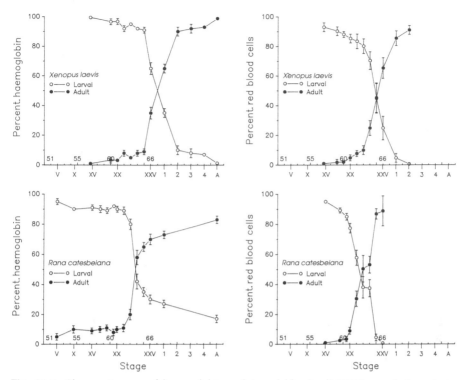

Fig. 12.5. Changes in type of haemoglobin and in red blood cell (RBC) populations during normal anuran larval metamorphosis. The percent haemoglobin concentration was determined by immunological methods for *Xenopus laevis* (Just, Schwager, Weber, Fey & Pfister 1980) and by electrophoretic methods for *Rana catesbeiana* (Just & Atkinson 1972). RBCs were separated in a continuous density gradient of Renografin 60 and quantified according to methods described by Forman & Just (1981). After quantification, the RBCs were lysed, and the nature of their haemoglobin was determined by either immunological or electrophoretic methods. The vertical bars represent ± 1 SEM. Abscissa as in Fig. 12.2; the A indicates data from adult animals. The Arabic numbers after T.K. stages refer to the weeks after climax.

22). As these new circulating cells age they cease their synthetic activity and become quiescent (Table 12.1, *X. laevis*, day 28; *R. catesbeiana*, day 60).

How do THs cause this haemoglobin transition? One obvious answer is that THs stimulate the production of adult RBCs (Fig. 12.5, Table 12.1). The rate of adult RBC production increases from essentially zero cells per day during pre- and prometamorphosis to about 5 million cells per day in *X. laevis* or about 10 million per day in *R. catesbeiana* during the climax stages (Table 12.2). When THs are used to induce metamorphic climax, the production of adult RBCs is less than during natural climax stages but much higher than during normal non-climax stages. Haemoglobin transition is also brought about by the massive and rapid destruction of larval RBCs during natural climax stages (Table 12.2). The methods used to induce climax stages produce slower rates of larval RBC destruction than occur during natural climax stages (Table 12.2). These rates of larval RBC destruction shorten the average life-span of larval

Table 12.1. Thyroid hormone induction of a red blood cell population containing adult haemoglobin in anuran larvae.

Species	Treatment[a]	Days	n	Percentages[b]		Amino acid incorporation d.p.m./10^9 cells[c]	
				Larval RBCs	Adult RBCs	Larval RBCs	Adult RBCs
Xenopus *laevis*	Th + T$_4$	0	6	91.0 ± 1.1^z	1.5 ± 0.4^z	0.01 ± 0.00^z	0.00 ± 0.00^z
	Th + T$_4$	7	6	85.3 ± 1.3^z	5.0 ± 2.6^z	0.01 ± 0.00^z	0.46 ± 0.05^y
	Th + T$_4$	14	6	65.7 ± 5.3^y	22.5 ± 6.2^y	0.02 ± 0.01^z	0.89 ± 0.25^x
	Th + T$_4$	21	6	20.4 ± 6.3^x	62.1 ± 7.5^x	0.01 ± 0.00^z	0.37 ± 0.13^y
	Th + T$_4$	28	6	5.1 ± 1.4^w	87.2 ± 2.1^w	0.00 ± 0.00^z	0.05 ± 0.04^z
	Th	0	6	90.1 ± 0.8^z	1.3 ± 0.2^z	0.01 ± 0.00^z	0.00 ± 0.00^z
	Th	7	6	89.3 ± 1.3^z	1.0 ± 0.6^z	0.01 ± 0.00^z	0.00 ± 0.00^z
	Th	14	6	87.0 ± 1.7^z	1.5 ± 0.7^z	0.02 ± 0.00^z	0.00 ± 0.00^z
	Th	21	6	88.7 ± 1.6^z	2.0 ± 1.1^z	0.01 ± 0.00^z	0.00 ± 0.00^z
	Th	28	6	85.4 ± 1.9^z	4.0 ± 2.1^z	0.01 ± 0.00^z	0.00 ± 0.00^z
Rana *catesbeiana*	T$_4$	16	6	84.7 ± 2.5^z	0.0	0.01 ± 0.10^z	0.97 ± 0.45^z
	T$_4$	22	20	78.7 ± 1.5^z	2.6 ± 1.9^z	0.01 ± 0.01^z	4.05 ± 1.35^y
	T$_4$	27	18	69.5 ± 3.3^y	4.7 ± 4.1^z	0.03 ± 0.01^z	2.65 ± 0.85^x
	T$_4$	30	19	69.4 ± 3.2^y	4.9 ± 3.6^z	0.02 ± 0.01^z	1.20 ± 0.30^z
	T$_4$	40	12	61.2 ± 2.8^y	25.3 ± 3.1^y	0.01 ± 0.00^z	0.90 ± 0.10^z
	T$_4$	45	5	65.0 ± 4.7^y	22.5 ± 5.4^y	0.02 ± 0.00^z	1.50 ± 0.60^z
	T$_4$	50	11	35.4 ± 4.8^x	53.8 ± 4.7^x	0.01 ± 0.00^z	0.20 ± 0.15^w
	T$_4$	60	6	18.1 ± 8.2^w	70.6 ± 8.5^w	0.01 ± 0.00^z	0.15 ± 0.00^w
	T$_4$	71	2	3.2 ± 3.2^v	92.1 ± 3.1^v	0.01 ± 0.00^z	0.00 ± 0.00^z
	H$_2$O	0	4	86.9 ± 2.1^z	00.0^z	0.00^z	0.00 ± 0.00^z
	H$_2$O	13	4	90.1 ± 1.5^z	00.0^z	0.00^z	0.00 ± 0.00^z
	H$_2$O	31	5	85.2 ± 1.0^z	00.0^z	0.01^z	0.00 ± 0.00^z
	H$_2$O	61	4	90.8 ± 1.8^z	00.0^z	0.01^z	0.00 ± 0.00^z

[a]*Xenopus* were placed in 0.04% thiourea at stage 55. One week later some of the animals were placed in 4×10^{-8} M T$_4$ in 0.04% thiourea. The remaining animals were maintained as controls in the 0.04% thiourea. *Rana* were immersed in 2.5×10^{-8} M T$_4$ as described by Forman & Just (1981).

[b]Cells were separated and quantified according to methods described by Forman & Just (1981). The type of haemoglobin present in cells was determined by immunological methods in *Xenopus* as described by Just, Schwager, Weber, Fey & Pfister (1980). The type of haemoglobin present in *Rana* RBCs was determined by electrophoretic methods described by Just & Atkinson (1972).

[c]Incorporation studies were performed by methods described by Forman & Just (1981).

z, y, x, w, v Numbers with the same letter in any data group not significantly different from each other. $P < 0.05$.

RBCs to less than 10 days during natural climax stages (Table 12.2), while the life-span of RBCs in pre- and prometamorphic larvae is about 100 days (Forman & Just 1976, 1981).

THs must have three separate and independent effects to cause the phenotypic switch in the RBC populations. THs must activate a system responsible for the selective destruction of larval RBCs in the presence of adult RBCs. This system may recognize the differences in the major histocompatibility proteins present in the cell membrane of larval and adult RBCs of *X. laevis* (see Flajnik & Du Pasquier 1988). The system might also use differences in the response of larval and adult RBCs to changes in the plasma osmotic pressure during normal metamorphosis of *R. catesbeiana* (for background see Burggren & Just 1992). THs must stimulate the organ or site within an organ where a new population of stem cells exists for the production of adult RBCs at climax. Finally, during climax stages, THs must inhibit cell division in the site containing the stem cells responsible for the larval RBC populations. Different origins or sites of larval and adult RBC production have been documented (Maeno, Todate & Katagiri 1985; Rollins-Smith & Blair 1990; Weber, Blum & Muller 1991). One cannot tell whether these three effects of THs on haemoglobin transition are direct or indirect on the RBCs.

Since THs greatly increase the basal metabolic rate of mammals, many investigators tried but failed to demonstrate a similar effect during amphibian metamorphosis. About a quarter of a century ago investigators turned their attention to the nucleus rather than the mitochondria as the site of TH action. Nuclear TH receptors have been demonstrated in the tail, intestine, liver, kidney and RBCs (for background see Burggren & Just 1992). The *X. laevis* genome contains four genes for TH receptors, two genes encode for the α TH receptors and two genes encode for the β TH receptors (Yaoita, Shi & Brown 1990). The number of TH receptors (particularly the β receptor) is very low during premetamorphosis, increases during prometamorphosis to high levels in early climax stages, and then declines (Yaoita & Brown 1990; Kawahara, Baker & Tata 1991). The TH receptor number increases in the larval RBC population until larval RBCs disappear. The adult RBC population also contains TH receptors, but the receptor numbers decline during haemoglobin transition (for background see Burggren & Just 1992).

TH administration to larvae increases the number of TH receptors in larval tissues. These results suggest that the rise in circulating TH concentrations during normal metamorphosis (Fig. 12.2) leads to an up-regulation of TH receptors during the course of natural prometamorphosis and early climax stages (Yaoita & Brown 1990; Kawahara *et al.* 1991; Schneider & Galton 1991; Kanamori & Brown 1992; Thomas, Drake & Frieden 1992). The up-regulation of RBC TH receptor number is observed during TH-induced metamorphosis of *R. catesbeiana* larvae and is dependent on both RNA and protein synthesis (Schneider & Galton 1991; Thomas, Drake & Frieden 1992). Thus THs do have a direct effect on RBCs, namely, the induction of TH receptors. No information is available on any other gene expressions that

Table 12.2. Kinetics of red blood cell populations during natural and thyroxine-induced climax stages in anuran larvae

Species	Type of metamorphosis	Adult RBC production rate	Larval RBC destruction rate	Half-life of larval RBCs (days)
Xenopus laevis[a]	Natural	0.53×10^7/day	1.17×10^7/day	6.5
	Induced	0.45×10^7/day	0.87×10^7/day	11.3
Rana catesbeiana[b]	Natural	1.08×10^7/day	2.09×10^7/day	7.6
	Induced	0.17×10^7/day	0.22×10^7/day	35.5

[a]Calculated from raw data in Fig. 12.5 and Table 12.1 by the methods of Forman & Just (1981).
[b]Data from Forman & Just (1981).

are altered when the RBC TH receptors are occupied by THs and how such changes in gene expression influence haemoglobin transition.

Haemoglobin transition apparently can be dissociated from circulating TH concentrations. When external metamorphosis is inhibited by chemical 'thyroidectomy' of *X. laevis*, haemoglobin transition still occurs, although the transition is delayed for many months (Just, Schwager & Weber 1977). Induction of complete anaemia during prometamorphosis causes premature appearance of adult haemoglobin in *X. laevis* (see Widmer, Hosbach & Weber 1983) but not in *R. catesbeiana* (see Maples, Palmer & Broyles 1986). Finally, restricted food availability during climax and the post-climax period greatly delays haemoglobin transition in *X. laevis* (Table 12.3). Availability of food during climax stages of *X. laevis* does not affect the rate of external metamorphosis (Table 12.3) although the size of the animals is affected by the end of climax and in the post-climax periods. If animals are well fed, haemoglobin transition is complete by two weeks after climax (96.1% adult haemoglobin), while animals that are not adequately nourished only begin haemoglobin transition by four weeks after climax (32.1% adult haemoglobin, Table 12.3). If young froglets (two weeks after climax) are given food, the animals increase adult haemoglobin production (Table 12.3). These observations (Table 12.3) may explain why normal haemoglobin transition has been reported to occur much later by others than by us (Fig. 12.5).

Even though these three conditions (dietary restrictions, anaemia and chemical 'thyroidectomy') appear to contradict the hypothesis that THs are involved in haemoglobin transition, such a conclusion is not warranted. Until the relative number of RBC TH receptors are known under these conditions, one cannot conclude that THs are not involved. Even if these three abnormal conditions can affect haemoglobin transition without the involvement of THs, that does not preclude the hypothesis that in normal metamorphosis THs cause haemoglobin transition.

Conclusion

There are numerous differences in anatomy, behaviour and ecological distribution of anuran larvae including the two genera *Xenopus* and *Rana*. This article presents evidence that the underlying physiological regulations, just like the molecular controls of the metamorphic process, are very similar. The physiological regulations included in this category are the following:

1. Early stages of larval development (premetamorphic stages) occur without THs; however, most developmental changes (prometamorphic and climax stages) are dependent on TH concentration.

2. Growth alone without the presence of THs does not result in morphological prometamorphosis or climax.

3. Maximum circulating concentration of THs occurs at or after the front legs emerge.

Table 12.3. Effects of food availability on rates of metamorphosis and haemoglobin transition in *Xenopus laevis*

Time (days)	High-calorie diet[a]			Low-calorie diet[b]		
	Average stage[c]	Larval haemoglobin[d] (%)	Adult haemoglobin[d] (%)	Average stage[c]	Larval haemoglobin[d] (%)	Adult haemoglobin[d] (%)
0	N.F 60.3^z	84.2 ± 2.1^z	15.8 ± 1.8^z	60.3^z	84.2 ± 2.1^z	15.8 ± 1.8^z
4	N.F 64.5^y	–	–	63.9^z	–	–
7	N.F 65.4^x	–	–	64.9^x	–	–
11	N.F 66.0^w	69.3 ± 4.1^y	30.7 ± 3.7^y	66.0^w	82.3 ± 3.3^z	17.7 ± 2.8^z
25	2 w.a.c.	4.9 ± 1.8^x	96.1 ± 1.9^x	2 w.a.c	78.7 ± 3.1^z	21.2 ± 2.1^z
39	4 w.a.c.	1.4 ± 1.2^x	98.6 ± 2.1^x	4 w.a.c.	67.9 ± 3.7^y	32.1 ± 3.2^y
39				4 w.a.c.^{e}	61.8 ± 4.2^y	38.2 ± 4.3^y
39				4 w.a.c.^{f}	31.7 ± 7.1^x	68.3 ± 6.8^x

[a]Larvae at stage 60–61 were placed into containers with *Tubifex* available *ad libitum*.
[b]Larvae at stages 60–61 were placed into containers without any food.
[c]The N.F stages are those of Nieuwkoop & Faber (1967); w.a.c. = weeks after climax is completed.
[d]The type and relative percentage of each type of haemoglobin was determined by methods described by Just, Schwager & Weber (1977).
[e]These animals were maintained until 2 w.a.c. without any *Tubifex* after which time they were fed one *Tubifex* per froglet per day.
[f]These animals were maintained until 2 w.a.c. without any *Tubifex* after which time they were fed *Tubifex ad libitum*.
z, y, x, w, v Numbers with the same letter in any data group are not significantly different from each other. $P < 0.05$.

4. Maximum release of the THs is dependent on a pituitary hormone (TSH) and on a hypothalamic hormone postulated to be CRH not TRH.

5. RBCs and numerous other tissues respond to THs during normal and induced metamorphosis because they have TH receptors.

6. The change from larval to adult haemoglobin involves a shift in RBC populations.

In some abnormal conditions (anaemia, shortage of food, chemical 'thyroidectomy'), haemoglobin transition can apparently be dissociated from TH levels in *Xenopus*; however, similar data do not exist for *Rana*.

Acknowledgement

We wish to dedicate this article to Professor Rudolf Weber of Bern, Switzerland, who first gave J.J.J. the opportunity to work with *Xenopus*. We express our gratitude to Gary Uglem, PhD, for his criticism and comments during the preparation of this manuscript. Partial support for original work came from NIH Biomedical Grant 2 S07RR07114–23.

References

Burggren, W.W. & Just, J.J. (1992). Developmental changes in physiological systems. In *Environmental physiology of the amphibians*: 467–530. (Eds Feder, M. & Burggren, W.). The University of Chicago Press, Chicago.

Denver, R.J. & Licht, P. (1989). Neuropeptide stimulation of thyrotropin secretion in the larval bullfrog: evidence for a common neuroregulator of thyroid and internal activity in metamorphosis. *J. exp. Zool.* **252**: 101–104.

Dodd, M.H.I. & Dodd, J.M. (1976). The biology of metamorphosis. In *Physiology of the Amphibia* 3: 467–599. (Ed. Lofts, B.). Academic Press, New York.

Dorn, A.R. & Broyles, R.H. (1982). Erythrocyte differentiation during the metamorphic hemoglobin switch of *Rana catesbeiana*. *Proc. natn. Acad. Sci. USA* **79**: 5592–5596.

Flajnik, M.F. & Du Pasquier, L. (1988). MHC Class I antigens as surface markers of adult erythrocytes during the metamorphosis of *Xenopus*. *Devl Biol.* **128**: 198–206.

Forman, L.J. & Just, J.J. (1976). The life span of red blood cells in the amphibian larva, *Rana catesbeiana*. *Devl Biol.* 50: 537–540.

Forman, L.J. & Just, J.J. (1981). Cellular quantitation of hemoglobin transition during natural and thyroid-hormone-induced metamorphosis of the bullfrog, *Rana catesbeiana*. *Gen. comp. Endocr.* 44: 1–12.

Garcia-Navarro, S., Malagon, M.M. & Garcia-Navarro, F. (1988). Immunohistochemical localization of thyrotropic cells during amphibian morphogenesis: a serological study. *Gen. comp. Endocr.* 71: 116–123.

Hanaoka, Y. (1967). The effects of posterior hypothalectomy upon the growth and metamorphosis of the tadpole of *Rana pipiens*. *Gen. comp. Endocr.* 71: 417–431.

Hemme, L. (1972). Die Differenzierungsgenese der TSH-Zellen von *Xenopus laevis* unter Normalbendingungen und nach Thiouracil Behandlung. *Z. Zellforsch. mikrosk. Anat.* **125**: 353–377.

Hsu, C.Y. & Yu, N.W. (1967). Growth of thyroidless tadpoles. *Bull. Inst. Zool. Acad. Sinica* **6**: 35–39.

Hsu, C.Y., Yu, N.W. & Liang, H.M. (1971). Age factor in the induced metamorphosis of thyroidectomized tadpoles. *J. Embryol. exp. Morph.* **25**: 331–338.

Just, J.J. (1972). Protein-bound iodine and protein concentration in plasma and pericardial fluid of metamorphosing anuran tadpoles. *Physiol. Zool.* **45**: 143–152.

Just, J.J. & Atkinson, B.G. (1972). Hemoglobin transitions in the bullfrog, *Rana catesbeiana*, during spontaneous and induced metamorphosis. *J. exp. Zool.* **182**: 271–280.

Just, J.J., Kraus-Just, J. & Check, D.A. (1981). Survey of chordate metamorphosis. In *Metamorphosis: a problem in developmental biology* (2nd edn): 265–326. (Eds Gilbert, L.I. & Frieden, E.). Plenum Press, New York.

Just, J.J., Schwager, J. & Weber, R. (1977). Hemoglobin transition in relation to metamorphosis in normal and isogenic *Xenopus*. *Roux's Archs devl Biol.* **183**: 307–323.

Just, J.J., Schwager, J., Weber, R., Fey, H. & Pfister, H. (1980). Immunological analysis of hemoglobin transition during metamorphosis of normal and isogenic *Xenopus*. *Roux's Archs devl Biol.* **188**: 75–80.

Kanamori, A. & Brown, D.D. (1992). The regulation of thyroid hormone receptor β genes by thyroid hormone in *Xenopus laevis*. *J. biol. Chem.* **267**: 739–745.

Kawahara, A., Baker, B.S. & Tata, J.R. (1991). Developmental and regional expression of thyroid hormone receptor genes during *Xenopus* metamorphosis. *Development* **112**: 933–943.

Maeno, M., Todate, A. & Katagiri, C. (1985). The localization of precursor cells for larval and adult hemopoietic cells of *Xenopus laevis* in two regions of embryos. *Dev. Growth Differ.* **27**: 137–148.

Maples, P.B., Palmer, J.C. & Broyles, R.H. (1986). *In vivo* regulation of hemoglobin phenotypes of developing *Rana catesbeiana*. *Devl Biol.* **117**: 337–341.

Moriceau-Hay, D., Doerr-Schott, J. & Dubois, M.P. (1982). Immunohistochemical demonstration of TSH-, LH- and ACTH-cells in the hypophysis of tadpoles of *Xenopus laevis* D. *Cell Tissue Res.* **225**: 57–64.

Nieuwkoop, P.D. & Faber, J. (1967). Normal table of *Xenopus laevis*. (2nd edn). North-Holland Publishing Company, Amsterdam, Netherlands.

Petranka, J.W., Just, J.J. & Crawford, E.C. (1982). Hatching of amphibian embryos: the physiological trigger. *Science* **217**: 257–259.

Rollins-Smith, L.A. & Blair, P. (1990). Contribution of ventral blood island mesoderm to hematopoiesis in postmetamorphic and metamorphosis-inhibited *Xenopus laevis*. *Devl Biol.* **142**: 178–183.

Schneider, M.J. & Galton, V.A. (1991). RNA species in tadpole erythrocytes by thyroid hormone. *Molec. cell. Endocr.* **5**: 201–208.

Streb, M. (1967). Experimentelle Untersuchungen über die Beziehung zwischen Schilddrüse and Hypophyse während der Larvalentwicklung und Metamorphose von *Xenopus laevis* Daudin. *Z. Zellforsch. mikrosk. Anat.* **82**: 407–433.

Tanaka, S., Sakai, M., Park, M.K. & Kurosumi, K. (1991). Differential appearance of the subunits of glycoprotein hormones (LH, FSH and TSH) in the pituitary of bullfrog (*Rana catesbeiana*) larvae during metamorphosis. *Gen. comp. Endocr.* **84**: 318–327.

Taylor, A.C. & Kollros, J.J. (1946). Stages in normal development of *Rana pipiens* larvae. *Anat. Rec.* **94**: 7–23.

Thomas, C.R., Drake, J. & Frieden, E. (1992). Thyroid hormone receptor induction by triiodothyronine in tadpole erythrocytes *in vivo* and *in vitro* and the effect of cycloheximide and actinomycin. *Gen. comp. Endocr.* **86**: 42–51.

Weber, R., Blum, B. & Müller, R.P. (1991). The switch from larval to adult globin gene expression in *Xenopus laevis* is mediated by erythroid cells from distinct compartments. *Development* **112**: 1021–1029.

Weber, R., Geiser, M., Müller, P., Sandmeier, E. & Wyler, T. (1989). The metamorphic switch in hemoglobin phenotype of *Xenopus laevis* involves erythroid cell replacement. *Roux's Archs devl Biol.* **198**: 57–64.

Widmer, H.J., Hosbach, H.A. & Weber, R. (1983). Globin gene expression in *Xenopus laevis*: anemia induces precocious globin transition and appearance of adult erythroblasts during metamorphosis. *Devl Biol.* **99**: 50–60.

Wilbur, H.M. & Collins, J.P. (1973). Ecological aspects of amphibian metamorphosis. *Science* **182**: 1305–1314.

Yaoita, Y. & Brown, D.D. (1990). A correlation of thyroid hormone receptor with amphibian metamorphosis. *Genes Dev.* **4**: 1917–1924.

Yaoita, Y., Shi, Y.B. & Brown, D.D. (1990). *Xenopus laevis* α and β thyroid hormone receptors. *Proc. natn. Acad. Sci. USA* **87**: 7090–7094.

Infections and defence

13 Parasites of *Xenopus*

R.C. TINSLEY

Synopsis

The *Xenopus* species carry a richer assemblage of parasites than most other anurans, with over 25 genera from seven invertebrate groups. This spectrum reflects a dual origin: some of the parasites are characteristic of anurans and represent 'heirlooms' from a common ancestry of parasites which have evolved with the anuran lineages. In addition, a component of the fauna is related to parasites which typically infect fishes: these represent ecological acquisitions and reflect the overlap, in terms of diet, habitats etc., between *Xenopus* and fish. For both these subsets of parasites, the representatives infecting *Xenopus* are highly distinctive, with life-history and other specializations which are often unique within the respective parasite groups. From a systematic point of view, they are best interpreted as having had a very distant origin from their nearest relatives and having now diverged to the point where almost all forms infecting *Xenopus* are placed in their own genera, families or higher taxonomic categories, strictly specific to *Xenopus*. Some of the parasites exploit *Xenopus* as an intermediate host and await passage to final hosts which are predators of *Xenopus* (and usually also of fish). These parasites may be pathogenic, infecting the heart, body musculature, eyes and lateral line system: by interfering with normal function, infection increases the risks of predation and thus facilitates completion of the parasite's life cycle. Other parasites employ *Xenopus* as a final host (in which their sexual reproduction occurs), and invade by a variety of routes (with the diet, by vector transfer, or by direct penetration of an active infective stage). For most of these parasites, pathogenic effects are difficult to detect and infection levels tend to occur below the point at which damage may be serious. The skin-infecting nematode, *Pseudocapillaroides xenopodis*, is exceptional in its pathogenicity, being responsible for lethal epidemics amongst laboratory-maintained *X. laevis*.

Introduction

A range of pathological conditions affecting *Xenopus*, including microbial infections and neoplasms, are well documented in the older literature, particularly by herpetologists who were medical practitioners (notably Dr E. Elkan). Detailed descriptions of tuberculosis affecting the skin and the respiratory and digestive tracts of *X. laevis* are provided by Reichenbach-Klinke & Elkan (1965). These and other accounts emphasize three related principles.

First, *Xenopus* may carry a range of pathogens without the development of disease until there is a physiological upset, caused by malnutrition, temperature change or other environmental stress; until this point, the condition may be kept in check and becomes overtly pathogenic only following the additional precipitating factor. Second, once disease has become established, it is striking that pathogenesis may develop to an extreme degree before *Xenopus* exhibits obvious ill-effects. Commonly, in tuberculosis of the lungs, liver or gut, a considerable part of the organ may be rendered non-functional before illness and death occur. Reichenbach-Klinke & Elkan (1965) observed that such 'crippling injuries and disease would have killed warm-blooded animals at a much earlier stage'. Third, in the often crowded conditions of laboratory maintenance, one animal may develop severe disease (tuberculosis, for instance) and die, whilst all others in the same aquarium remain unaffected. These observations suggest that immune defences are normally highly effective in the control of microbial disease (see, in this volume, Du Pasquier, Wilson & Robert pp. 301–313; Horton, Horton & Ritchie pp. 279–299; Kreil pp. 263–277).

In contrast to these chronic conditions, a variety of microbial pathogens may cause epidemic disease leading rapidly to high mortality. These diseases include the severe generalized sepsis attributable to haemolytic bacteria, particularly *Aeromonas hydrophila*. Little is known of the influence of these infections in wild *Xenopus* populations: virtually all experience derives from lethal outbreaks which may originate and develop under the unnatural conditions of laboratory maintenance.

The scope of this account is restricted to the infections of *Xenopus* included within the remit of parasitology: the protozoan, platyhelminth, nematode and other metazoan parasites which represent associations acquired in nature and forming part of the natural 'biology of *Xenopus*'. For most of these agents, the association with *Xenopus* probably extends through the greater part of the evolutionary history of the host group. The reciprocal influences form the basis of a host–parasite relationship involving complex ecological, physiological and evolutionary interactions.

The parasite fauna

Diversity

The parasite fauna of *Xenopus* is characterized by its extraordinary richness. Based on records from one of the species, *X. laevis* (on which most information is available), the fauna associated with *Xenopus* includes over 25 genera from seven major invertebrate groups (Table 13.1). The majority of these parasites are highly distinctive and possess specializations which are found nowhere else amongst their respective groups, indeed in some cases their adaptations are unique within the animal kingdom. This distinctiveness is illustrated at a taxonomic level by the fact that, for those parasites which have been formally named, every one is represented by a separate species specific to *Xenopus*, that almost all (excluding the larval digeneans) are represented by distinct genera, and that many belong to a separate *Xenopus*-specific higher taxon (tribe, sub-family or even family). As a group, therefore, this parasite fauna stands apart from almost all its respective relatives.

One parasite phylum has not, so far, been reliably recorded from any *Xenopus* species—the Acanthocephala[1]. The apparent absence of acanthocephalans is noteworthy since this group parasitizes many other anurans, and is transmitted via crustacean intermediate hosts which form an important component of the diet of *Xenopus*.

Sites of infestation

Virtually all the organ systems of *Xenopus* provide habitats for parasites. Those sites which are in direct communication with the outside, principally the alimentary canal and organs opening off it (nasal and eustachian passages, gall bladder, urinary bladder, kidneys) have no physical barrier to entry; these harbour parasites which typically reach sexual maturity in *Xenopus* and there is a straightforward exit for their eggs or larval stages. Other habitats are enclosed and access to the site requires active tissue penetration. Regions including the pericardial and peritoneal cavities, lateral line system, eyelids and general musculature, harbour larval stages of cestodes, nematodes and digeneans which employ *Xenopus* as an intermediate host. There is no direct exit from these regions and the parasites depend on passive transfer to another host, usually by predation, before further development and maturation can occur. The blood stream is also a totally enclosed habitat and may be infected by protozoans and nematode microfilariae; in these cases transmission from host to host is dependent on a blood-feeding vector.

One potential site appears to be largely unexploited: the lungs of *Xenopus* are very well developed and vascularized; they have the major role in respiratory exchange (cutaneous respiration is relatively less important). Exceptionally, Macnae, Rock & Makowski (1973) found one *X. laevis* (out of a large but unspecified number examined) infected with strigeiform metacercariae in cysts on the lungs, and Cosgrove & Jared (1974) reported one *X. laevis* (out of 435) with a nematode identified as *Rhabdias* sp. in the lung. Otherwise, this niche appears to be vacant. This, like the absence of an acanthocephalan parasite in *Xenopus*, is surprising: in other anurans, the lungs are very frequently occupied by digeneans and nematodes.

Although *Xenopus* is more or less continuously exposed to water-borne infection, the external skin surface is generally not infected by permanent parasites. These may be precluded by the occasions when the skin dries, during

---

[1]Although Leadley-Brown (1970) reported an acanthocephalan in the rectum of *X. laevis*, her illustration indicates that this was actually the digenean *Progonimodiscus doyeri* in which the posterior acetabulum is mistaken for the acanthocephalan proboscis.

Table 13.1. The parasite fauna of *Xenopus laevis* based on my records (1967 to the present) from *X. laevis laevis* at the Cape, Republic of South Africa, with the addition of metacercariae recorded by Porter (1938) (*Cercaria, Opisthioglyphe* spp.) and Macnae *et al.* (1973) (*Clinostomum* sp.), and some protozoans recorded by Thurston (1970) (from *X. l. victorianus* in Uganda). Other unidentified digenean metacercariae and larval cestodes and nematodes recorded at the Cape are omitted. Parasites recorded in *X. laevis* elsewhere include *Pleurogenes cystolobatus* in Zaire (Manter & Pritchard 1964). References provide the most recent comprehensive accounts (including some life cycles, systematics) with bibliographies listing earlier publications.

Parasite	Site	References
MONOGENEA		
Protopolystoma xenopodis	Urinary bladder, kidneys	Tinsley & Owen (1975)
Gyrdicotylus gallieni	Mouth	Harris & Tinsley (1987); Jackson & Tinsley (1994)
DIGENEA		
Adults		
Dollfuschella rodhaini	Stomach	Macnae *et al.* (1973)
Oligolecithus elianae	Intestine	Macnae *et al.* (1973)
Xenopodistomum xenopodis	Gall bladder	Tinsley & Owen (1979)
Progonimodiscus doyeri	Rectum	Macnae *et al.* (1973)
Metacercariae		
Diplostomum (Tylodelphys) xenopodis	Pericardium	Tinsley & Sweeting (1974)
Neascus sp.	Lateral line	Elkan & Murray (1952)
Echinostomum xenopodis	Eyelids	Porter (1938)
Clinostomum sp.	Body cavity	Macnae *et al.* (1973)
Cercaria xenopodis	Lateral line	Porter (1938)
Opisthioglyphe xenopodis	Dermis	Porter (1938)
CESTODA		
Cephalochlamys namaquensis	Intestine	Thurston (1967); Ferguson & Appleton (1988)
NEMATODA		
Camallanus kaapstaadi	Oesophagus	Thurston (1970); Jackson & Tinsley (1995a)
Camallanus xenopodis	Intestine	Jackson & Tinsley (1995a)
Batrachocamallanus slomei	Stomach	Jackson & Tinsley (1995b)
Pseudocapillaroides xenopodis	Epidermis	Cosgrove & Jared (1974); Wade (1982)
Microfilariae	Blood	Thurston (1970)
ACARI		
Xenopacarus africanus	Nostrils, eustachian passages	Fain *et al.* (1969)
HIRUDINEA		
Marsupiobdella africana	External skin	Van der Lande & Tinsley (1976)
PROTOZOA		
Balantidium xenopodis	Rectum	Thurston (1970)
Nyctotherus sp.	Rectum	Thurston (1970)
Protoopalina xenopodus	Rectum	Thurston (1970)
Hexamita intestinalis	Rectum	Thurston (1970)
Chilomastix caulleryi	Rectum	Fantham (1922)
Entamoeba sp.	Intestine	Thurston (1970)
Trichodina xenopodos	Urinary bladder	Kruger *et al.* (1991)
Trypanosoma sp.	Blood	Thurston (1970)
Cryptobia sp.	Blood	Thurston (1970)

overland migration and during aestivation, and by the periodic moulting of the skin. In laboratory-maintained *X. laevis*, Tinsley & Whitear (1980) recorded two protozoan epibionts which were able to avoid the effects of moulting. Highly mobile *Trichodina* sp. occurred on the skin surface: it is likely that many would be eaten by the host each time the slough is cast; however, some could be thrown off and then re-establish on the same or other hosts, especially in confined aquarium conditions. In contrast, the sessile ciliate *Epistylis* sp. forms colonies confined to the claws which provide a more stable habitat because they are not involved in the regular moulting process. Tinsley & Whitear (1980) also recorded bacteria attached to the claws, particularly in association with epibiont colonies, but scanning electronmicroscopy revealed few bacteria on the skin surface. This freedom from external microbial infection may be linked with specific peptide skin secretions (reviewed by Kreil, this volume pp. 263–277). The leech *Marsupiobdella africana* does inhabit the external body surface, but individuals remain on the host for only about three weeks, during feeding. Their response to skin moulting has not been observed. Other permanent parasites of *Xenopus*, whose closest relatives are skin parasites of fish, have adopted alternative habitats which avoid the hazards of life on the external skin: the representatives infecting *Xenopus* are mesoparasites. *Gyrdicotylus gallieni* occurs in the oral cavity and *Trichodina xenopodos* infects the urinary bladder (Table 13.1).

Host specificity

The great majority of parasites listed in Table 13.1 are strictly host-specific to *Xenopus*. Only three are reliably recorded from other anurans. *Cephalochlamys* is a common parasite of the *Xenopus* taxa more or less throughout their geographical ranges; there are two instances of its occurrence in other anurans, in *Rana angolensis* in Zimbabwe (Mettrick 1963) and in *Dicroglossus occipitalis* in Gabon (Dollfus 1968). Since infection requires ingestion of copepod intermediate hosts, transmission to non-pipids must be rare but both these ranid frogs occur frequently in water and could feed on prey which has recently ingested (or become contaminated with) infected copepods.[2] *Marsupiobdella africana* has been found once on *Rana vertebralis* in addition to *X. laevis* (see Moore 1958). However, Van der Lande & Tinsley (1976) showed experimentally that although the leeches may temporarily attach to other anurans, they feed and complete their life history only on *Xenopus* species. *Progonimodiscus doyeri* has been recorded once in the ranid *Conraua crassipes* in Cameroon by Gassmann (1975). In addition to these exceptions concerning Table 13.1, the digenean *Pleurogenes cystolobatus* occurs in *Xenopus laevis* and *Bufo regularis* in Zaire (Manter &

[2]Infection of the North African *Pleurodeles poireti* by *C. namaquensis*, recorded in an aquarium by Dollfus (1968), should be regarded as unnatural, particularly since there is no geographical overlap between this urodele and *Xenopus*.

Pritchard 1964), and *Diplodiscus magnus* infects *X. tropicalis* and species of
Dicroglossus, Hylarana and *Kassina* in Ghana (Fischthal & Thomas 1968).
Unusually, the life cycles of these digeneans could permit transmission both
to aquatic *Xenopus* and to other anurans. Infection with diplodiscine flukes
may occur through ingestion of infected snails (by *Xenopus* when underwater
or by other anurans when snails emerge onto vegetation). Alternatively, the
cercariae may encyst on skin surfaces leading to infection when moulted skin
is eaten (Vercammen-Grandjean 1960), and this route could involve a variety
of anurans as in the case of *D. magnus*.

Nevertheless, these instances of cross-infection are extremely rare: the
general principle emerges that parasites forming the highly characteristic
fauna associated with *Xenopus* do not normally infect other host groups
and, reciprocally, that parasites from other definitive hosts, both fish and
other anurans, infect *Xenopus* only exceptionally.

Origins and relationships

The parasites of *Xenopus* can be divided, on the basis of their systematic
relationships and host specificity, into two assemblages. A part of the parasite
fauna has affinities with that represented in other anuran amphibians. Thus,
Protopolystoma is related to the genus *Polystoma* which occurs in the urinary
bladders of a wide range of anurans. *Progonimodiscus* is equivalent to
Diplodiscus which occurs in the corresponding site—the rectum—in other
frogs and toads. The rectal opalinid of *Xenopus, Protoopalina*, is represented
in other anurans by species of *Opalina*. The nearest relatives of the other gut
trematodes *Dollfuschella* and *Oligolecithus* also occur in non-pipid anurans,
and the same is true of the rectal flagellates *Nyctotherus* and *Balantidium*.
These links reflect the common ancestry of the parasites which probably
infected the early anuran stock and then evolved with their respective host
groups. However, the *Xenopus* parasites are generally quite distinct from those
in other anurans: this is indicated by their isolated taxonomic position. For
instance, *Xenopacarus* belongs to a group of trombidiform mites which all
infect the nasal cavities of anurans, but the *Xenopus* parasite is separated from
the other members into its own genus and tribe within the Lawrencarinae (see
Fain, Baker & Tinsley 1969). *Xenopodistomum* belongs to the Plagiorchidae
which includes many intestinal parasites of anurans; however, the *Xenopus*
parasite has no obvious close links with other genera in this group (Tinsley
& Owen 1979).

The other major group of *Xenopus* parasites is related to forms which are
more commonly represented on fishes. *Gyrdicotylus*, in the buccal cavity of
Xenopus, is similar to the genus *Gyrodactylus* which occurs on the skin,
fins and gills of a wide range of fish. The camallanid nematodes which are
found in the pharynx, oesophagus, stomach and intestine of *Xenopus* are
typically gut parasites of fish. Strigeatoid digeneans invade the vertebrate
second intermediate host by a skin-penetrating cercaria; their hosts are

therefore characteristically aquatic and most commonly fish. *Xenopus* is infected by three of the four strigeatoid groups: *Diplostomulum, Neascus* and *Tetracotyle*. *Trichodina* occurs in the urinary bladder of *Xenopus* and nearest relatives are frequent ectoparasites of fish. This general affinity reflects the ecological link, the sharing of habitats and diets, between the respective host groups. The *Xenopus* parasites are nevertheless not recent transfers from fish hosts: the representatives infecting *Xenopus* are morphologically distinct and taxonomically isolated, reflecting a long association with *Xenopus*, to which they are now strictly host-specific. Thus, the pseudophyllidean cestodes have life cycles involving copepod intermediate hosts and this determines transmission to aquatic vertebrates, principally fish. The representative in *Xenopus, Cephalochlamys*, follows this pattern but is assigned to a family of its own, distinct from other pseudophyllideans.

The camallanid nematodes illustrate how *Xenopus* may have been 'captured' by fish parasites which utilize copepod crustaceans as intermediate hosts and which would therefore be transmitted to a range of planktivorous predators. Three camallanid species occur in South African *X. laevis* (Table 13.1). The closest relatives of *Batrachocamallanus* are found amongst *Procamallanus* species infecting a range of African freshwater fishes. However, these lineages have now diverged to the extent that the forms infecting *Xenopus* are assigned to a separate host-specific genus (Jackson & Tinsley 1995b). Taxonomic analysis of the *Camallanus* species infecting *Xenopus* indicates that they may be derived from at least two independent colonizations by different parasite lineages occurring in teleost fishes (Jackson & Tinsley 1995a). There are no recorded instances of present-day species of camallanids from fishes transferring to *Xenopus*; equally the *Xenopus* camallanids are now strictly host-specific and do not infect their ancestral hosts.

Phylogenetic affinities

Amongst the diverse groups associated with *Xenopus*, including protozoan, helminth, annelid and arthropod representatives, a significant number of examples retain very primitive phylogenetic features. The monogenean *Gyrdicotylus gallieni* possesses the highly distinctive features of the Gyrodactylidae, including a unique form of viviparity. Its excretory system (an important phylogenetic character) cannot be derived from that of any known gyrodactylid genus, suggesting that *Gyrdicotylus* has been distinct from a very early stage in the evolution of the family; this is corroborated by the distinctive armature of the penis which is unlike that of other gyrodactylids (Harris & Tinsley 1987). The mite *Xenopacarus africanus* is characterized by the most primitive organization of the ereynetal organ found within its family, the Ereynetidae. This primitive development has provided an explanation of the origin of the more advanced structures found in mites infecting other anurans, including *Bufo* species (Fain *et al.* 1969). *Protopolystoma xenopodis* exhibits a series of characters (presence of two pairs of haptoral hamuli and 16 penis hooks, absence of vaginae and a uterus) none of which occurs elsewhere amongst anuran polystomatids. Further consideration of larval characters indicates that *Protopolystoma* fully deserves the status suggested by its name and has previously unrecognized links with polystomatids infecting chelonian reptiles (Tinsley 1981).

Other examples from the *Xenopus* fauna also exhibit characters which reveal interrelationships between groups otherwise considered to be entirely separate. The leech *Marsupiobdella africana* possesses a series of exceptional features including a velum, which represents an intermediate condition between the proboscis typical of rhynchobdellids and the pharynx of hirudids. This novel structure demonstrates a link between the two orders of the Hirudinea, the Rhynchobdellida and the Gnathobdellida (Van der Lande & Tinsley 1976). The nematode *Pseudocapillaroides xenopodis* has morphological features which place it within the Capillariidae; however, the parasite is distinguished by its site of infection (the host epidermis) and by its production of embryonated eggs *in utero*, characteristics which demonstrate a close affinity to a separate family, the Trichosomoididae (Moravec & Cosgrove 1982).

These characters which unite otherwise distinct systematic groups suggest that the parasites in *Xenopus* may have very ancient origins within their respective lineages. Superimposed on this complement of primitive characters are others which are highly specialized, often found nowhere else amongst related parasites. *Marsupiobdella africana* possesses a permanent brood pouch, unique amongst the annelid Clitellata (all oligochaetes and leeches), and this permits the most advanced reproductive behaviour found amongst glossiphoniid leeches (Van der Lande & Tinsley 1976). The digenean *Xenopodistomum xenopodis* shows extreme morphological specialization to its habitat within the gall bladder of *Xenopus*; although the changes accompanying ontogeny demonstrate how this unusual body form evolved from the more typical digenean plan, no other relatives infecting anurans show comparable adaptation (Tinsley & Owen 1979). The attachment mechanism of the monogenean *Gyrdicotylus gallieni* is unique for a gyrodactylid, with a highly specialized suctorial mechanism adapted to the soft tissue substrate of the oral cavity.

Thus, the highly distinctive fauna associated with *Xenopus* has a mixture of primitive and specialized features. These characteristics seem best interpreted in terms of, first, the early separation of these parasites from their respective evolutionary lines and, second, their subsequent prolonged isolation, facilitating adaptation to their specific host and life styles.

Life-cycle patterns

The life cycles of the 29 parasites associated with *Xenopus* represent a major part of the diversity shown within the entire field of parasitology. As a host, *Xenopus* is involved in a complex web of interactions with other organisms—with its parasites, its prey and its predators.

Indirect life cycles

Digeneans

Digenean platyhelminths have life cycles involving two or more hosts of which the first is almost always a mollusc. The adult parasite is generally found in the alimentary tract and associated organs of a vertebrate final host and eggs pass out with the faeces. A miracidium hatches, usually a swimming ciliated larva, and this infects the snail within which a succession of stages multiply asexually and produce another swimming stage, the cercaria. Commonly, this penetrates into the tissues of a second intermediate host within which it encysts. The parasite, now a metacercaria, awaits transfer to the final host which is typically next in a predator–prey sequence. There are a great many variations on this basic pattern, but it serves as a general framework to examine the interactions of the digeneans infecting *Xenopus*.

Xenopus as intermediate host. Six of the digeneans infecting *Xenopus* occur as metacercariae (Table 13.1): in these cases, invasion involves direct cutaneous penetration by a cercaria, and it follows that *Xenopus*—now the second intermediate host—must be eaten by a predator for the parasites to complete their life cycles. In some cases the life cycle requires four hosts: the invading cercaria becomes a mesocercaria in *Xenopus*, transforms into a metacercaria in a predator of *Xenopus* and only becomes an adult in a predator of the third intermediate host.

These life-cycle patterns most commonly involve fishes as hosts of the mesocercariae or metacercariae which are ultimately transmitted to predators of fish. The aquatic ecology of *Xenopus* leads to involvement with several life cycles, but none of these parasites infects fishes as alternative hosts: all are host-specific to *Xenopus*. The metacercariae occur in species-specific locations: those in the tissues (muscles, lateral line organs, eyelids) are enclosed in a cyst which may be laid down both by the parasite (to protect itself) and by the host (to enclose the parasite); those in the body cavities (pericardium, peritoneum) are generally unencysted and mobile. The life cycle of one of these has recently been determined (P.H. King & J.G. Van As unpubl.). *Diplostomum (Tylodelphys) xenopodis* infects the freshwater snail *Bulinus tropicus*; the cercariae are equipped with large penetration glands for invasion of *Xenopus* and migrate to the pericardial sac; in experimental trials, metacercariae transformed into adults after ingestion by the darter (*Anhinga melanogaster*), and natural infections were also found in the duodenum of the reed cormorant (*Phalacrocorax africanus*).

Other species have been recorded as larval stages without evidence of the final host but predators of *Xenopus* such as the black-headed heron (*Ardea melanocephala*) and the hammerhead stork (*Scopus umbretta*), together with egrets and cormorants, are suspected as final hosts of some of the metacercariae (Macnae *et al.* 1973).

Natural infections in snails have been reported to release cercariae which infect *Xenopus laevis*: Porter (1938) observed that *Cercaria xenopodis* from *Physopsis africanus* invaded and encysted in the neuromast organs around the eyes and elsewhere in the lateral line system; *Echinostomum xenopodis* from *Physopsis africanus* and *Lymnaea natalensis* infected the skin around the eyes and lateral line organs; encysted *Opisthioglyphe xenopodis* were recovered from both *Lymnaea natalensis* and tadpoles of *X. laevis*. (Final hosts of *Opisthioglyphe* spp. do not normally include birds and another predator of tadpoles may be involved.) Other parasites are known only from the metacercarial stage in *Xenopus*: *Clinostomum* sp. occurs in the peritoneal cavity (Macnae *et al.* 1973; Prudhoe & Bray 1982); *Neascus* sp. encysts in the dermis below the lateral line sense organs (Elkan & Murray 1952).

Given its exposure to water-borne infection, it is not surprising that *Xenopus* is invaded by a variety of cercariae, and there are reports of other infections which have not been identified. Macnae *et al.* (1973) recorded strigeid metacercariae encysted in groups of 3–5 worms on the lungs of one *X. laevis*.

***Xenopus* as final host.** Four other digenean genera exploit southern African *Xenopus laevis* as final host in the life cycle. These belong to groups which are typical parasites of anurans (in contrast to the metacercariae which are almost all allied with groups infecting fish). Generally, these life cycles involve arthropods or even the metamorphosed young of the anuran as second intermediate hosts (which are, presumably, regular items in the diet). For the digeneans of most anurans, the second intermediate host is often infected in water and then eaten after metamorphosis into a terrestrial stage. *Xenopus*, on the other hand, is likely to become infected by ingestion of aquatic prey. Thus, the first intermediate host of *Oligolecithus elianae* is *Lymnaea natalensis* from which cercariae emerge and penetrate larvae of *Culex pipiens* and *Aedes aegypti*; *Xenopus* acquires infection through predation on mosquito larvae (Vercammen-Grandjean 1960). The miracidia of *Progonimodiscus doyeri* penetrate *Biomphalaria pfeifferi*, and the cercariae which emerge encyst on a variety of surfaces including the shells of snail hosts and amphibian skin: *Xenopus* becomes infected by feeding on snails and possibly also when moulted skin is ingested (Vercammen-Grandjean 1960). Miracidia of *Dollfuschella rodhaini* infect *Lymnaea natalensis*, but the subsequent course of transmission has not been determined (Vercammen-Grandjean 1960). The life cycle of *Xenopodistomum xenopodis* is entirely unknown, but this digenean has been recorded only in South Africa (Macnae *et al.* 1973; Tinsley & Owen 1979) and it is possible that one or more of its intermediate hosts has a restricted geographical range. In contrast, the other digeneans which mature in *X. laevis* employ relatively ubiquitous species of snail and arthropod intermediate hosts and, correspondingly, the parasites are widely distributed throughout the range of *Xenopus* species.

From this outline, it emerges that the two groups of digenean parasites associated with *Xenopus* represent ecological indicators of predator–prey interactions: the metacercarial infections rely on specific predators of *Xenopus*, the infections of adult digeneans reflect specific prey species in the diet of *Xenopus*.

Cestodes and nematodes

Xenopus **as intermediate host.** Encysted larval stages of tapeworms and nematodes occur in the tissues of *Xenopus*, including the wall of the gut and the peritoneal membranes (Cosgrove & Jared 1974; Macnae *et al.* 1973; Thurston 1970). There is no information on the identity of these and final hosts must be assumed to be predators of *Xenopus*.

Xenopus **as final host.** In contrast to the ecological complexity of the digeneans, the cestode and camallanid nematode parasites of *Xenopus* employ almost identical two-host life cycles characteristic of related parasites in many aquatic vertebrates. The species of *Cephalochlamys*, *Camallanus* and *Batrachocamallanus* occur as adults in the alimentary tract of *Xenopus*. Larval stages are infective when passed in the faeces and are eaten by copepods; they penetrate into the haemocoel of the copepod, develop to an infective stage, and transmission relies on predation of copepods by *Xenopus* (Thurston 1967, 1970).

Nothing is known of the life cycle of the filarial nematode whose infective microfilariae were found in the blood of *Xenopus* by Thurston (1970). Adults, normally located in the internal tissues, were not found. A blood-feeding vector must be involved in host-to-host transmission.

Direct life cycles

There are four groups of metazoan parasites which have direct (single-host) life cycles.

Monogeneans

Monogeneans are predominantly ectoparasites of fishes and almost all are transmitted by a swimming infective stage. Typically, the adult parasite deposits eggs into the external environment; these develop on the substratum and produce ciliated larvae—the oncomiracidia—which hatch and locate the next host. One group of monogeneans, the Polystomatidae, has radiated onto tetrapod vertebrates, a colonization which probably accompanied the evolution of early land vertebrates. The original external habitats of the fish-parasitic monogeneans, the skin and gills, would have been precluded in this evolutionary step, and the lineages which successfully evolved with the proto-amphibians moved into equivalent 'aquatic' habitats at either end of the alimentary tract—the oral cavity or the urinary bladder. The Polystomatidae retain the characteristic aquatic transmission route but show a series of remarkable adaptations for exploiting anurans which only enter water periodically (Tinsley 1983, 1990). However, *Protopolystoma xenopodis* has a host which is essentially fish-like in habits and the life-cycle pattern is abbreviated in comparison with that of other anuran polystomatids. The adult, in the urinary bladder, manufactures eggs continuously (mean: nine eggs per parasite per day at 20 °C); these are expelled and develop to produce an oncomiracidium (three weeks at 20 °C), and the hatched infective stage has 12–24 h of swimming life to locate another *Xenopus*, invading via the cloaca. The juvenile worms develop in the kidneys for 2–3 months and then migrate to the bladder where, as reproducing adults, they may live for up to two years (Thurston 1964; Tinsley & Owen 1975; Jackson & Tinsley 1988a).

Whereas the close relatives of *Protopolystoma* all infect tetrapods, especially anurans, the second monogenean parasite of *Xenopus* belongs to a group in which virtually all other species are fish parasites. The gyrodactylids have a method of reproduction unique within the Animal Kingdom. The worm is viviparous and the offspring, which develops in the large uterus of the parent, itself contains a developing offspring in its uterus, and this in turn may contain a further offspring *in utero*. The resulting series of generations, boxed like 'Russian dolls', confers an exceptional capacity for rapid population increase within infected hosts. On fish, gyrodactylids occur on the skin and gills, but the representative on *Xenopus* occupies the oral cavity, avoiding the risks of desiccation (during aestivation or occasional overland migration by the host) and of periodic skin sloughing (Tinsley & Whitear 1980). Transmission to new hosts involves exit of established worms from one host and active transfer to the next, but all stages are already adult from birth and invading worms can continue reproduction on the next host soon after re-establishment (Harris & Tinsley 1987).

Nematodes and mites
Two other direct-life-cycle parasites of *Xenopus* show multiplication within the host, the nematode *Pseudocapillaroides xenopodis* and the mite *Xenopacarus africanus*. *P. xenopodis* creates tunnels in the epidermis within which embryonated eggs and all developmental stages occur (Moravec & Cosgrove 1982; Wade 1982); although the life cycle has not been described, it is likely that transfer is 'contagious', with nematodes invading the skin either during host–host contact or after release onto the substratum. The mites also increase in numbers in individual hosts, and adults, larvae and nymphs occur side-by-side in the nostrils and eustachian passages (Fain *et al.* 1969).

Leeches
Marsupiobdella africana is one of very few external parasites of *Xenopus*, and all stages are able to transfer from host to host, either directly or via temporary attachment to the substrate. Reproduction involves adaptations more specialized than in any other leech. *M. africana* has a unique internal brood pouch within which up to 50 offspring develop to an infective stage. The parent worm carries this brood internally until it locates a *Xenopus*, when the offspring are discharged more or less explosively onto its skin. The young

leeches remain attached and feed on blood for about three weeks. However, they detach just before sexual maturity and this serves both to distribute each brood in the external environment and to promote out-breeding (Van der Lande & Tinsley 1976).

Protozoans

The rectal protozoans, species of *Balantidium*, *Nyctotherus* and *Protoopalina*, belong respectively to groups which are common parasites of anurans although, where determined, the species found in *Xenopus* are specific to this host. *Entamoeba* sp. has been recorded in the intestine (Thurston 1970), and flagellates including *Hexamita intestinalis* (see Thurston 1970) and *Chilomastix caulleryi* (see Fantham 1922) in the rectum. These are all transmitted by cysts passed in the faeces and probably accidentally ingested by *Xenopus* when the bottom mud of ponds is disturbed during prey capture. The related intestinal ciliates of other anurans, *Nyctotherus*, *Balantidium* and *Opalina* spp., have transmission cycles synchronized with host reproduction (see Smyth & Smyth 1980). This feature correlates with the fact that only the tadpole stage of their hosts, species of *Rana, Bufo* and other semi-terrestrial anurans, is detritivorous. Therefore, production and release of cysts is focused into the period when the adult anuran enters water to breed, invasion is almost certainly confined to ingestion of cysts by tadpoles, and hosts cannot be infected after metamorphosis. On the other hand, because all stages of *Xenopus* feed in water, it is likely that there is continuous re-infection by the host-specific *Xenopus* protozoans throughout life.

Protozoans recorded from the blood of *Xenopus* include *Trypanosoma* and *Cryptobia* spp. free in the bloodstream and a haemogregarine within red blood cells (Thurston 1970). Entry and exit from this habitat must, by analogy with related parasites, be accomplished via a blood-feeding vector. *Marsupiobdella africana* could fulfil this role but since this leech is apparently restricted to South Africa there must be alternative vectors elsewhere. *M. africana* is more or less host-specific to *Xenopus* (apart from a phoretic association with freshwater crabs), but other leeches and perhaps bugs with a wider host range could transfer blood parasites between *Xenopus*.

Trichodinid ciliates are common ectoparasites of fishes; they are highly mobile and transfer from host to host usually involves active swimming. *Trichodina xenopodos* inhabits the urinary bladder and transmission must involve passage in the urine. Kruger, Basson & Van As (1991) showed that the parasites can survive in water for a short period, and Professor J. Green (pers. comm.) has recovered very large numbers of *Trichodina* in plankton samples taken from ponds inhabited by *Xenopus laevis* which proved to be heavily infected. Kruger *et al.* (1991) suggested that *T. xenopodos* transfers during mating, but it is likely that recently-eliminated parasites could also penetrate via the cloaca into the urinary bladder at other times, particularly when *Xenopus* are densely aggregated (in confined ponds during drought, for instance).

Mechanisms of parasite transmission

In all the life-cycle patterns described above (including those of parasites related to species in other anurans), the principles of transmission are typical of parasites of fish. Transmission is water-borne and, across the spectrum of parasite taxonomic groups, *Xenopus* becomes infected in one of three ways. First, invasion may be active, involving an infective stage which is specifically equipped for locomotion, host location and recognition, and for migration to the preferred site of infection within the host's body. Second, entry to the host may be passive, generally with ingested food items, and the parasite has no control over events during transmission except that emergence within the host's gut may be triggered by specific cues, followed by active migration to the final habitat. Third, host-to-host transfer may be mediated by a vector, most commonly feeding on the blood of successive hosts and incidentally transferring parasites which may or may not require a period of development in the vector. In better-studied systems, these blood parasites may have specific behavioural traits which optimize uptake from the bloodstream by the vector and which facilitate entry (often by inoculation) into the next host.

Protopolystoma and the six digenean metacercariae employ swimming infective stages specialized for active host-finding and invasion. These have locomotor structures (the monogenean oncomiracidium with cilia, the digenean cercariae with contractile tails) required for a few hours only and lost immediately after successful entry into the host's body. The energy reserves are fixed at the time of release into the external environment and the larva dies if the host is not located within a limited period. The oncomiracidium of *P. xenopodis* is specialized to locate a specific portal of entry, the cloacal opening, and then to migrate along the urinary ducts to the kidneys. A chemical trail such as a concentration gradient of excretory products would be a likely factor, but this has not been studied. Certain sense organs on the surface of the oncomiracidium disappear soon after host infection and this suggests that these sensilla are responsible specifically for host recognition (Tinsley & Owen 1975).

The cercariae infecting *Xenopus* are equipped for skin penetration with well-developed glands secreting histolytic enzymes. The cercarial tail is shed during penetration into the subcutaneous tissues and cystogenous glands discharge secretions which form a cyst wall enclosing the larva. It is not known whether metacercariae such as *Neascus* sp. which encyst in relatively superficial sites (tissues around the lateral line organs) penetrate preferentially above these organs. Porter (1920, 1938) described attraction of *Echinostomum xenopodis* and *Cercaria xenopodis* to skin adjacent to the eyes and lateral line organs, but also suggested that this correlation may arise because, in her experiments, the surface-swimming larvae came into contact first with these areas as the host floated at the surface. In some fish parasites with equivalent life cycles, cercariae penetrate anywhere on the body and then follow specific migration routes to sometimes distant sites: thus, *Diplostomum spathaceum* migrates along the course of blood vessels to the preferred site (the eye) and this is completed within about 36 h post-penetration (Erasmus 1972). The final site for *D. xenopodis*, the pericardial cavity, must also require very precise behavioural adaptations but the migration cues are unknown.

Three further metazoan parasites with direct life cycles rely on transmission involving active locomotion and recognition: the leech, gyrodactylid monogenean and mite. However, in these cases infection may be accomplished by any development stage of the parasite, and these must therefore all retain sensory and locomotor abilities for host identification and invasion. Accidental detachment may be an important element in host-to-host transfer.

For *Marsupiobdella africana*, offspring released from the brood pouch or detached from a host are able to locate and re-attach to another *Xenopus*, but their range of travel and presumably energy reserves are relatively limited. The adaptations for brooding enable the adult leech to have complete control of transmission. During development of the offspring, the parent remains quiescent but its behaviour changes once the young become infective: the gravid parent responds to vibrations in the water and actively moves towards potential hosts. At the moment of contact, the offspring are discharged more or less explosively onto the host, and recognition of the skin secretions of *Xenopus* by the parent leech appears to provide the cue both for host identification and for release of offspring. Juveniles can be carried for up to two months by the parent, greatly extending their infectivity, but in addition to the protective role of the brood pouch the adaptations enable the parent to exercise discrimination in the timing of transmission. The host-finding ability of *M. africana* may also benefit from a phoretic association between the leech and various freshwater crabs: the crabs regularly share habitats with *X. laevis* and temporary attachment of *M. africana* to the mobile carrier host may result in transport to ponds occupied by *Xenopus* (see Van der Lande & Tinsley 1976).

The mite, *Xenopacarus africanus*, and the monogenean, *Gyrdicotylus gallieni*, share a common route of entry into the host: via the nostrils. *G. gallieni* detached from one host can survive for up to 48 h on the substrate (at 20 °C); the worms are mobile and, like *M. africana*, may be sensitive to specific skin chemicals of *Xenopus*. When they make contact with a submerged *Xenopus* they attach and travel rapidly over the body surface until they locate the nostrils. Recognition of the invasion route into the oral cavity appears to exploit the nostril olfactory currents (described by Elepfandt, this volume pp. 97–120): the parasite responds to water currents by rapid reaching movements with which it gains entry into the nostril chamber (Harris & Tinsley 1987). The mites are markedly hydrophobic and those which emerge from the host float on the water surface. Invasion occurs when potential hosts surface to breathe and protrude the snout and nostrils through the water film (unpubl. obs.).

Little is known of the transmission biology of *Pseudocapillaroides xenopodis*. Because of its mode of reproduction, involving the deposition of embryonated eggs, unique amongst capillariids, larvae hatch within the epidermal tunnels made by the adults. This must contribute to a progressive increase in worm burdens in infected hosts. No other members of the Capillariidae infect host skin (related forms in amphibians are parasites of the alimentary tract), but it must be assumed that host-to-host transfer occurs by skin penetration, either through existing abrasions or through intact skin.

For all the remaining metazoan parasites of *Xenopus*, host invasion is achieved passively and there is no possibility of discrimination on the part of individual infective stages. Instead, transmission depends on selection determined through evolution. The digeneans which mature in *Xenopus*, the cestode and the camallanid nematodes all employ transfer through the food chain as a means of entry, and their respective intermediate hosts must have been incorporated into the life cycle for the reasonably high probability that they form regular items in the diet of *Xenopus*. Indeed, some of the transmission routes provide additional insight into the feeding habits of *Xenopus*. Thus, heavy infections of the tapeworm *Cephalochlamys namaquensis* in adult hosts indicate that copepods remain a significant component of the diet throughout life, even for large *Xenopus*. Porter (1938) speculated that adult *X. laevis* may be the final hosts of *Opisthioglyphe xenopodis* whose metacercariae were recorded in *Xenopus* tadpoles. This transfer mechanism would depend on cannibalism which is a routine occurrence in many *Xenopus* populations (Tinsley, Loumont & Kobel, this volume pp. 35–59). The transmission of *Progonimodiscus doyeri* metacercariae encysted on the skin of *Xenopus* exploits the fact that moulted skin is eaten by the host.

Although the protozoan parasites of *Xenopus* are poorly studied, they illustrate a spectrum of host–parasite interactions including varied transmission mechanisms. *Trichodina xenopodos* transfers to new hosts by active invasion; the intestinal protozoans rely on passive transmission by means of cysts passed with the faeces of one host and ingested with food by the next; the protozoans (and the nematode microfilariae) found in the bloodstream are transmitted by a vector.

Pathology

Amongst the helminth parasites of *Xenopus*, one distinct group has major potential for pathogenic effects: these are the parasites which employ *Xenopus* as an intermediate host in the life cycle and reach sexual maturity only when the host is eaten by an appropriate predator. In this group, host death is an essential part of transmission and any parasite-induced effects which predispose the infected individual to attack by the final host predator would have selective advantage in parasite evolution. Parasite life cycles in which infection may lead to specific alteration of host behaviour are reviewed critically by Barnard & Behnke (1990), including the metacercarial infections of fishes which are reported to increase vulnerability to predation. The parasites occur in delicate organs including the eyes and brain, where they may alter escape responses, and in the skin where they may disrupt camouflage patterns and increase conspicuousness. The metacercarial infections of *Xenopus* appear to follow these characteristics. In particular, the encystment of several species, *Cercaria xenopodis, Echinostomum xenopodis* and *Neascus* sp., in the lateral line organs could disrupt the sensory input documented by Elepfandt (this volume pp. 97–120). One effect of *Neascus* noted by Elkan & Murray (1952) was the reversal of normal colour pattern: infected lateral line plaques became darkened and the rest of the body surface was pale, destroying camouflage; affected animals responded abnormally, became paralysed and died soon after the symptoms developed. Histological examination revealed extensive degeneration of the lateral line sense organs (Elkan & Murray 1952). Tinsley & Sweeting (1974) recorded that metacercariae of *Diplostomum xenopodis* occurred in burdens of up to 3000 worms per host and that in heavy infections the pericardial sac was distended with fluid containing parasites and sediment. The effects could have severe consequences for cardiac function (see Nigrelli & Maraventano 1944).

However, despite logical predictions that these infections should be highly pathogenic, it is difficult to establish evidence that this potential is realized. The *X. laevis* studied by Elkan & Murray (1952) which demonstrated the lateral line infection had been imported from South Africa to England over a year before the symptoms developed. Clearly, the hosts had carried the same parasite burden throughout this period (since further cercarial invasion was prevented during laboratory maintenance by the absence of snail intermediate hosts). It is unexplained how such a destructive infection could remain asymptomatic for over one year and what circumstances were responsible for precipitating the final effects. Tinsley & Sweeting (1974) recorded that *X. laevis* carrying very heavy infections of *D. xenopodis* survived in good condition throughout over three years of laboratory maintenance. Of course, in the laboratory, hosts are protected from a range of environmental stresses under which parasite infections might become more significant. However, Tinsley & Sweeting (1974) recorded no mortality associated with this infection during laboratory conditions simulating stresses which *Xenopus* might experience naturally (including artificial aestivation, starvation, hormone stimulation). Nigrelli & Maraventano (1944) attributed the deaths of *X. laevis* to burdens of 25–150 metacercariae, but 26% of hosts in the samples studied by Tinsley & Sweeting carried 100–3000 worms and no fatalities attributable to infection were recorded. It still seems common sense to conclude that, under natural conditions, the extra demands of infection could result in heavily parasitized animals lacking the stamina or escape responses to avoid predation as effectively as those unparasitized. Nevertheless, demonstration of these effects remains elusive. An important factor in this system, noted in connection with bacterial diseases, may be the remarkable tolerance of *Xenopus* to a range of dysfunction under normal circumstances. Elepfandt's observations (this volume pp. 97–120) help to explain the apparent lack of ill-effects from extensive lateral line damage in *Neascus*-infected hosts (see above): *Xenopus* can comprehensively monitor environmental stimuli with just a few functioning lateral line plaques. Perhaps other independent effects on condition (including malnutrition) trigger lethal effects in *Xenopus* which might otherwise cope with pre-existing disease.

Porter (1920, 1938) described severe pathological consequences following experimental exposure of *X. laevis* to cercariae of *Echinostomum xenopodis*. The infective dose was not recorded but invasion led to extreme irritation of the eyes, inflammation of blood vessels and oedema. One young *Xenopus* died 18 days post-infection, an older one died after 65 days. Parker (1932) recorded an echinostome infection in natural *X. laevis bunyoniensis* populations in Uganda which caused extreme swelling of the lower eyelids. Elepfandt (this volume pp. 97–120) has shown that loss of sight would not interfere with prey capture in *Xenopus* but that a major role of the upwardly-directed eyes concerns predator avoidance.

One helminth parasite of *Xenopus* is clearly associated with serious pathogenic effects: *Pseudocapillaroides xenopodis* has been responsible for lethal epidemics in laboratory populations. Affected animals developed roughened skin, became anorexic and emaciated, and died. Once the infection became apparent, no spontaneous cures were noted. The outbreaks described by Moravec & Cosgrove (1982) and Cohen *et al.* (1984) occurred in laboratory-raised *X. laevis*, and that described by Wade (1982) occurred in *X. laevis* imported from South Africa at least 18 months before the condition developed. As in the other pathogenic infections of *Xenopus*, there remains a question of the circumstances precipitating the epidemic or the development of symptoms in already infected animals. The implication of Wade's report is that the potential of *P. xenopodis* for *in situ* population growth was kept in check for over a year before breakdown. One outbreak amongst my own laboratory-maintained *Xenopus*, including both laboratory-raised and imported animals, occurred after the temperature, normally constant at 22 °C, fell to 15 °C for 10 days. This temperature decrease may have affected immune competence, permitting a pre-existing low-level infection to become epidemic. Cohen *et al.* (1984) provided evidence that a thymus-dependent immune response normally provides protection against the development of this disease (see below).

There is little critical information on pathology induced by the other helminths in *Xenopus*. Parasite diet and habitat may provide some indication of the potential for pathogenic effects. Several of the gut parasites feed on epithelial tissues (including *Gyrdicotylus gallieni*, *Oligolecithus elianae*); *Xenopodistomum xenopodis* located in the gall bladder can extend its long neck down the bile duct to feed in the intestine (Tinsley & Owen 1979). There are five parasite genera infecting *Xenopus* for which host blood makes the major component of the diet: the monogenean *Protopolystoma xenopodis*, the mite *Xenopacarus africanus*, the leech *Marsupiobdella africana*, and the nematodes *Camallanus* and *Batrachocamallanus* species. Precise measurement of the cost of parasite-induced blood loss has been obtained recently for the pelobatid *Scaphiopus couchii* (see Tocque & Tinsley 1992, 1994) but there are no equivalent data for *Xenopus*. However, a key factor influencing tolerance of blood and other tissue loss is worm burden and this is discussed further below.

Parasite population biology

Population dynamics

The helminth infections of *Xenopus* fall into two categories with respect to the dynamics of infection levels.

The interaction between *Xenopus* and its metacercarial infections represents a closed system. Exposure to aquatic invasion is more or less continuous, worm burdens tend to accumulate throughout life without loss, and the resting stages of the parasites probably survive as long as the intermediate host—until the chance for onward transfer to the final host. For *Diplostomum xenopodis*, Tinsley & Sweeting (1974) found no decline in infection levels in the absence of re-infection, and metacercariae in heavy burdens remained viable throughout a study period exceeding three and a half years. In infected populations there is likely to be a positive correlation between worm burden and host age, although the relationship may be swamped by stochastic effects, including occasional chance encounters with local, abnormally high cercarial densities. The dispersal and mixing of *Xenopus* populations may also be responsible for the presence of a few heavily infected individuals within otherwise lightly infected populations. In a sample of 55 *X. laevis*, Tinsley & Sweeting (1974) recorded one host with 2000 metacercariae, alongside 39 uninfected and 15 with only 1–50 (mean 18) worms per host. This seems best explained by host dispersal and migration which, during the life of *Xenopus*, would tend to produce a more even distribution of infected animals. Of course, this tendency towards steady accumulation of worm burdens will be limited if the infection prejudices host survival. However, if the infection becomes lethal only above a certain threshold then this would guarantee transfer of maximum numbers of parasites to potential final hosts which, in turn, would then contaminate the ponds with infective stages for further transfer through snail and *Xenopus* populations.

In complete contrast, the majority of metazoan parasites infecting *Xenopus* occupy open organ systems: parasite infrapopulation size will be determined by the balance between recruitment and loss, and a dynamic flux will be dependent on interaction between parasite survival and invasion rates.

For the parasites which multiply within the host, continued reproduction could theoretically enable infrapopulations to persist indefinitely in infected hosts. However, despite this potential there is now very good documentation for one species, at least, that the survival of infrapopulations is strictly limited. Gyrodactylid monogeneans such as *Gyrdicotylus gallieni* have a capacity for continuous *in situ* production of offspring quite unlike all other platyhelminth parasites (Harris & Tinsley 1987); nevertheless, infections are terminated by a host reaction and individual infrapopulations become extinct after 2–5 months (Jackson & Tinsley 1994). These *Xenopus* then remain refractory to re-infection for an undetermined period, probably several weeks or months (see below).

An encounter between *Xenopus* and a single gravid *Marsupiobdella africana* can result in instantaneous infection by up to 50 juvenile leeches. These cause obvious irritation and tend to accumulate in the cloacal region where they avoid the host's attempts to remove them with its clawed feet. Each infection is of fixed duration because the leeches drop off after 2–3 weeks (at 22 °C), before reproducing (Van der Lande & Tinsley 1976).

The important implication of these data is that worm burdens recorded in samples of host populations are temporary and associated pathogenic effects may be correspondingly short-term.

Distribution of infection in host populations

It is now well recognized that infections of macroparasites commonly exhibit an overdispersed (negative binomial) distribution within host populations. Characteristically, the majority of individuals within a host population carry low worm burdens or are uninfected, whilst only a small minority carry high burdens. Features generally assumed responsible for this pattern include variations in susceptibility to infection, heterogeneity in the distribution of infective stages in the environment, and differences (including host behavioural traits) in exposure to infection. Available records of parasite infection levels in natural *Xenopus* populations (mostly *X. laevis*) generally conform to this familiar pattern (Fig. 13.1). The important outcome is that, within a natural host population, a relatively small proportion of hosts would experience potential ill-effects of heavy infections; the great majority would be unaffected.

Population data for both categories of parasite, in closed and open systems, show similar patterns of distribution, but with markedly different scales. *Diplostomum xenopodis* worm burdens actually represent lifetime accumulation, providing a vertical record through time, whereas those of most other parasites represent a 'snapshot' showing the horizontal distribution of worm burdens at one point in time.

Records for *D. xenopodis* are provided by Tinsley & Sweeting (1974). Based on 410 *X. laevis* sampled over four years, the prevalence of infection was 60%. The distribution of infection levels was markedly skewed (Fig. 13.1a): 25% of infected animals carried very low burdens (1–10 worms per host), but 26% carried 100 worms or more, and 2% carried 1000 or more up to a maximum of 3000 worms per host.

The tapeworm *Cephalochlamys namaquensis* provides an example of the dynamic flux between recruitment and loss to which most helminth populations may be subject. Thurston (1967) showed that worms are mature after

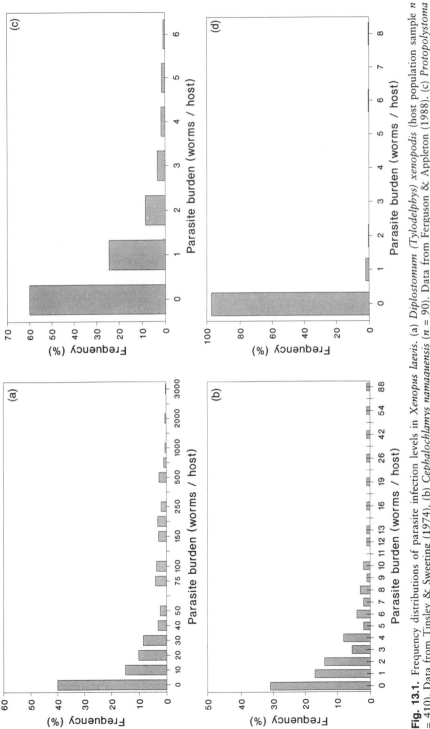

Fig. 13.1. Frequency distributions of parasite infection levels in *Xenopus laevis*. (a) *Diplostomum* (*Tylodelphys*) *xenopodis* (host population sample *n* = 410). Data from Tinsley & Sweeting (1974). (b) *Cephalochlamys namaquensis* (*n* = 90). Data from Ferguson & Appleton (1988). (c) *Protopolystoma xenopodis* (*n* = 1200). Data from Tinsley (1972). (d) *Xenopodistomum xenopodis* (*n* = 396). Data from Tinsley & Owen (1979).

about nine days post-infection. One population examined by Ferguson & Appleton (1988) (during May–July) showed a high prevalence of infection: 69%. The ratio of immature to mature worms was 3.2 : 1 indicating either recent intense recruitment or significant post-infection mortality or a combination of the two. Forty-five percent of the population carried only one or two worms, 6% carried more than 25 worms up to a maximum of 88 worms per host (Fig. 13.1b). Tapeworm size (reflecting absorptive area) is likely to be a major determinant in the removal of nutrients from the host's food intake and hence the potential for negative effects. Size is influenced by age but also potentially by host nutrition and parasite intraspecific competition. Ferguson & Appleton (1988) showed that the majority of tapeworms (around 60%) were very small (<4 mm long), representing the smallest drain on host resources, and few were large (4% were >24 mm).

Studies of natural infections of *Protopolystoma xenopodis* have consistently recorded a relatively restricted range of intensities of infection. Amongst large samples of South African *X. l. laevis* (*n* = 1200) examined by Tinsley (1972), the maximum burden of adult worms was six (mean 1.8 per host). In *X. vestitus*, Tinsley (1973) recorded a maximum of six (mean 2.4) (*n* = 241). Further *X. l. laevis* samples studied by Jackson & Tinsley (1988a) confirmed a normal maximum of six adult worms per host (96% of infections) but recorded rare higher burdens (one instance each of seven, 11 and 15 worms). Jackson & Tinsley (1988b) also recorded exceptional cases of eight and 12 adult worms per host. Thurston (1970) and Macnae *et al.* (1973) reported up to eight (mean 2.0) and nine adult worms per host respectively. Significantly, within this restricted range of worm burdens, at least 50% of *Xenopus* in natural populations carry no adult *P. xenopodis* and, of those infected, over 60% carry only a single worm. A typical frequency distribution of infection levels (from the data employed by Tinsley 1972) is shown in Fig. 13.1c. This restricted range of infection levels could be caused either by poor invasion success or by poor parasite survival following successful invasion.

There is relatively comprehensive documentation of the population biology of polystomatid monogeneans: data for several host–parasite systems suggest that transmission is usually highly efficient, but established parasite infrapopulations are subject to strong regulation which leads to low numbers of adult worms per host (Tinsley 1993). In the case of *Protopolystoma xenopodis*, host populations usually show evidence of more or less continuous invasion: recently-invaded worms have an overall prevalence of 40% in samples of *X. l. laevis* from the Cape. Developing juvenile worms in the kidneys reflect invasions during the preceding 2–3 months, and older worms in the urinary bladder represent the accumulated adult parasite population, aged three months to two years post-infection. Tinsley (1972) recorded a major change in the numbers of parasites in these two age classes. The intensity of infection by juvenile parasites (range 1–59, mean 7.1 worms per host) was four times that of adult parasites (range 1–6, mean 1.8 worms per host), and this difference is even more marked when the relative durations of these stages are considered. A significant component of the loss of parasites is attributable to elimination of high worm burdens in the period before maturation: 30% of hosts infected by juvenile parasites carried more than six worms, exceeding the maximum recorded amongst adult parasites. Despite the capacity in this life cycle for continuous re-infection of individual hosts by parasites capable of surviving up to two years, 82% of infected *X. l. laevis* carry only one or two adult *P. xenopodis*. Further insight into the relatively high turnover within established parasite populations is provided by data from a host population sample (158 *X. l. laevis*) maintained in the laboratory without opportunity for re-infection: during a period of nine months after capture, over 75% of the original parasite population was lost (Tinsley 1972). This evidence suggests that, in the case of *P. xenopodis*, the potential for parasite-induced pathogenic effects is greatly reduced by continuous elimination of established worms leading to relatively low burdens in individual hosts.

The laboratory experimental studies of Jackson & Tinsley (1994) showed that infrapopulations of *Gyrdicotylus gallieni* can increase exponentially to maximum burdens exceeding 200 worms per host. However, since these peaks are transient, with the infection subsequently eliminated by a host response, records of natural infection levels can only provide 'snapshots' of various stages in the population cycle. It is intriguing, therefore, that the wild-caught host samples (*n* = 483) examined by Harris & Tinsley (1987) revealed an overall prevalence of only 3.1% and a maximum burden of only seven worms (mean 1.9 per infected *Xenopus*). During laboratory maintenance, infection levels increase: Thurston's (1970) data provide a glimpse of this potential with a record of 77 *G. gallieni* in one *Xenopus* kept for six weeks after capture, but Thurston found no infections in other *Xenopus* samples examined. The experimentally infected hosts (*n* = 117) which formed the basis of Jackson & Tinsley's study did not show signs of irritation or stress, and no pathological consequences were observed. Clearly, any effects would have only a temporary influence on individual hosts, and the data on infection levels indicate that effects at the host population level would probably be negligible.

There is little published information on the infection levels of the digenean species which mature in the gut and associated organs, but the brief comments accompanying taxonomic descriptions indicate a patchy occurrence of infections and generally low burdens (rarely more than 'a few' worms per host). More detailed data for *Xenopodistomum xenopodis*, based on 396 *X. l. laevis* examined over a 10-year period, revealed an overall prevalence of 2.8% (Tinsley & Owen 1979); amongst the 11 infected *Xenopus*, eight carried single worms, and the other three carried two, six and eight worms respectively (Fig. 13.1d). Clearly, unless such an infection might be rarely observed because of rapidly lethal effects, its extreme rarity would suggest a very minor influence on host populations. Moreover, both Elkan (1960) and Tinsley & Owen (1979) observed no pathological effects, even in hosts which harboured infections during six months of laboratory maintenance.

Factors regulating infection levels

In most host–parasite relationships, the potential for pathogenic effects is related to the size of the parasite infrapopulations within individual hosts. Infection levels are determined by interaction of parasite characteristics (including transmission biology, discussed above, and parasite longevity), ecological factors (including seasonal changes, temperature, and variations in transmission rates with habitat availability and host density) and host physiological factors (including immunity).

Parasite longevity

The most commonly encountered parasites in imported Cape *X. laevis* are *Protopolystoma xenopodis*, *Cephalochlamys namaquensis*, the camallanid nematodes and *Diplostomum xenopodis*, all occurring in more than 50% of host population samples. Each of these has been found to survive more than one year in laboratory-maintained hosts (in the absence of re-infection) (R.C. Tinsley unpubl.). More comprehensive data for *Protopolystoma xenopodis* demonstrate maximum survival for two years (Jackson & Tinsley 1988a). Data are less complete for the digeneans occurring in the alimentary tract of *Xenopus* since infection levels are commonly low and worm expulsion soon after host capture may accompany stress, altered diet, etc. *Xenopodistomum xenopodis* has been found to survive up to six months in laboratory-maintained *Xenopus*, but the prevalence of this fluke in imported *X. laevis* was only 3% in the samples studied by Tinsley & Owen (1979), so the chance of finding infections in hosts maintained for longer periods would in any case be low. Digenean metacercariae commonly survive for several years: *Tetracotyle* sp. encysted in the body musculature of *X. vestitus* remained viable for over five years in laboratory-maintained hosts (Tinsley & Sweeting 1974) (cf. *D. xenopodis*, above).

Ecological factors

There is little information on parasite population ecology derived from field-based studies. Tinsley (1972) reported seasonal changes in *Protopolystoma xenopodis* populations based on samples of *X. laevis* exported from the Cape immediately after capture. In late summer and autumn, densely aggregated host populations experience heavy invasion (prevalence of recently invaded parasites over 50%, mean intensity 18.0 worms per host). During winter rainfall, host populations disperse and there is greatly reduced transmission resulting in minimum levels of larval infection (prevalence 11%, mean intensity 2.0 worms per host). The parasites which invaded in autumn reach maturity during the following spring and produce the highest adult worm burdens (prevalence 53%, mean intensity 2.3), but there is reduced recruitment into the adult parasite population during summer and autumn, corresponding with the reduced invasion rates during the preceding winter. As a result, burdens of adult parasites decline to a minimum in winter (prevalence 11%, mean intensity 1.5) before the maturation of the autumn invasions contributes to augmented adult infections again in the following spring. These seasonal cycles are determined by two environmental factors operating in unison: high summer temperatures, which increase rates of parasite egg production, egg development and maturation, coincide with low rainfall when hosts occur at maximum population densities in confined habitats; low winter temperatures result in decreased parasite egg production etc. at a time of high rainfall when host populations disperse in flooded habitats. The peaks in the abundance of recently-invaded larvae and subsequently of adult parasites are staggered by around 5–6 months in winter (because of reduced developmental rates at low temperatures) and around 3–4 months in summer (at highest temperatures). Significantly, these seasonal cycles also show that the peak in worm burdens following maximum (autumn) invasion is relatively transient: continuous worm losses contribute to declining parasite populations until the next autumn's recruitment.

Host immunity

Although there is now considerable information on the functioning of the immune system of *Xenopus* (Horton, Horton & Ritchie, this volume pp. 279–299; Du Pasquier, Wilson & Robert, this volume pp. 301–313), there is only sketchy, circumstantial evidence for the operation of host immunity against parasite infection.

In isolated, naive *X. laevis* (never previously infected), experimental infections of *Gyrdicotylus gallieni* are eliminated after 2–5 months. However, lineages of *G. gallieni* can be maintained for up to 10 months—the maximum duration of the experiments conducted by Jackson & Tinsley (1994)—by transfer of parasites to successive naive hosts. Infections can also persist for up to 11 months in aquarium populations of 20–30 *Xenopus*: this accords with a cycling of infection amongst a host group heterogeneous with respect to short-term immunity where parasite infrapopulations survive temporarily on individuals whose protective response has waned (Harris & Tinsley 1987). This host reaction is better documented in the gyrodactylid infections of fishes where the previously infected host remains refractory to re-infection for a period of weeks or months. Some indication of the significance of immunity in natural *Xenopus* populations is provided by a series of experimental exposures carried out by Jackson & Tinsley (1994) using a sample of 43 wild-caught *X. laevis*. Only two (4.7%) could be infected with *G. gallieni* compared with a success rate of over 60% using naive *X. laevis*. This high level of presumed natural resistance accords with the very low prevalence of *G. gallieni* infection (3.1%) recorded in host population samples examined by Harris & Tinsley (1987) (see above). The immunological nature of the protective response is unknown.

The parasite of *Xenopus* for which most population data are available, *Protopolystoma xenopodis*, shows highly restricted infection levels of adult worms. As outlined above, despite continuous re-infection and a capacity for extended survival (two years maximum longevity), adult infrapopulations rarely exceed six parasites, over 80% of infections comprise only one or two worms, and 40% of individuals in host populations are parasite-free (Tinsley 1972). Similar ecological and experimental data for a related polystomatid, *Pseudodiplorchis americanus*, have been interpreted in terms of regulation by host immunity (Tinsley 1993), but the evidence remains circumstantial. Post-invasion stages of the blood-feeding *P. xenopodis* infect the kidneys for a period of two months, causing tissue damage. There is a major reduction in worm burdens before maturation and it is possible that immunological responses are triggered during this renal phase in the host–parasite interaction.

The most exciting indication that host immunity may control helminth infection in *Xenopus* comes from a study of *Pseudocapillaroides xenopodis* by Cohen *et al.* (1984). Laboratory-raised *Xenopus* which developed disease symptoms had been thymectomized as tadpoles. Implantation with thymuses from MHC-compatible donors led to reversal of the course of disease and significant recovery. The possibility that *Xenopus* normally mounts a protective thymus-dependent immune response against *P. xenopodis* provides an explanation for sudden epidemic outbreaks in laboratory populations, especially where temperature decrease may have been a precipitating factor (see above).

Concluding comment

The origins and relationships of the parasites of *Xenopus* reveal two important influences on the evolution of this rich and distinctive parasite fauna. First, the spectrum of parasites reflects the specialization of pipids for a fully aquatic life, and particularly the fact that these are virtually the only anurans which feed underwater. This contributes to an ecological isolation of *Xenopus* from other anurans and an overlap with fishes. Second, the phylogenetic affinities and strict host-specificity of the parasites are best interpreted in terms of the distant evolutionary relationship between pipids and other anuran groups. A consequence of these influences is that *Xenopus* contains a dual parasite fauna: one component represents the host transfers acquired from parasite lineages which typically infect fishes; the second component comprises 'heirlooms' inherited from the parasite communities characteristic of anurans. Taxonomic interpretation is unambiguous: all evidence points to a long association of host and parasites within an isolated evolutionary unit.

This close evolutionary interrelationship could provide a model system for evaluating evidence of host–parasite co-evolution. There is considerable diversity in the nature of the interactions involving more than 25 invertebrate genera associated with *Xenopus*. These interactions are complex, determined by a variety of factors including the roles of partners—prey, predators, other hosts—involved in the life cycles. There is clear evidence of the high degree of specialization of the parasites, adapted to specific facets of host biology. Evidence of a reciprocal basis in co-evolution could be reflected in the outcome of potential pathogenic interactions. Here the information is still sketchy and circumstantial but it is intriguing, that, for most parasites of *Xenopus*, infection levels tend to occur below the threshold at which damage may be significant. In other better-studied host–parasite systems—principally the diseases of medical and veterinary importance in mammals—part of this regulation is mediated by the host immune response. Further investigation of the immune control of parasite infections in *Xenopus* is now needed: this would provide documentation which is relatively rare for entirely natural host–parasite systems.

References

Barnard, C.J. & Behnke, J.M. (Eds) (1990). *Parasitism and host behaviour*. Taylor & Francis, London.

Cohen, N., Effrige, N.J., Parsons, S.C.V., Rollins-Smith, L.A., Nagata, S. & Albright, D. (1984). Identification and treatment of a lethal nematode (*Capillaria xenopodis*) infestation in the South African frog, *Xenopus laevis*. *Devl comp. Immunol.* **8**: 739–741.

Cosgrove, G.E. & Jared, D.W. (1974). Diseases and parasites of *Xenopus*, the clawed toad. In *Gulf Coast regional symposium on diseases of aquatic animals*: 225–242. (Eds Amborski, R.L., Hood, M.A. & Miller, R.R.). Center for Wetlands Research, Louisiana State University, Baton Rouge.

Dollfus, R. Ph. (1968). Présence insolite chez un urodele et en Afrique du nord d'un *Cephalochlamys* (Cestoda, Pseudophyllidea). *Bull. Mus. natn. Hist. nat. Paris* **39**: 1192–1201.

Elkan, E. (1960). Some interesting pathological cases in amphibians. *Proc. zool. Soc. Lond.* **134**: 275–296.

Elkan, E. & Murray, R.W. (1952). A larval trematode infection of the lateral line system of the toad *Xenopus laevis* (Daudin). *Proc. zool. Soc. Lond.* **122**: 121–126.

Erasmus, D.A. (1972). *The biology of trematodes*. Edward Arnold, London.

Fain, A., Baker, R.A. & Tinsley, R.C. (1969). Notes on a mite *Xenopacarus africanus* n.g., n.sp. parasitic in the nasal cavities of the African clawed frog *Xenopus laevis* (Ereynetidae: Trombidiformes). *Rev. Zool. Bot. afr.* **80**: 340–345.

Fantham, H.B. (1922). Some parasitic Protozoa found in S. Africa. V. *S. Afr. J. Sci.* **19**: 332–339.

Ferguson, R.R. & Appleton, C.C. (1988). Some aspects of the morphology, population structure and larval biology of *Cephalochlamys namaquensis* (Cestoda: Diphyllidea), a parasite of the clawed toad, *Xenopus laevis*. *S. Afr. J. Zool.* **23**: 117–123.

Fischthal, J.H. & Thomas, J.D. (1968). Digenetic trematodes of amphibians and reptiles from Ghana. *Proc. helminth. Soc. Wash.* **35**: 1–15.

Gassmann, M. (1975). Contribution à l'étude des trématodes d'amphibiens du Cameroun. *Annls Parasit. hum. comp.* **50**: 559–577.

Harris, P.D. & Tinsley, R.C. (1987). The biology of *Gyrdicotylus gallieni* (Gyrodactylidea), an unusual viviparous monogenean from the African clawed toad, *Xenopus laevis*. *J. Zool., Lond.* **212**: 325–346.

Jackson, H.C. & Tinsley, R.C. (1988a). Environmental influences on egg production by the monogenean *Protopolystoma xenopodis*. *Parasitology* **97**: 115–128.

Jackson, H.C. & Tinsley, R.C. (1988b). The capacity for viable egg production by the monogenean *Protopolystoma xenopodis* in single and multiple infections. *Int. J. Parasit.* **18**: 585–589.

Jackson, J.A. & Tinsley, R.C. (1994). Infrapopulation dynamics of *Gyrdicotylus gallieni* (Monogenea: Gyrodactylidae). *Parasitology* **108**: 447–452.

Jackson, J.A. & Tinsley, R.C. (1995a). Evolutionary relationships, host range and geographical distribution of *Camallanus* Railliet & Henry, 1915 species (Nematoda: Camallaninae) from clawed toads of the genus *Xenopus* (Anura: Pipidae). *Syst. Parasit.* **32**: 1–21.

Jackson, J.A. & Tinsley R.C. (1995b). Representatives of *Batrachocamallanus* n. gen. (Nematoda: Procamallaninae) from *Xenopus* species (Anura: Pipidae): geographical distribution, host range and evolutionary relationships. *Syst. Parasit.* **31**: 159–188.

Kruger, J. Basson, L. & Van As, J.G. (1991). Redescription of *Trichodina xenopodos* Fantham, 1924 (Ciliophora: Peritrichida), a urinary bladder parasite of *Xenopus laevis laevis* Daudin, 1802, with notes on transmission. *Syst. Parasit.* **19**: 43–50.

Leadley-Brown, A. (1970). *The African clawed toad,* Xenopus laevis: *a guide for laboratory practical work*. Butterworth, London.

Macnae, W., Rock, L. & Makowski, M. (1973). Platyhelminths from the South African clawed toad, or platanna *(Xenopus laevis)*. *J. Helminth*. **47**: 199–235.

Manter, H.W. & Pritchard, M.H. (1964). Mission de zoologie médicale au Maniema (Congo, Léopoldville) (P.L.G. Benoit, 1959). 5. Vermes—Trematoda. *Annls Mus. r. Afr. centr. (Sér. 8vo) (Sci. zool.)* No. 132: 75–101.

Mettrick, D.F. (1963). Some cestodes of reptiles and amphibians from the Rhodesias. *Proc. zool. Soc. Lond.* **141**: 239–250.

Moore, J.P. (1958). The leeches (Hirudinea) in the collection of the Natal Museum. *Ann. Natal Mus.* **14**: 303–340.

Moravec, F. & Cosgrove, G.E. (1982). *Pseudocapillaroides xenopi* gen. et sp. nov. from the skin of the South African clawed frog, *Xenopus laevis* Daud. (Nematoda: Capillariidae). *Rev. Zool. afr.* **96**: 129–137.

Nigrelli, R.F. & Maraventano, L.W. (1944). Pericarditis in *Xenopus laevis* caused by *Diplostomulum xenopi* sp. nov., a larval strigeid. *J. Parasit.* **30**: 184–190.

Parker, H.W. (1932). Scientific results of the Cambridge expedition to the East African lakes, 1930–1931. 5. Reptiles and amphibians. *J. Linn. Soc.* (Zool.) **38**: 213–229.

Porter, A. (1920). The experimental determination of the vertebrate hosts of some South African cercariae from the molluscs *Physopsis africana* and *Limnaea natalensis*. *Med. Jl S. Afr.* **15**: 128–133.

Porter, A. (1938). The larval trematodes found in certain South African Mollusca with special reference to schistosomiasis (bilharziasis). *Publs S. Afr. Inst. med. Res.* **42**: 1–492.

Prudhoe, S. & Bray, R.A. (1982). *Platyhelminth parasites of the Amphibia*. British Museum (Natural History) London and Oxford University Press, Oxford.

Reichenbach-Klinke, H. & Elkan, E. (1965). *The principal diseases of lower vertebrates*. Academic Press, London & New York.

Smyth, J.D. & Smyth, M.M. (1980). *Frogs as host-parasite systems.* 1. Macmillan, London.

Thurston, J.P. (1964). The morphology and life cycle of *Protopolystoma xenopi* (Price) Bychovsky in Uganda. *Parasitology* 54: 441–450.

Thurston, J.P. (1967). The morphology and life-cycle of *Cephalochlamys namaquensis* (Cohn, 1906) (Cestoda: Pseudophyllidea) from *Xenopus muelleri* and *X. laevis. Parasitology* 57: 187–200.

Thurston, J.P. (1970). Studies on some Protozoa and helminth parasites of *Xenopus*, the African clawed toad. *Rev. Zool. Bot. afr.* 82: 349–369.

Tinsley, R.C. (1972). The adaptation for attachment by the Polystomatidae (Monogenoidea). *C.r. Multicolloque eur. Parasit.* 1: 65–68.

Tinsley, R.C. (1973). Observations on Polystomatidae (Monogenoidea) from East Africa with a description of *Polystoma makereri* n. sp. *Z. Parasitenk.* 42: 251–263.

Tinsley, R.C. (1981). The evidence from parasite relationships for the evolutionary status of *Xenopus* (Anura: Pipidae). *Monit. zool. ital. (N.S.) (Suppl.)* 15: 367–385.

Tinsley, R.C. (1983). Ovoviviparity in platyhelminth life-cycles. *Parasitology* 86: 161–196.

Tinsley, R.C. (1990). Host behaviour and opportunism in parasite life cycles. In *Parasitism and host behaviour*: 158–192. (Eds Barnard, C.J. & Behnke, J.M.). Taylor & Francis, London.

Tinsley, R.C. (1993). The population biology of polystomatid monogeneans. *Bull. fr. Pêche Piscic.* 238: 120–136.

Tinsley, R.C. & Owen, R.W. (1975). Studies on the biology of *Protopolystoma xenopodis* (Monogenoidea): the oncomiracidium and life-cycle. *Parasitology* 71: 445–463.

Tinsley, R.C. & Owen, R.W. (1979). The morphology and biology of *Xenopodistomum xenopodis* from the gall bladder of the African clawed toad, *Xenopus laevis. J. Helminth.* 53: 307–316.

Tinsley, R.C. & Sweeting, R.A. (1974). Studies on the biology and taxonomy of *Diplostomulum (Tylodelphylus) xenopodis* from the African clawed toad, *Xenopus laevis. J. Helminth.* 48: 247–263.

Tinsley, R.C. & Whitear, M. (1980). The surface fauna of *Xenopus* skin. *Proc. R. Soc. Edinb.* 79B: 127–129.

Tocque, K. & Tinsley, R.C. (1992). Ingestion of host blood by the monogenean *Pseudodiplorchis americanus*: a quantitative analysis. *Parasitology* 104: 283–289.

Tocque, K. & Tinsley, R.C. (1994). The relationship between *Pseudodiplorchis americanus* (Monogenea) density and host resources under controlled environmental conditions. *Parasitology* 108: 175–183.

Van der Lande, V.M. & Tinsley, R.C. (1976). Studies on the anatomy, life history and behaviour of *Marsupiobdella africana* (Hirudinea: Glossiphoniidae). *J. Zool., Lond.* 180: 537–563.

Vercammen-Grandjean, P.H. (1960). Les trématodes du lac Kivu-Sud (Vermes). *Ann. Mus. r. Afr. Cent. (Sér. 4to) (Zool.)* No. 5: 1–171.

Wade, S.E. (1982). *Capillaria xenopodis* sp. n. (Nematoda: Trichuroidea) from the epidermis of the South African clawed frog (*Xenopus laevis* Daudin). *Proc. helminth. Soc. Wash.* 49: 86–92.

14 Skin secretions of *Xenopus laevis*

GÜNTHER KREIL

Synopsis

A variety of peptides, enzymes and other proteins have been isolated from skin secretions of Amphibia. Among the peptides at least three different groups can be discerned: (1) those with structural and functional homology to mammalian hormones and neurotransmitters; (2) antimicrobial peptides of different types, one being characterized by the propensity to form amphipathic helices; and (3) opioid peptides containing a D-amino acid essential for biological activity, for which no mammalian counterparts have so far been found.

Of the first group, thyrotropin releasing hormone, caerulein (related to mammalian cholecystokinin) and xenopsin have been isolated from skin secretions of *X. laevis*. cDNA cloning has been used to elucidate the structure of the precursors of these peptides. Moreover, the antimicrobial peptides PGS/magainins I and II, PGLa and several others have been studied in detail.

Several enzymes involved in the processing of peptide precursors to the final product have also been isolated from skin secretions of *X. laevis*.

Recent studies with skin secretions of frogs from other families of amphibian species have yielded similar results. These have also been found to contain numerous peptides; in particular, a large number of peptides with antimicrobial properties have been discovered in the past years.

Introduction

FIRST WITCH. Round about the cauldron go;
  In the poison'd entrails throw.—
  Toad that under cold stone
  Days and nights has thirty-one
  Swelter'd venom, sleeping got,
  Boil thou first i'th' charmed pot.
    (Shakespeare: Macbeth, Act IV)

The skin of Amphibia has for centuries been used in folk medicine and witch-craft. With the advent of modern science, organic chemists, pharmacologists, physiologists and others started to investigate the structure and biological function of the many compounds present in skin secretions of frogs and toads. The pioneering studies of Erspamer and his colleagues have shown that amphibian skin is a rich source of a variety of peptides with diverse functions (Erspamer & Melchiorri 1983; Erspamer, Falconieri Erspamer & Cei 1986; Bevins & Zasloff 1990). Many of these peptides are similar or identical to peptide hormones present in mammalian brain and other parts of the nervous system or in the gastrointestinal tract. This 'brain–gut–skin triangle' (Erspamer, Melchiorri, Brocardo *et al.* 1981) has presented an intriguing problem for many years. A second class of peptides present in the skin of amphibians possesses antimicrobial activity. This was first suggested for bombinin, a peptide isolated from skin secretion of *Bombina variegata* (see Csordas & Michl 1970). More recently, additional peptides with similar action have been found in *Bombina* (Simmaco, Barra *et al.* 1991; Gibson, Tang *et al.* 1991) as well as in other species (Zasloff 1987; Giovannini *et al.* 1987; Mor, Nguyen *et al.* 1991). It is notable that many of these antimicrobial peptides have a basic net charge and can form an amphipathic helix. A third group of peptides is characterized by the presence of a D-amino acid in the second position. These peptides, to date only found in skin secretions of Phyllomedusinae, exhibit high affinity and selectivity for opiate receptors (Montecucchi *et al.* 1981; Kreil *et al.* 1989; Mor, Delfour *et al.* 1989; Lazarus *et al.* 1989; Erspamer, Melchiorri, Falconieri Erspamer, Negri *et al.* 1989; Mignogna, Severini *et al.* 1992; Negri *et al.* 1992). Lastly, a diverse group of peptides has been detected in the skin of different Amphibia and for these no function has so far been assigned (see e.g. Erspamer, Falconieri Erspamer & Cei 1986).

In certain cases, the amount of a given peptide present in one gram of skin may be measured in milligrams. The biosynthesis of these peptides should thus be amenable to experimental analysis. We therefore started, more than ten years ago, to use recombinant DNA techniques in attempts to characterize the precursors of amphibian skin peptides (Hoffmann, Bach *et al.* 1983; Hoffmann, Richter & Kreil 1983). All these studies, in our laboratory and others (Sures & Crippa 1984; Zasloff 1987; Poulter *et al.* 1988), were originally performed using *Xenopus laevis*; however, similar experiments have been carried out more recently using amphibian species from different families, e.g. *Bombina variegata* (Richter, Egger & Kreil 1990; Simmaco, Barra *et al.* 1991), *B. orientalis* (Spindel *et al.* 1990; Gibson, Tang *et al.* 1991), *Phyllomedusa sauvagei* (Richter, Egger & Kreil 1987), *Ph. bicolor* (Richter, Egger, Negri *et al.* 1990), and *Rana pipiens* (Krane *et al.* 1988).

Dermal glands of *X. laevis*

The skin of X. *laevis* contains numerous glands which have a rather unusual structure. Each consists of a syncytial secretory compartment filled with elliptical 'storage granules'. The nuclei are located at the periphery of the syncytium and the entire gland is surrounded by myoepithelial cells. Contraction of these cells triggered by adrenergic stimulation forces the granules and the entire contents of the syncytial compartment through a gland duct to the skin surface (Dockray & Hopkins 1975; Flucher *et al.* 1986). The granules are highly unstable, yet they can partly be isolated from the secretion; they dissociate rapidly at low ionic strength. Current evidence suggests that these granules are not surrounded by a phospholipid bilayer and thus differ from typical secretory vesicles. Electron microscopy reveals the presence of a banded structure suggesting the presence of a scaffold, probably made up of proteins. From a cDNA expression library, two types of clones that appear to encode such scaffold proteins have been isolated. One codes for a small family of 26–28 kDa proteins characterized by a very basic COOH-terminal region rich in histidine (Berger & Kreil 1989). A second group of cDNAs encode a protein with a highly repetitive structure. As the repeat unit is composed of Ala, Pro, Glu and Gly, this was termed the APEG protein, using the single letter abbreviations for these amino acids (Gmachl *et al.* 1990). At the COOH-end, this protein contains a sequence of about 60 amino acids which shows homology to pS2, a peptide earlier isolated from human breast cancer cells (Rio *et al.* 1988), and to porcine pancreatic spasmolytic polypeptide (Thim 1989). Another protein which contains two copies of a similar peptide is present in the mucous skin glands of X. *laevis* (Hoffmann 1988).

Besides these proteins which may be part of a scaffold, the elliptical granules apparently contain all the constituents of the skin secretion, i.e. peptides and biogenic amines. In addition, processing enzymes required to liberate the peptides from their respective precursors are also present (see below).

The contents of the granular glands are released by stress. After injection of adrenaline or electrical stimulation, almost total discharge of the glands occurs within one minute. The secreted mixture acts as a potent deterrent to predators (Barthalmus & Zielinski 1988). Subsequently, the regeneration of these glands proceeds through morphologically distinct stages over a period of several weeks (see Flucher *et al.* 1986).

Peptides from skin secretion of *X. laevis*

Using biochemical and pharmacological techniques, three peptides were shown to be present in the skin secretion of X. *laevis*: caerulein, thyrotropin releasing hormone, and xenopsin. The structure of their precursor polypeptides has been elucidated via cDNA cloning. These studies, as well as independent investigations by Gibson, Williams and their colleagues (Gibson, Poulter *et al.* 1986; Giovannini *et al.* 1987; Poulter *et al.* 1988), led to the discovery of a large number of additional peptides present in the skin secretion of these frogs.

Caerulein

The decapeptide caerulein was originally discovered in skin of *Litoria caerulea* and subsequently shown also to be present in species from several other families including *X. laevis* (Erspamer 1971; Erspamer, Falconieri Erspamer, Mazzanti *et al.* 1984). Close relatives of this peptide, namely phyllocaerulein and [Asn-2, Leu-5] caerulein, were detected in the skin of other species (Erspamer, Falconieri Erspamer, Mazzanti *et al.* 1984; Erspamer, Melchiorri, Falconieri Erspamer, Montecucchi *et al.* 1985). Caerulein is homologous to cholecystokinin and it possesses the same biological activities as its mammalian counterpart. Using recombinant DNA techniques, several cDNAs encoding precursors of caerulein have been isolated from the skin of *X. laevis* (Hoffmann, Bach *et al.* 1983; Wakabayashi, Kato & Tachibana 1985; Richter, Egger & Kreil 1986). The predicted precursor polypeptides contained one, three or four copies of the end-product, separated by 'spacer' peptides which showed a high degree of similarity, both within a given precursor and between different ones (see later). Subsequent analysis of two of the genes encoding these precursors suggests that they originate through duplication and/or deletion of small exons (Vlasak *et al.* 1987). Interestingly, in one of the genes an exon was discovered containing the genetic information for a new peptide related to caerulein. It is not known at present whether, through alternative splicing, this exon may also be expressed in *X. laevis*.

Thyrotropin releasing hormone

In 1976, Guillemin and Schally received the Nobel prize for their work on thyrotropin releasing hormone (TRH). In a major effort it was possible to purify this hormone from several hundred thousand mammalian hypothalami and establish its sequence. In hindsight, it is ironic that the skin of a few dozen frogs would have sufficed to obtain milligram quantities of TRH. This was first demonstrated for *Bombina orientalis* (see Yasuhara & Nakajima 1975) and subsequently for other species including *X. laevis* (Jackson & Reichlin 1977; Bennett *et al.* 1982).

These observations prompted us to start working with amphibian skin. The biosynthesis of a simple tripeptide, pGlu-His-Pro. amide, with a very high biological activity, was of obvious interest. Moreover, it had been proposed that TRH was synthesized by soluble enzymes via a ribosome-independent mechanism; however, this could not be confirmed by others (reviewed by Jackson 1989). Using recombinant DNA techniques, we were able to demonstrate that in the skin of *X. laevis*, TRH was liberated from larger precursor polypeptides (Richter, Kawashima *et al.* 1984; Kuchler, Richter *et al.* 1990). Two similar cDNAs from skin were found to encode precursors each containing seven copies of the end product. Subsequently, an additional cDNA encoding a somewhat different TRH precursor was isolated from brain of *X. laevis* (see Bulant *et al.* 1992). Interestingly, the TRH precursor from mammalian brain shows, except for the region encoding the five copies of the end-product, no similarity to its amphibian counterparts (Lechan *et al.* 1986).

Xenopsin

From skin extracts of *X. laevis*, a peptide was isolated which was shown to have potent contractile activity on some smooth muscle preparations (Araki *et al.* 1973). This peptide, termed xenopsin, is homologous to neurotensin present in mammalian brain (Carraway & Leeman 1973). Peptides resembling xenopsin more closely than neurotensin have since been found in mammals as well (Carraway, Mitra & Muraki 1990; Carraway & Mitra 1990). Amphibian xenopsin is derived from a small precursor the sequence of which was elucidated by recombinant DNA techniques (Sures & Crippa 1984). The pre- and pro-regions of the xenopsin precursor show a striking homology to the corresponding parts of prepro-laevitide (Poulter *et al.* 1988) as well as to those of caerulein and PGLa (Kuchler, Kreil & Sures 1989).

Antimicrobial peptides

In the course of sequence analysis of the caerulein precursors, it was noted that they contained regions encoding a set of homologous basic peptides which could form amphipathic helices (Hoffmann, Bach *et al.* 1983; Richter, Egger & Kreil 1986). A similar segment was found to be present in the xenopsin precursors (Sures & Crippa 1984). By chance, several cDNAs encoding the precursor of yet another basic peptide were isolated. This peptide was termed PGLa (*p*eptide with amino-terminal *g*lycine and carboxy-terminal *l*eucine *a*mide; Hoffmann, Richter & Kreil 1983). An independent analysis by Gibson, Williams and their colleagues (Gibson, Poulter *et al.* 1986; Giovannini *et al.* 1987) as well as some of our own work (Andreu *et al.* 1985; Richter, Aschauer & Kreil 1985) demonstrated that all these basic peptides predicted from the sequence of cDNA clones were in fact present in the skin secretions of *X. laevis*. Further analysis of skin secretions, as well as cDNA cloning experiments, subsequently led to the discovery of additional peptides. One was laevitide, a peptide of unknown function, the precursor of which also contains one of these basic peptides (Poulter *et al.* 1988), the other being the PGS/magainin peptides (Giovannini *et al.* 1987; Zasloff 1987). These latter peptides are derived from precursors containing several copies of the two types of PGS/magainin entities.

By analogy to bombinin, isolated more than ten years earlier from skin secretion of *B. variegata*, it was obvious to everybody working in this field that the basic peptides from skin secretion would probably also be antimicrobial and/or possess haemolytic activity. Moreover, a number of similar peptides from insect haemolymph have been characterized by Boman and his colleagues

(Boman & Hultmark 1987; Boman 1991). The extensive press coverage accompanying the finding that magainins also had antibacterial activity (Zasloff 1987) thus seemed rather startling (see review by Gibson 1991). Subsequent studies have in fact confirmed that, as far as tested, all the other peptides, i.e. PGLa and the fragments derived from the precursors of caerulein and xenopsin, show antimicrobial activity (see e.g. Soravia, Martini & Zasloff 1988). These peptides can apparently interact directly with the membranes of bacteria thereby dissipating the electrochemical gradient (Westerhoff *et al.* 1989). Moreover, synergistic interaction between, e.g., PGLa and magainins has been observed (Williams *et al.* 1990). The fact that the corresponding peptides composed only of D-amino acids have the same activity as the L-isomers demonstrates that they do not interact with a chiral centre (Wade *et al.* 1990).

Processing enzymes

In general, secreted peptides are derived from larger precursors through a variety of enzymatic reactions. This involves first the cleavage of signal or pre-peptide on transfer through the membrane of the endoplasmic reticulum. Subsequently, in the lumen of the endoplasmic reticulum, in the Golgi stacks, and finally in secretory granules, the peptide precursors are modified and hydrolysed by a variety of enzymes ultimately to yield the end-product. In view of the large number of different peptides secreted from amphibian skin, it is not surprising that a wide variety of processing enzymes, apparently involved in liberating these peptides from their respective precursors, are also present in this tissue. Moreover, many precursors of frog skin peptides are unusually complex in that several copies of peptides with different functions may be present in one polypeptide chain.

Like other secreted proteins, the intact precursors of peptides may be modified in the endoplasmic reticulum and the Golgi apparatus. This can involve hydroxylation of proline residues and the formation of tyrosine sulphate. Dermorphin containing hydroxyproline has been isolated from skin of *Phyllomedusa rhodei* (see Erspamer, Melchiorri, Falconieri Erspamer, Montecucchi *et al.* 1985), while tyrosine sulphate is present in caeruleins present in several amphibian species (Erspamer 1971). After exit from the trans-Golgi network, a number of endoproteases, exoproteases and other enzymes act on the precursor polypeptides to yield the end products. From the structure of the precursors of caerulein (Richter, Egger & Kreil 1986), one can conclude that the following reactions must take place to yield the final products: (1) cleavage at pairs of basic amino acids, like Arg–Arg; (2) cleavage after single arginine residues; (3) hydrolysis of the COOH-terminal arginines by a carboxypeptidase; (4) formation of terminal amides involving COOH-terminal glycine residues exposed after the previous endo- and exoproteolytic reactions; (5) stepwise cleavage of dipeptides from the amino end by a dipeptidyl-aminopeptidase; (6) formation of pyroglutamic acid from amino terminal glutamine residues.

It was noted some time ago that several and possibly all of these processing enzymes are present in the skin secretions of *X. laevis*. Some of them have actually been isolated from this source. Two enzymes have been shown to be endoproteases cleaving after single arginine residues. It was first noted by Kuks *et al.* (1989) that during the processing of several precursors analysed from skin of *X. laevis*, hydrolysis must take place after the second arginine in the conserved sequence Arg–X–Val–Arg–Gly. An endoprotease highly specific for this sequence has been purified from skin secretion of *X. laevis* (see Kuks *et al.* 1989). A second endopeptidase with a different specificity has been purified from the same source by Darby, Lackes & Smyth (1991). Two enzymes acting at the termini of processing intermediates have also been investigated. These were the amidating enzyme (Mizuno *et al.* 1986; Mollay, Wichta & Kreil 1986; Ohsuye *et al.* 1988; Suzuki *et al.* 1990) and a dipeptidyl aminopeptidase (Mollay, Vilas *et al.* 1986; Kreil 1990).

Are skin peptides also present in amphibian brain and gut?

In view of the similarity of amphibian skin peptides to hormones and neurotransmitters isolated from mammals and other vertebrates, two obvious questions arise: are these skin peptides identical to peptides present in amphibian brain and gastrointestinal tract and are they derived from the same precursors? This has been the subject of few studies but with the tools at hand it is a topic that can now be investigated in more detail.

In the case of TRH, it was shown early on that the skin peptide had the same structure as the hormone isolated from mammalian hypothalamus (Yasuhara & Nakajima 1975; Jackson & Reichlin 1977). Two mRNAs encoding the TRH precursor have been detected in *Xenopus* brain, one of which is clearly different from the mRNAs present in skin (Bulant *et al.* 1992). The second brain mRNA, which co-migrates with one of the skin mRNAs, has not been analysed.

The question whether the caerulein present in skin corresponds to cholecystokinin (CCK) and/or gastrin in amphibian brain and gastrointestinal tract has also been investigated. Studies with specific antibodies suggested that caerulein-like peptides do occur in the intestine but not in the brain of *X. laevis*, while CCK-like peptides have been detected in both tissues (Dimaline 1983). In an investigation of *Rana catesbeiana* (see Johnsen & Rehfeld 1992), it was shown that gastrointestinal cholecystokinin is derived from a precursor that is not related to the caerulein precursors from *X. laevis* skin. On the other hand, several antimicrobial skin peptides, including fragments of the precursors of caerulein and xenopsin, are present in the stomach of *X. laevis* (see Moore *et al.* 1991).

Similar studies have been performed with the skin peptides bombesin and ranatensin and their mammalian counterparts, gastrin releasing peptide (GRP) and neuromedin B. In the case of *Bombina orientalis*, it could be shown that the bombesin-like peptides are derived from two types of precursors, one being present in brain and stomach only, while the other is present also in the skin (Nagalla *et al*. 1992). The former was found to be more homologous to the mammalian GRP precursor than to the bombesin precursor isolated from the same species. From the brain of *X. laevis*, a cloned cDNA encoding the precursor of a peptide closely resembling mammalian neuromedin B could be isolated. The corresponding mRNA was also detected in gut, ovaries and early embryos, but not in the skin of these frogs (Wechselberger, Kreil & Richter 1992).

These limited studies indicate that the genes expressed in skin may also be active in other tissues of amphibians. On the other hand, separate genes apparently encode the precursors of hormones and neurotransmitters typically found in vertebrate brain and gastrointestinal tract.

Concluding remarks

It has been known for some time that amphibian skin contains a wide variety of biologically active peptides. A thorough analysis of the skin secretion of one species, *Xenopus laevis*, has revealed an unexpected complexity. While only a few constituents were detected by classical pharmacological tests, dozens of additional peptides were detected by a combination of cDNA cloning and systematic analysis of individual components. It seems likely that this would also be the case for the skin secretions of other species. The work on amphibian skin has in addition yielded interesting results about the enzymes involved in the processing of peptide precursors.

It was noted early on that many skin peptides were homologous to mammalian hormones and neurotransmitters. However, the types of peptides present in amphibian skin vary widely. For example, some contain tachykinins, others bradykinins, still others caeruleins etc. Yet it is apparent that each species must contain a complete set of peptides typical for vertebrate species in brain, intestine and other organs. Some studies have recently been performed to test whether peptides present in amphibian skin are products of the same genes as the neurotransmitters, hormones etc. functioning in other parts of the organisms. The limited data available so far indicate that genes encoding skin peptides are very different from their counterparts in endocrine and exocrine glands and thus have a long history of independent evolution. Yet some of the genes expressed in dermal glands of amphibians are clearly also active in other parts of these animals. The current picture can therefore be summarized in the following way: as with all other vertebrates, Amphibia have a set of genes encoding peptide hormones, neurotransmitters and other biologically active peptides. A subset of these genes, which differs from species to species, has been duplicated and the genes appear to have evolved separately over many millions of years. These genes, while sometimes still active in the same cells as the original copies, are now expressed principally in the dermal glands of amphibians. Future studies will show to what extent this simple picture holds true.

Postscript

Since this manuscript was first submitted, a number of papers have been published dealing with skin secretions of Amphibia. In two excellent reviews the current knowledge on peptides from frog skin (Lazarus & Attila 1993) and on opioid peptides from Phyllomedusinae (Erspamer 1993) has been presented.

In skin secretions of *Xenopus laevis*, a new family of peptides termed xenoxins has been discovered (Kolbe *et al.* 1993). Xenoxins are homologous to neurotoxins and cytotoxins from snake venom. However, they are low in general toxicity as tested in mice; their biological activity is currently not known. It seems likely that a common ancestor of xenoxins and the snake venom toxins was present early in the evolution of vertebrates and that this ancestral peptide had a low toxicity.

Evidence has been presented that skin secretions of *X. laevis* contain, in addition to processing enzymes, enzymes which inactivate peptide hormones and antibacterial peptides. Two metallo-endoproteases with different specificities have been characterized (Resnick *et al.* 1991; Joudiou *et al.* 1993).

Recent studies with other amphibian species have shown that *X. laevis* is by no means an exception. Skin secretions of *Rana* sp. contain numerous antimicrobial peptides with a common sequence motif (Morikawa, Hagiwara & Nakajima 1992; Simmaco, Mignogna *et al.* 1993, 1994; Clark *et al.* 1994), at least five different dermaseptins are present in the skin of *Phyllomedusa sauvagei* (Daly *et al.* 1992; Mor & Nicolas 1994), and besides numerous bombinins (Simmaco, Barra *et al.* 1991) a second family of peptides termed bombinins H have been discovered in *Bombina variegata* (see Mignogna, Simmaco *et al.* 1993). Some of these latter peptides contain a D-amino acid, D-allo-isoleucine, at the second position. Besides the opioid peptides from Phyllomedusinae, this represents the second example of amphibian skin peptides containing a D-amino acid (Kreil 1994).

These frogs are representatives of only four of about 20 families of amphibian species. In each family, specific antimicrobial peptides have been found which are not related to those of other groups. Indeed, studies with two hylid frogs from Australia led to the discovery of almost two dozen new peptides of unknown function (Waugh *et al.* 1993). It can thus be expected that a large variety of additional peptides exist in skin secretions of different frogs and that sequence comparisons of these components might be useful for taxonomic studies.

References

Andreu, D., Aschauer, H., Kreil, G. & Merrifield, R.B. (1985). Solid-phase synthesis of PYLa and isolation of its natural counterpart, PGLa [PYLa (4–24)] from skin secretion of X. laevis. Eur. J. Biochem. 149: 531–535.

Araki, K., Tachibana, S., Uchiyama, M., Nakajima, T. & Yasuhara, T. (1973). Isolation and structure of a new peptide "xenopsin" from skin of X. laevis active on the smooth muscle. Chem. pharm. Bull., Tokyo 21: 2801–2804.

Barthalmus, G.T. & Zielinski, W.J. (1988). Xenopus skin mucus induces oral dyskinesias that promote escape from snakes. Pharmacol. Biochem. Behav. 30: 957–959.

Bennett, G.W., Marsden, C.A., Clothier, R.M., Waters, A.D. & Balls, M. (1982). Co-existence of TRH and 5-hydroxytryptamine in the skin of X. laevis. Comp. Physiol. Biochem. (C) 72: 257–261.

Berger, H. & Kreil, G. (1989). Secretory granules in the dermal glands of Xenopus laevis: structure of a basic polypeptide deduced from cloned cDNA. FEBS Lett. 249: 293–296.

Bevins, C.L. & Zasloff, M. (1990). Peptides from frog skin. A. Rev. Biochem. 59: 395–414.

Boman, H.G. (1991). Antibacterial peptides: key components needed in immunity. Cell 65: 205–207.

Boman, H.G. & Hultmark, D. (1987). Cell-free immunity in insects. A. Rev. Microbiol. 41: 103–126.

Bulant, M., Richter, K., Kuchler, K. & Kreil, G. (1992). A cDNA from brain of Xenopus laevis coding for a new precursor of thyrotropin-releasing hormone. FEBS Lett. 296: 292–296.

Carraway, R.E. & Leeman, S.E. (1973). Isolation of a new hypotensive peptide, neurotensin, from bovine hypothalami. J. biol. Chem. 248: 6854–6861.

Carraway, R.E. & Mitra, S.P. (1990). Isolation and sequence of canine xenopsin and an extended fragment from its precursor. Peptides 11: 747–752.

Carraway, R.E., Mitra, S.P. & Muraki, K. (1990). Isolation and structures of xenopsin-related peptides from rat stomach, liver and brain. Regul. Peptides 29: 229–239.

Clark, D.P., Durell, S., Maloy, W.L. & Zasloff, M. (1994). Ranalexin. A novel antimicrobial peptide from bullfrog (Rana catesbeiana) skin, structurally related to the bacterial antibiotic polymyxin. J. biol. Chem. 269: 10849–10855.

Csordas, A. & Michl, H. (1970). Isolation and structure of a hemolytic polypeptide from the defensive secretion of European Bombina species. Monatsh. Chem. 101: 182–189.

Daly, W.J., Caceres, J., Moni, W.R., Gusovski, F., Moos, M., Seamon, B.K., Milton, K. & Myers, W.C. (1992). Frog secretion and hunting magic in the upper Amazon: identification of a peptide that interacts with an adenosine receptor. Proc. natn. Acad. Sci. USA 89: 10960–10963.

Darby, N.J., Lackes, D.B. & Smyth, D.G. (1991). Purification of a cysteine endopeptidase which is secreted with bioactive peptides from the epidermal glands of *Xenopus laevis. Eur. J. Biochem.* **195**: 65–70.

Dimaline, R. (1983). Is caerulein amphibian CCK? *Peptides* **4**: 457–462.

Dockray, G.J. & Hopkins, C.R. (1975). Caerulein secretion by dermal glands of *Xenopus laevis. J. Cell Biol.* **64**: 724–733.

Erspamer, V. (1971). Biogenic amines and active polypeptides of the amphibian skin. *A. Rev. Pharmacol.* **11**: 327–350.

Erspamer, V. (1993). The opioid peptides of the amphibian skin. *Int. J. devl Neurosci.* **10**: 3–30.

Erspamer, V., Falconieri Erspamer, G. & Cei, J.M. (1986). Active peptides in the skins of two hundred and thirty American amphibian species. *Comp. Biochem. Physiol. (C)* **85**: 125–137.

Erspamer, V., Falconieri Erspamer, G., Mazzanti, G. & Endean, R. (1984). Active peptides in the skin of one hundred amphibian species from Australia and Papua New Guinea. *Comp. Biochem. Physiol. (C)* **77**: 99–108.

Erspamer, V. & Melchiorri, P. (1983). Actions of amphibian skin peptides on the central nervous system and the anterior pituitary. *Neuroendocr. Perspect.* **2**: 37–106.

Erspamer, V., Melchiorri, P., Broccardo, M., Falconieri Erspamer, G., Falaschi, P., Improta, G., Negri, L. & Renda, T. (1981). The brain-gut-skin triangle: new peptides. *Peptides* **2** Suppl. 2: 7–16.

Erspamer, V., Melchiorri, P., Falconieri Erspamer, G., Montecucchi, P.C. & De Castiglione, R. (1985). *Phyllomedusa* skin: a huge factory and store-house of a variety of active peptides. *Peptides* **6** Suppl. 3: 7–12.

Erspamer, V., Melchiorri, P., Falconieri-Erspamer, G., Negri, L., Corsi, M.R., Severini, C., Barra, D., Simmaco, M. & Kreil, G. (1989). Deltorphins—a family of naturally occurring peptides with high affinity and selectivity for delta-opioid binding sites. *Proc. natn. Acad. Sci. USA* **86**: 5188–5192.

Flucher, B.E., Lenglachner-Bachinger, C., Pohlhammer, K., Adam, H. & Mollay, C. (1986). Skin peptides in *Xenopus laevis*: morphological requirements for precursor processing in developing and regenerating granular skin glands. *J. Cell Biol.* **103**: 2299–2309.

Gibson, B.W. (1991). Lytic peptides from skin secretions of *Xenopus laevis*: a personal perspective. *ACS Symp. Ser.* **444**: 222–236.

Gibson, B.W., Poulter, L., Williams, D.H. & Maggio, J.E. (1986). Novel peptide fragments originating from PGLa and the caerulein and xenopsin precursors from *Xenopus laevis. J. biol. Chem.* **261**: 5341–5349.

Gibson, B.W., Tang, D., Mandrell, R., Kelly, M. & Spindel, E.R. (1991). Bombinin-like peptides with antimicrobial activity from skin secretions of the Asian toad, *Bombina orientalis. J. biol. Chem.* **266**: 23103–23111.

Giovannini, M.G., Poulter, L., Gibson, B.W. & Williams, D.H. (1987). Biosynthesis and degradation of peptides derived from *Xenopus laevis* prohormones. *Biochem. J.* **243**: 113–120.

Gmachl, M., Berger, H., Thalhammer, J. & Kreil, G. (1990). Dermal glands of *Xenopus laevis* contain a polypeptide with a highly repetitive amino acid sequence. *FEBS Lett.* **260**: 145–148.

Hoffmann, W. (1988). A new repetitive protein from *Xenopus laevis* skin highly homologous to pancreatic spasmolytic polypeptide. *J. biol. Chem.* **263**: 7686–7690.

Hoffmann, W., Bach, T.C., Seliger, H. & Kreil, G. (1983). Biosynthesis of caerulein in the skin of *Xenopus laevis*: partial sequences of precursors as deduced from cDNA clones. *EMBO J.* **2**: 111–114.

Hoffmann, W., Richter, K. & Kreil, G. (1983). A novel peptide designated PYLa and its precursors as predicted from cloned mRNA of *Xenopus laevis* skin. *EMBO J.* **2**: 711–714.

Jackson, I.M.D. (1989). Controversies in TRH biosynthesis and strategies towards the identification of a TRH precursor. *Ann. N.Y. Acad. Sci.* **553**: 7–13.

Jackson, I.M.D. & Reichlin, S. (1977). TRH: abundance in the skin of the frog *Rana pipiens*. *Science* **198**: 414–415.

Johnsen, A.H. & Rehfeld, J.F. (1992). Identification of cholecystokinin/gastrin peptides in frog and turtle: evidence that cholecystokinin is phylogenetically older than gastrin. *Eur. J. Biochem.* **207**: 419–428.

Joudiou, C., Carvalho, K.M., Camarao, G., Boussetta, H. & Cohen, P. (1993). Characterization of the thermolysin-like cleavage of biologically active peptides by *Xenopus laevis* peptide hormone inactivating enzyme. *Biochemistry* **32**: 5959–5966.

Kolbe, H.V.J., Huber, A., Cordier, P., Rasmussen, U.B., Bouchon, B., Jaquinod, M., Vlasak, R., Délot, E.C. & Kreil, G. (1993). Xenoxins, a family of peptides from dorsal gland secretion of *Xenopus laevis* related to snake venom cytotoxins and neurotoxins. *J. biol. Chem.* **268**: 16458–16464.

Krane, I.M., Naylor, S.I., Helin-Davis, D., Chin, W.W. & Spindel, E.R. (1988). Molecular cloning of cDNAs encoding the human bombesin-like peptide neuromedin B and its amphibian homolog ranatensin. *J. biol. Chem.* **263**: 13317–13323.

Kreil, G. (1990). Processing of precursors by dipeptidyl-aminopeptidases: a case of molecular ticketing. *Trends biochem. Sci.* **15**: 23–26.

Kreil, G. (1994). Peptides containing a D-amino acid from frogs and molluscs (minireview). *J. biol. Chem.* **269**: 10967–10970.

Kreil, G., Barra, D., Simmaco, M., Erspamer, V., Falconieri Erspamer, G., Negri, L., Severini, C., Corsi, R. & Melchiorri, P. (1989). Deltorphin, a novel amphibian skin peptide with high selectivity and affinity for δ-opioid receptors. *Eur. J. Pharmacol.* **162**: 123–128.

Kuchler, K., Kreil, G. & Sures, I. (1989). The genes for the frog skin peptides GLa, xenopsin, levitide and caerulein contain a homologous export exon encoding a signal sequence and part of an amphiphilic peptide. *Eur. J. Biochem.* **179**: 281–285.

Kuchler, K., Richter, K., Trnovsky, J., Egger, R. & Kreil, G. (1990). Two precursors of TRH from skin of *X. laevis*: each contains seven copies of the end-product. *J. biol. Chem.* **265**: 11731–11733.

Kuks, P.F.M., Creminon, C., Leseney, A.-M., Bourdais, J., Morel, A. & Cohen, P. (1989). *Xenopus laevis* skin Arg-Xaa-Val-Arg-Gly-endoprotease: a highly specific protease cleaving after a single arginine of a consensus sequence of peptide hormone precursors. *J. biol. Chem.* **264**: 14609–14612.

Lazarus, L.H. & Attila, M. (1993). The toad, ugly and venomous, wears yet a precious jewel in his skin. *Progr. Neurobiol.* **41**: 473–507.

Lazarus, L.H., Wilson, W.E., de Castiglione, R. & Guglietta, A. (1989). Dermorphin gene sequence peptide with high affinity and selectivity for delta-opioid receptors. *J. biol. Chem.* **264**: 3047–3050.

Lechan, R.M., Wu, P., Jackson, I.M.D., Wolf, H., Cooperman, S., Mandel, G. & Goodman, R.H. (1986). TRH precursor: characterization in rat brain. *Science* **231**: 159–161.

Mignogna, G., Severini, C., Simmaco, M., Negri, L., Falconieri Erspamer, G., Kreil, G.

& Barra, D. (1992). Identification and characterization of two dermorphins from skin extracts of the Amazonian frog *Phyllomedusa bicolor*. *FEBS Lett.* **302**: 151–154.

Mignogna, G., Simmaco, M., Kreil, G. & Barra, D. (1993). Antibacterial and haemolytic peptides containing D-alloisoleucine from the skin of *Bombina variegata*. *EMBO J.* **12**: 4829–4832.

Mizuno, K., Sakata, J., Kojima, M., Kangawa, K. & Matsuo, H. (1986). Peptide C-terminal alpha-amidating enzyme purified to homogeneity from *Xenopus laevis* skin. *Biochem. biophys. Res. Commun.* **137**: 984–991.

Mollay, C., Vilas, U., Hutticher, A. & Kreil, G. (1986). Isolation of a dipeptidyl aminopeptidase, a putative processing enzyme, from skin secretion of *X. laevis*. *Eur. J. Biochem.* **160**: 31–35.

Mollay, C., Wichta, J. & Kreil, G. (1986). Detection and partial characterization of an amidating enzyme in skin secretion of *X. laevis*. *FEBS Lett.* **202**: 251–254.

Montecucchi, P.C., De Castiglione, R., Piani, S., Gozzini, L. & Erspamer, V. (1981). Amino acid composition and sequence of dermorphin, a novel opiate-like peptide from skin of *Phyllomedusa sauvagei*. *Int. J. Peptide Protein Res.* **17**: 275–283.

Moore, K.S., Bevins, C.L., Brasseur, M.M., Tomassini, N., Turner, K., Eck, H. & Zasloff, M. (1991). Antimicrobial peptides in stomach of *Xenopus laevis*. *J. biol. Chem.* **266**: 19851–19857.

Mor, A., Delfour, A., Sagan, S., Amiche, M., Pradelles, P., Rossier, J. & Nicolas, P. (1989). Isolation of dermenkephalin from amphibian skin, a high affinity delta-selective opioid heptapeptide containing a D-amino acid. *FEBS Lett.* **255**: 269–274.

Mor, A., Nguyen, V.H., Delfour, A., Migliore-Samour, D. & Nicolas, P. (1991). Isolation, amino acid sequence, and synthesis of dermaseptin, a novel antimicrobial peptide of amphibian skin. *Biochemistry* **30**: 8824–8830.

Mor, A. & Nicolas, P. (1994). Isolation and structure of novel defensive peptides from frog skin. *Eur. J. Biochem.* **219**: 145–154.

Morikawa, N., Hagiwara, K. & Nakajima, T. (1992). Brevinin-1 and -2, unique antimicrobial peptides from the skin of the frog *Rana brevipoda porsa*. *Biochem. biophys. Res. Commun.* **189**: 184–190.

Nagalla, S.R., Gibson, B.W., Tang, D., Reeve, J.R. Jr. & Spindel, E.R. (1992). Gastrin-releasing peptide (GRP) is not mammalian bombesin: identification and molecular cloning of a true amphibian GRP distinct from amphibian bombesin in *Bombina orientalis*. *J. biol. Chem.* **267**: 6916–6922.

Negri, L., Falconieri Erspamer, G., Severini, C., Potenza, R.L., Melchiorri, P. & Erspamer, V. (1992). Dermorphin-related peptides from the skin of *Phyllomedusa bicolor* and their amidated analogs activate two µ-opioid receptor subtypes that modulate antinociception and catalepsy in the rat. *Proc. natn. Acad. Sci. USA* **89**: 7203–7202.

Ohsuye, K., Kitano, K., Wada, Y., Fuchimura, K. & Tanaka, S. (1988). Cloning of cDNA encoding a new peptide C-terminal alpha-amidating enzyme having a putative membrane-spanning domain from *Xenopus laevis* skin. *Biochem. biophys. Res. Commun.* **150**: 1275–1281.

Poulter, L., Terry, A.S., Williams, D.H., Giovannini, M.G., Moore, C.H. & Gibson, B.W. (1988). Laevitide, a neurohormone-like peptide from the skin of *Xenopus laevis*. *J. biol. Chem.* **263**: 3279–3283.

Resnick, N.M., Maloy, W.L., Guy, H.R. & Zasloff, M. (1991). A novel endopeptidase from *Xenopus* that recognizes alpha-helical secondary structure. *Cell* **66**: 541–554.

Richter, K., Aschauer, H. & Kreil, G. (1985). Biosynthesis of peptides in the skin of *Xenopus laevis*: isolation of novel peptides predicted from the sequence of cloned cDNAs. *Peptides* 6: 17–21.

Richter, K., Egger, R. & Kreil, G. (1986). Sequence of preprocaerulein cDNAs cloned from skin of *Xenopus laevis*, a small family of precursors containing one, three or four copies of the final product. *J. biol. Chem.* 261: 3676–3680.

Richter, K., Egger, R. & Kreil, G. (1987). D-alanine in the frog skin peptide dermorphin is derived from L-alanine in the precursor. *Science* 238: 200–202.

Richter, K., Egger, R. & Kreil, G. (1990). Molecular cloning of a cDNA encoding the bombesin precursor in skin of *Bombina variegata*. *FEBS Lett.* 262: 353–355.

Richter, K., Egger, R., Negri, L., Corsi, R., Severini, C. & Kreil, G. (1990). cDNAs encoding [D-Ala-2] deltorphin precursors from skin of *Phyllomedusa bicolor* also contain genetic information for three dermorphin-related opioid peptides. *Proc. natn. Acad. Sci. USA* 87: 4836–4839.

Richter, K., Kawashima, E., Egger, R. & Kreil, G. (1984). Biosynthesis of TRH in the skin of *Xenopus laevis*: partial sequence of the precursor deduced from cloned cDNA. *EMBO J.* 3: 617–621.

Rio, M.C., Bellocq, J.P., Daniel, J.Y., Tomasetto, C., Lathe, R., Chenard, M.P., Batzenschlager, A. & Chambon, P. (1988). Breast cancer-associated pS2 protein: synthesis and secretion by normal stomach mucosa. *Science* 241: 705–708.

Simmaco, M., Barra, D., Chiarini, F., Noviello, L., Melchiorri, P., Kreil, G. & Richter, K. (1991). A family of bombinin-related peptides from the skin of *Bombina variegata*. *Eur. J. Biochem.* 199: 217–222.

Simmaco, M., Mignogna, G., Barra, D. & Bossa, F. (1993). Novel antimicrobial peptides from skin secretion of the European frog *Rana esculenta*. *FEBS Lett.* 324: 159–161.

Simmaco, M., Mignogna, G., Barra, D. & Bossa, F. (1994). Antimicrobial peptides from skin secretions of *Rana esculenta*. Molecular cloning of cDNAs encoding esculentin and brevinins and isolation of new active peptides. *J. biol. Chem.* 269: 11956–11961.

Soravia, E., Martini, G. & Zasloff, M. (1988). Anti-microbial properties of peptides from *Xenopus* granular gland secretions. *FEBS Lett.* 228: 337–340.

Spindel, E.R., Gibson, B.W., Reeve, J.R. & Kelly, M. (1990). Cloning of cDNAs encoding amphibian bombesin. *Proc. natn. Acad. Sci. USA* 87: 9813–9817.

Sures, I. & Crippa, M. (1984). Xenopsin: the neurotensin-like octapeptide from *Xenopus* skin at the carboxyl terminus of its precursor. *Proc. natn. Acad. Sci. USA* 81: 380–384.

Suzuki, K., Shimoi, H., Iwasaki, Y., Kawahara, T., Matsuura, Y. & Nishikawa, Y. (1990). Elucidation of the amidating reaction mechanism by frog amidating enzyme, peptidylglycine alpha-hydroxylating monooxygenase, expressed in insect cell culture. *EMBO J.* 9: 4259–4265.

Thim, L. (1989). A new family of growth factor-like peptides: "trefoil" disulphide loop structures as a common feature in breast cancer associated peptide, pancreatic spasmolytic polypeptide and frog skin peptides. *FEBS Lett.* 250: 85–90.

Vlasak, R., Wiborg, O., Richter, K., Burgschwaiger, S., Vuust, J. & Kreil, G. (1987). Conserved exon-intron organization in two different caerulein precursor genes of *Xenopus laevis*: additional detection of an exon potentially coding for a new peptide. *Eur. J. Biochem.* 169: 53–58.

Wade, D., Boman, A., Wahlin, B., Drain, C.M., Andreu, D., Boman, H.G. & Merrifield,

R.B. (1990). All D-amino acid-containing channel forming antibiotic peptides. *Proc. natn. Acad. Sci. USA* **87**: 4761–4765.

Wakabayashi, T., Kato, H. & Tachibana, S. (1985). Complete nucleotide sequence of mRNA for caerulein precursor from *X. laevis. Nucleic Acids Res.* **13**: 1817–1821.

Waugh, R.J., Stone, D.J.M., Bowie, J.H., Wallace, J.C. & Tyler, M.J. (1993). Peptides from Australian frogs. Structures of caeridins from *Litoria caerulea. J. chem. Soc. Perkin Trans.* **1993**: 573–576.

Wechselberger, C., Kreil, G. & Richter, K. (1992). Isolation and sequence of a cDNA encoding the precursor of a bombesin-like peptide from brain and early embryos of *Xenopus laevis. Proc. natn. Acad. Sci. USA* **89**: 9819–9822.

Westerhoff, H.V., Juretic, D., Hendler, R.W. & Zasloff, M. (1989). Magainins and the disruption of membrane-linked free-energy transduction. *Proc. natn. Acad. Sci. USA* **86**: 6597–6601.

Williams, R.W., Starman, R., Taylor, K.P.M., Gable, K., Beeler, T., Zasloff, M. & Covell, D. (1990). Raman spectroscopy of synthetic antimicrobial peptides magainin 2a and PGLa. *Biochemistry* **29**: 4490–4496.

Yasuhara, T. & Nakajima, T. (1975). Isolation of TRH from skin of *Bombina orientalis. Chem. pharm. Bull., Tokyo* **23**: 3301–3303.

Zasloff, M. (1987). Magainins—antimicrobial peptides from *Xenopus* skin: isolation, characterization of two active forms, and partial cDNA sequence of a precursor. *Proc. natn. Acad. Sci. USA* **84**: 5449–5453.

15 Immune system of *Xenopus*: T cell biology

JOHN D. HORTON, TRUDY L. HORTON and PAMELA RITCHIE

Synopsis

Evidence for T and B lymphocyte populations in *Xenopus* is reviewed, and attention is focused on the recent creation of monoclonal antibodies that identify T cell surface markers. A second section summarizes evidence for T helper and T cytotoxic lymphocytes and their genetic restriction, discusses the role of T cells in transplantation immunity, and outlines studies on T cell-derived cytokines. A third section considers the biochemistry, cell distribution and ontogeny of *Xenopus* major histocompatibility complex (MHC) proteins; implications for T cell functioning of the differences in MHC antigen expression between larvae and adults are discussed. The role of the thymus in T cell development is then dealt with in detail. After considering the embryological origins of thymic lymphocytes, describing thymus histogenesis and discussing T cell changes during ontogeny, the possibility of extrathymic development of some 'T-like' cells is discussed in the light of recent flow cytometric studies on early-thymectomized (Tx) *Xenopus*. This fourth section also describes experiments probing the role of the thymus in establishing negative and positive selection of T-lineage lymphocytes. The final section deals with establishment and characterization of peripheral tolerance, following grafting of larvae with allogeneic tissues, and discusses immunoregulatory events occurring during metamorphosis.

Introduction

The immune system of *Xenopus* has been studied in depth for some 25 years and is now well characterized at the organismal and cellular levels: dissection of the molecular and genetic basis of immunity is in progress (see reviews by Du Pasquier, Schwager & Flajnik 1989; Horton 1994). This paper highlights studies pertaining to *Xenopus* T cell biology, with special focus on T cell development: B cell biology is reviewed by Du Pasquier (this volume pp. 301–313).

Identification of T cells

T/B cell dichotomy

Thymectomy of *Xenopus* from 4–8 days of age has clearly demonstrated the existence of thymus-dependent and T-independent components of immunity (see reviews by Manning & Horton 1982; Katagiri & Tochinai 1987). Such early thymectomy abrogates acute allograft rejection and mixed leucocyte culture (MLC) reactivity; responsiveness to T cell mitogens is severely diminished. Cellular and humoral responses to T cell-dependent erythrocyte antigens are abolished, as is the low molecular weight (IgY) antibody response to haptenated keyhole limpet haemocyanin. *In vivo* IgM antibody and *in vitro* proliferative responses to T cell-independent antigens/mitogens are not affected by early thymectomy. T-dependent and T-independent components of immunity have also been readily demonstrated by use of N-methyl-N-nitrosourea (NMU). This lymphotoxic agent permanently removes the thymic cortex and also the T-dependent lymphoid areas in peripheral lymphoid organs (Clothier, Balls & Ruben 1989).

T cell surface markers

Thymus-dependent and T-independent regions of the spleen have been shown by immunocytochemistry to represent zones rich in T and B lymphocytes respectively. Thus the T-independent white pulp follicles are rich in B cells (Bleicher & Cohen 1981; Obara, Tochinai & Katagiri 1982), as shown by the selective staining of these regions with anti-immunoglobulin monoclonal antibodies (mAbs). In contrast, splenic T cells, which can be shown by flow cytometry to be surface immunoglobulin-negative (sIg⁻ᵛᵉ) (see Fig. 15.1), are found in the perifollicular (marginal) zone and scattered in the red pulp. We have demonstrated this (J.D. Horton, T.L. Horton & P. Ritchie unpubl.) by using mAbs that identify *Xenopus* T cell sub-populations, such as XT-1, which recognizes a 120 kDa molecule on a T cell subset (Nagata 1988), and AM22 (Flajnik, Ferrone *et al.* 1990) or F17 (Ibrahim *et al.* 1991), both of which recognize a 35 kDa, CD8-like molecule. Flow cytometry reveals that the antigen recognized by XT-1 (called XTLA-1) is expressed by many *Xenopus* species, but not by *X. tropicalis* (see Varley & Horton 1991). The majority of XT-1⁺ᵛᵉ cells co-express the putative CD8 antigen (our unpublished observations—see Fig. 15.1).

Recently new anti-*Xenopus* T cell monoclonal antibodies have been created (Ibrahim *et al.* 1991) that are especially useful for probing aspects of T cell development. The mAb 2B1 recognizes a single chain glycoprotein (71–82 kDa CD5 homologue) which is expressed constitutively on *all Xenopus* T cells (M.D. Cooper pers. comm.). D4–3 recognizes a major subpopulation of thymocytes and peripheral T cells of *Xenopus*, the antigen recognized being a 110 kDa heterodimer—a putative *Xenopus* αβ T cell receptor (TCR) candidate. The mAb D12–2 recognizes only a minor thymocyte and splenocyte

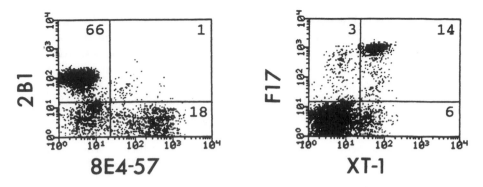

Fig. 15.1. Flow cytometric analysis of mAb-stained *Xenopus* splenocytes.

Splenocytes from a six-month-old control *X. laevis* were adjusted to 1×10^6 leucocytes/ml. To provide background fluorescence levels, one aliquot was stained first with CT3 (anti-chicken thymocyte mAb), followed by FITC-conjugated anti-mouse Ig, then finally stained with PE-conjugated mouse Ig. Markers were set using this preparation to place 98–99% background fluorescence in lower left quadrant (data not shown). One experimental aliquot was stained with XT-1 (mAb staining a T cell subset—see text), followed by FITC-αMIg and then double-stained with a second anti-*Xenopus* mAb—PE-conjugated F17 (anti-CD8). The second experimental aliquot was first stained with 8E4:57 (anti-*X. laevis* IgM mAb), followed by FITC-αMIg, and double-stained with PE-2B1 (pan T cell mAb). Ten thousand splenocytes from each sample were analysed on a FACSCAN. Fluorescence parameters shown pertain to lymphocytes, gated by forward/side scatter. The x and y axes show four decade log fluorescence intensities of FITC- and PE-stained cells respectively. Percentages of single- or double-labelled cells are shown in appropriate quadrants. (Data from ongoing unpublished studies being conducted in our laboratory.)

population, the surface antigen being a 75 kDa heterodimer, possibly the γδ TCR homologue (Ibrahim *et al.* 1991).

T cell functions

T lymphocyte subsets

Specific helper, cytotoxic and suppressor T cell activities exist in *Xenopus*. Whether these various functions reside in separate subpopulations is not yet established, but may soon be answered in view of the increasing array of anti-*Xenopus* T cell antibodies noted above. *In vitro* collaboration between splenic or peripheral blood-derived T cells (identified by their non-adherence to nylon wool, their lack of surface Ig staining and their thymus dependence) and B cells in secondary IgY antibody production has been shown to be major histocompatibility complex (MHC)-restricted in *Xenopus* (Bernard *et al.* 1981). It has recently been shown that both antigen-specific and allogeneic MHC-specific secondary T cell proliferation responses are restricted by MHC class II molecules (Harding 1990). The generation of cytotoxic T lymphocytes (CTL) requires *in vivo* priming and restimulation of responders in MLC:

no lysis is effected when target cells differ from stimulators by two MHC haplotypes (Bernard *et al.* 1979). There is evidence indicating that *Xenopus* CTL can be generated against targets expressing class I or class II MHC, and also against minor histocompatibility antigens (Watkins, Harding & Cohen 1988; T.L Horton, Horton & Varley 1989). Thymic T cells are able to suppress *in vitro* antibody production in *Xenopus* (Hsu, Julius & Du Pasquier 1983) and this suppressive ability has been shown to be MHC-unrestricted and sensitive to cyclophosphamide and irradiation (Clothier, Last *et al.* 1989). Thymus T cells are poor helpers and graft-versus-host (GVH) inducers, whereas splenic T cells are much stronger in these respects (Hsu *et al.* 1983; Du Pasquier *et al.* 1989). GVH reactivity, often inducing recipient death, of implanted allogeneic spleens has been described in *Xenopus* (Nakamura 1985).

T cell-mediated immunity

The relative contribution made by T and B cells to skin allo- and xenograft destruction has been assessed. Thymus-dependency of rapid graft rejection appears less evident when xenograft rejection is studied, especially when donor and host are distantly related species (Clothier, Ali *et al.* 1989; J.D. Horton, Horton, Ritchie & Varley 1992). Skin xenograft (*X. tropicalis*) destruction in control *X. laevis* is associated with B cell accumulation under the foreign skin, rather than to a heavy T cell infiltration that occurs within allografts (J.D. Horton, Horton, Ritchie & Varley 1992). Since T cells are selected in the thymus to respond preferentially with antigens presented by self MHC proteins, it seems likely that graft antigens presented by donor cells expressing xenogeneic MHC appear 'too different' to be directly recognized by the host's cytotoxic T cells. Perhaps donor xenoantigens can only be successfully presented (to T cells) by host antigen-presenting cells (APCs), i.e. in the context of the host's MHC proteins, which would tend to lead to antibody responses, rather than to T cell cytotoxicity.

Lymphokine-activated, NK-like cells, rather than cytotoxic T lymphocytes, appear to be associated with the cellular response to a herpes-like virus in *Xenopus* (see Watkins 1985). T-dependent reactivity towards nematodes of the genus *Capillaria*, which can live in the skin of *Xenopus*, causing 'flaky skin' disease, have been described (Cohen, Effrige *et al.* 1984). *Xenopus* naturally infected with a cocco-bacilloid micro-organism possess activated splenic T cells which constitutively secrete cytokines (Haynes, Harding *et al.* 1992).

Cytokines

'Interleukin-2-like' material (T cell growth factor—TCGF) has been identified from culture supernatants of T cell mitogen-stimulated or MLC-activated splenocytes from control, but not from early-thymectomized (Tx) *Xenopus* (Watkins & Cohen 1987a; Turner *et al.* 1991). Purification of *Xenopus*

TCGF indicates a protein of molecular mass 16 kDa (Watkins & Cohen 1987b; Haynes & Cohen 1993). Since crude supernatants from mitogen- or alloantigen-activated splenocytes undoubtedly contain a mix of cytokines, it is not surprising that these supernatants promote proliferation not only of activated T cells, but also of surface Ig^{+ve} (B) cells (enriched by panning with anti-Ig mAbs) from control *Xenopus* (Cohen & Haynes 1991) and surface Ig-negative splenocytes (enriched by cell sorting) from Tx animals (Turner *et al.* 1991).

Larval *Xenopus* thymocytes (contrast larval splenocytes) are unresponsive to T cell mitogens, but can be costimulated by addition of T cell mitogen together with either TCGF or IL-1-rich supernatant (Watkins, Parsons & Cohen 1988). This indicates that larval thymic cell populations are unable to produce, but can respond to, IL-1 and Il-2-like material.

Xenopus TCGF and mammalian IL-2 are not functionally cross-reactive *in vitro* (Watkins & Cohen 1987a). In contrast, there is evidence that human recombinant IL-2 (rIL-2) can modulate a variety of *in vivo* immune reactivities in *Xenopus* (see Ruben 1986). *Xenopus* splenocytes can bind an anti-human IL-2 receptor antibody (against the P55 Tac peptide), an interaction blocked by preincubation of cells with the rIL-2 (Langeberg *et al.* 1987). However, at present the physiological significance of these findings remains controversial (compare Ruben, Langeberg *et al.* 1990 with Haynes, Moynihan & Cohen 1991). Molecular characterization of the receptor for homologous IL-2 in *Xenopus* is now in progress (Cohen & Haynes 1991) and will be aided by the development of novel monoclonal antibodies against this receptor (M. Flajnik pers. comm.).

MHC proteins

Characterization and cell distribution

In mammals, the MHC codes for polymorphic histocompatibility antigens, the class I and II MHC proteins. The latter play a crucial role in presentation of antigenic peptides (derived intra- or extracellularly respectively) to T cell receptor molecules. The *Xenopus* MHC has been defined by functional criteria and is being characterized by biochemical, molecular and genetic techniques (Flajnik & Du Pasquier 1990a; Kaufman, Flajnik & Du Pasquier 1991; Flajnik, Canel *et al.* 1991a). In *Xenopus*, phenomena such as T–B cell collaboration, the generation of alloantigen-specific cytotoxic responses and thymic education of T-lineage cells are all under control of its MHC, a single chromosomal region called the XLA (by analogy with human HLA). Sato *et al.* (1993) have shown that the MHC remains functionally diploid in X. *laevis*, a tetraploid species with many duplicated gene loci; they argue that too many copies of the MHC gene complex would be detrimental to T cell function.

Molecular cloning of *Xenopus* class I and II MHC is in progress. The original class I cDNA clone isolated (Flajnik, Canel *et al.* 1991a), now known to relate to a large family of non-MHC-linked class I molecules (Flajnik, Kasahara *et al.* 1993) predicted amino acid sequences similar to MHC class I molecules of higher vertebrates. However, the *Xenopus* α-3 domain is more similar to Ig-like domains of mammalian class II β chains than to those of mammalian class I molecules (Shum *et al.* 1993).

Xenopus class I alpha chains are 40–44 kDa molecules which are non-covalently bound to non-MHC-coded β_2 microglobulin (Flajnik, Kaufman *et al.* 1984). The predicted amino acid sequences of class I peptide-binding domains (α1 and α2) indicate structural similarities to certain peptide-binding heat shock proteins (Flajnik, Canel *et al.* 1991b). Class I molecules are polymorphic and expressed on the surfaces of all adult cells, highest expression being on haemopoietic cells. Class I molecules expressed by erythrocytes (where the α chain is not associated with β_2 microglobulin) and leucocytes appear to be distinct (Flajnik & Du Pasquier 1990a). Several reagents specific for class I antigens of *Xenopus* are now available; the mAb TB17 is especially useful since it can recognize cell surface class I molecules from several, but not all, strains of *X. laevis* and also LG15 clonal *Xenopus* (Flajnik, Taylor *et al.* 1991).

Xenopus class II molecules are composed of MHC-encoded α and β chains, both of which are 30–35 kDa transmembrane glycoproteins (Kaufman, Flajnik, Du Pasquier & Riegert 1985). cDNA clones for *X. laevis* MHC class II β chain genes have now been isolated (Sato *et al.* 1993). Amino acid sequence identity with representative mammalian MHC class II β chains is predicted to be nearly 50%. An invariant chain is found transiently associated with class II during biosynthesis (Kaufman, Flajnik & Du Pasquier 1991). Studies employing anti-class II mAbs have shown that the polymorphic class II proteins are expressed constitutively on only a limited range of adult cells, including thymocytes, B *and* T cells (contrast mammals) and various APCs that include putative 'Langerhans-like' cells of skin epidermis (Du Pasquier & Flajnik 1990). These dendritic cells may present antigens (in association with MHC class II) entering the skin to T cells in the vicinity; thus the inner epidermal layer contains XT-1[+ve] cells (J.D. Horton, Horton, Ritchie & Varley 1992).

Du Pasquier & Flajnik (1990) have shown that larval T cells are MHC class II[-ve], whereas larval B cells are class II[+ve]. They also reveal that class II expression is associated with the epithelial surfaces of the larval gills and gut, in addition to the skin, suggesting the possibility that class II MHC proteins may be used by diverse cell types to present conventional antigens to T cells. They suggest that such epithelial-associated MHC class II may also be important in binding native bacterial toxins and that the widespread distribution of class II in tadpoles might represent the way in which antigen was presented in a more primitive immune system.

Ontogeny

One particularly interesting feature of MHC expression in *Xenopus* is that, in contrast to the early larval appearance of class II molecules, class I proteins have not been found on the surface of any cell surface prior to metamorphosis (Flajnik & Du Pasquier 1990b). Thus class I expression is not essential for early development or for functioning of the larval immune system. Lack of surface class I on larval thymus stromal cells would appear to preclude the selection of class I-restricted cytotoxic T cells prior to metamorphosis. Flajnik & Du Pasquier (1990b) suggest that this may well be advantageous to an animal which must become immunocompetent when still possessing few lymphocytes. The larval T cell repertoire can therefore be dedicated to interacting with class II-expressing B cells (and other APCs) and also effecting class II MHC-restricted killing. It should be noted, however, that antibodies cross-reacting with class I MHC α chains are found intracytoplasmically in thymus epithelial cells of young tadpoles (Flajnik & Du Pasquier 1990a, b) and that the putative CD8 (Tc-cell 'specific'?) molecule is expressed on some larval T cells (Du Pasquier & Flajnik 1990). It has been postulated that since class I expression is absent from haemopoietic cells of the larval thymus, some T cells could retain the capacity to destroy class I^{+ve} self cells (including those whose class I proteins present class II peptides). In this way tadpole tissues expressing both class I and II molecules at metamorphosis (such as pharynx, gut and skin) may be targeted for destruction (Flajnik & Du Pasquier 1990b).

The thymus and T cell development

Thymocyte origins and thymus histogenesis

The thymus arises dorsally from the second pharyngeal pouch around day 3 after fertilization (see Manning & Horton 1982). By seven days, two major cell types are found in the thymus, the epithelial cells and the immigrant lymphoid cells. These immigrants colonize the thymus on day 4 (Tochinai 1978), having originated from the embryonic lateral plate mesoderm (Turpen *et al.* 1982).

The embryonic location of stem cells that migrate to the thymus has been probed by transplanting stem cell primordia between ploidy-distinct embryos. Flow cytometry of propidium iodide-stained nuclei (Smith, Flajnik & Turpen 1989), in addition to microdensitometry of Feulgen-stained cells and microscopic analysis of cells in metaphase, have been used to identify the ploidy of the differentiated cell populations. The *X. borealis* fluorescence cell marker has also been employed by some workers (Maeno, Tochinai & Katagiri 1985). It has emerged that two areas within the lateral plate mesoderm, one located dorsally and the other ventrally, give rise to haemopoietic cells in *Xenopus*.

The relative contribution of the dorsal stem cell compartment (DSC) and ventral mesoderm (the ventral blood island (VBI) comparable to the avian and mammalian yolk sac) to thymus lymphopoiesis (and to peripheral T cell populations) is beginning to emerge. Maeno, Tochinai *et al.* (1985) indicated that early larval thymocytes were derived from VBI, whereas DSC-derived cells were not detectable within the thymus until late larval and perimetamorphic life. In contrast, Bechtold, Smith & Turpen (1992) have recently revealed that both VBI- and DSC-derived cells enter the thymus four days after fertilization and that the thymus glands of five-week-old larvae and young adults contain a mixture of cells derived from both regions. Temporal differences in the VBI and DSC contributions to the developing larval thymus indicated by Bechtold *et al.* (1992) suggest heterogeneity within the thymocyte population, associated with the embryonic origin of the colonizing stem cells. These authors reveal that both VBI- and DSC-derived precursors contribute to the larval and adult peripheral T cell pool.

During metamorphosis the thymus is translocated to the ear region, and at the end of transition is temporarily depleted of its lymphocytes (Du Pasquier 1982). It has been shown that a wave of stem cell entry into the thymus (involving both VBI- and DSC-derived components—Bechtold *et al.* 1992) occurs during metamorphosis (between 38 and 57 days after fertilization) (Turpen & Smith 1989). Presumably this precursor immigration and subsequent wave of T cell development just after metamorphosis allows these lymphocytes to mature (be 'educated') in an environment where adult-specific antigens are expressed.

In the adult, the rapidly-proliferating cortical lymphocytes of the thymus are particularly sensitive to γ-irradiation, whereas various stromal cell types are radiation-resistant (Russ & Horton 1987). Diverse cell types containing granules, together with secretory cells, are also found in the thymus, especially in the adult medulla (Clothier & Balls 1985). IgM-secreting plasma cells have also been identified in the *Xenopus* thymus, although this organ appears not to be involved with the production of these cells (Du Pasquier *et al.* 1989). The thymus undergoes regression at sexual maturity.

Use of mAbs to monitor T cell development

Monoclonal antibodies against leucocyte-specific antigens have recently been described in *Xenopus* (Ohinata, Tochinai & Katagiri 1989; Smith & Turpen 1991), which will be useful for exploring leucocyte differentiation (e.g. see Ohinata, Tochinai & Katagiri 1990) and leucocyte interactions.

Anti-*Xenopus* MHC class II mAbs reveal that surface class II expression on thymus epithelium begins at seven days after fertilization, indicating that class II is not required to attract stem cells to the thymus (Du Pasquier & Flajnik 1990). Larval thymic lymphoid cells do not express class II MHC proteins. At seven days, the T cell differentiation antigen (XTLA-1) also begins to appear on thymic lymphoid cells (Nagata 1986): this occurs before the emergence of XT-1[+ve] cells in the periphery.

Although some larval T cells may persist beyond metamorphosis, this early T cell population is significantly diluted by expanding 'adult' populations

(Rollins-Smith, Blair & Davis 1992). The newly-formed 'adult' T lymphocytes (in thymus and spleen) now express readily-detectable surface class II MHC, although the immunological significance of this class II expression is unclear. MHC class II expression on T cells, but not the immigration of thymocyte precursors to the thymus, is dependent upon completion of metamorphosis, since expression fails to occur in larvae prevented from metamorphosing (Rollins-Smith & Blair 1990). (Such inhibition is achieved by addition of sodium perchlorate to the aquarium, which blocks thyroxine output from the thyroid.)

Sequential development of T cell functions

Thymectomy at sequential times during larval development suggests that the thymus needs to be present for only a short period to establish normal (*in vivo* and *in vitro*) alloimmune responses, whereas it is required for a considerably longer larval period to establish antibody production to T-dependent antigens and reactivity to T cell mitogens (see Manning & Horton 1982). Studies on the ontogeny of T cell-dependent immunity in intact *Xenopus* reveal that alloimmune reactivity, together with the ability of splenocytes to respond to T cell mitogens, develops early in life (when the lymphoid system contains only 0.5×10^6 lymphocytes), whereas good IgY primary antibody responses are only seen in the froglet (see discussion in Du Pasquier 1982).

Extrathymic development of 'T-like' cells?

Early thymectomy of *Xenopus* larvae dramatically impairs, but does not always eliminate, T-dependent cell functions such as allograft rejection and T cell mitogen reactivity (see review by Cohen & Turpen 1978; Manning & Horton 1982; Du Pasquier 1982). One possible explanation for such residual T cell function in Tx *Xenopus* is the early exodus of lymphocytes from the thymus prior to thymectomy, followed by subsequent expansion of these T lineage cells in the periphery. This explanation seems unlikely in view of evidence based on sequential thymectomy (Manning & Collie 1977) and because of the earliness of thymectomy in some studies. For example, *Xenopus* Tx at only four days of age (i.e. when stem cell *immigration* to the thymus is first occurring) can, as froglets, sometimes chronically reject MHC-disparate skin grafts: furthermore, following graft rejection, splenocytes of such Tx animals become able to display a non-donor-specific MLC reactivity, but are still unable to respond normally to the T cell mitogen PHA (Nagata & Cohen 1983). Therefore, a more favoured explanation (see Manning & Collie 1977) for residual 'T cell' function in Tx *Xenopus* is the existence of an alternative, extrathymic maturation pathway. Such 'T-independent T-like' cells (which often express γδ T cell receptors) are frequently found in aged nude ('thymusless') mammals and are currently receiving considerable attention by developmental immunologists (see Bell 1989).

The possibility that early removal of the thymus in anurans activates

Table 15.1. Ontogenetic changes in mAb-defined T and B cell populations in the spleen: comparison of control and thymectomized *X. laevis*

Experimental group	Age (months)	Mean (± SD) percentage of mAb-stained positive cells[a]			
		2B1	F17/AM22	XT-1	8E4:57
Controls	1 (larva)	37±7(2)[b]	17(1)	13±7(2)	54±3(2)
	2–5	47±6(9)	23±4(7)	23±7(4)	38±8(4)
	6–9	63±8(5)	25±9(7)	25±5(5)	20±5(6)
	10–14	62±3(7)	26±3(5)	31(1)	29±4(2)
Tx (5–7 day)	1 (larva)	4±1(4)	4±1(4)	3±1(4)	76±7(3)
	2–5	4±1(4)	3±2(4)	3±1(4)	72±5(3)
	6–9	8±3(3)	5±2(7)	6±3(7)	48±3(4)
	10–14	17±2(3)	8±1(2)	N.D.	55(1)

Splenocytes stained and analysed as described in Fig. 1, using either a FACSCAN or Coulter Epics cytometer. For mAB details – see text.
[a]Marker set to exclude 98–99% background staining.
[b]Nos. in parentheses show number of experiments. Two to three spleens used for each experiment on larvae or 2–5 month froglets, single spleen used in each experiment on older animals.
For source of data – see Fig. 1.

alternative pathways of lymphopoiesis was supported some years ago by experiments employing VBI-grafted, intact and Tx *Rana pipiens* (Turpen & Cohen 1976; Cohen & Turpen 1980). Whether such putative extrathymic lymphopoietic pathways include the differentiation of 'T-like' cells is now being explored in Tx *Xenopus*, by utilizing a panel of anti-T cell mAbs. Our preliminary experiments (Table 1, Fig. 15.2) revealed that the spleen of both larval and young (2–5-month-old) post-metamorphic thymectomized (Tx at 5–7 days) *Xenopus* had virtually undetectable levels of 2B1[+ve], XT-1[+ve] or CD8[+ve] T cells as judged by flow cytometric analysis, whereas sIg[+ve] B lymphocytes were plentiful. In contrast, a small but distinct population of splenocytes that express T cell markers emerged in six-month-old Tx *X. laevis* (Table 15.1, Fig. 15.2). Although somewhat higher percentages of such 'T-like' cells were evident by 10–14 months of age (Table 15.1) in these preliminary experiments, we have more recently in dual colour flow cytometric studies found no evidence for T cell markers in spleen or gut following thymectomy (J.D. Horton & M.D. Cooper in prep.).

The emergence of a small population of mAb-defined 'T-like' lymphocytes in the spleen of some six-month-old Tx animals was coincident with the first appearance of residual splenocyte reactivity (stimulation indices of <10% of those found with control cells) towards T cell mitogens (Ho 1992). Twelve-month-old thymectomized animals that lack T cell marker expression fail to respond to T cell mitogens. Thus although T-like lymphocytes can arise in some thymectomized *Xenopus*, one scenario following early thymic ablation is absence of any T cell development.

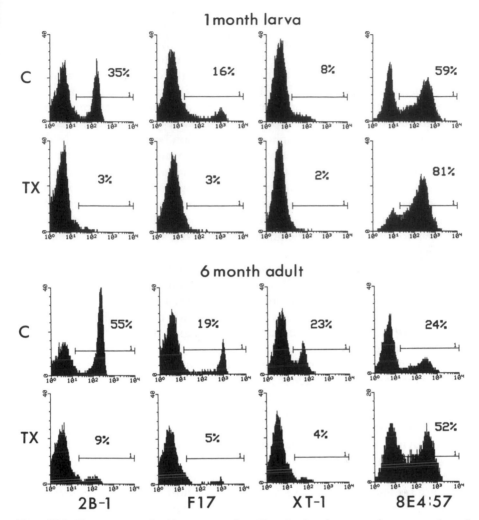

Fig. 15.2. Flow cytometric histograms of mAb-stained splenocytes from control and thymectomized *Xenopus*: comparison of larval and adult profiles.

Analyses were performed on mAb-stained splenic lymphocytes from pooled, one-month-old larvae, and individual six-month-old adults. Profiles of control and thymectomized (Tx at 5 days) *X. laevis* are compared. Note that, in these preliminary experiments, 'T-like' cells emerge in six-month-old Tx animals. These cells are seen as distinct peaks with the 2B1, F17 and XT-1 mAbs, with fluorescence intensities similar to those found with control T cells. Splenocytes stained and analysed as described in Fig. 1. The percentage of positive cells in region 1 (that excludes 98–99% background staining) is given for each individual mAb. x axis, four-decade log fluorescence intensity; y axis, relative cell number. (For source of data, see Fig. 1.)

Thymus 'education' of T cells

Implantation of thymus, which may be either lymphoid or lymphocyte-depleted following γ-irradiation, to early-thymectomized (larval or adult) *Xenopus* reveals that foreign thymus grafts can promote the differentiation of host precursor cells along a T cell pathway (Nagata & Cohen 1984). Use of xenogeneic donor/host combinations demonstrates the persistence of a range of donor thymic stromal cell types in the implant (J.D. Horton, Russ *et al.* 1987), whereas the thymus lymphocyte population becomes host-derived. The *in vivo* construction of thymuses with MHC-disparate epithelial and lymphoid compartments can readily be achieved by a different surgical approach that is feasible with the amphibian embryo. This approach involves joining the anterior part of one 24-h *Xenopus* embryo, containing the thymic epithelial buds, to the posterior portion of an MHC-incompatible embryo, from which the haemopoietic stem cells, including lymphocytes, arise (Flajnik, Du Pasquier & Cohen 1985).

These two experimental approaches have been used to explore the role which is played by thymic stromal cells in establishing tolerance (negative selection) of T-lineage lymphocytes to self- and allo-antigens, and in restricting the MHC antigen specificities (positive selection) with which helper and effector T cell populations preferentially interact during recognition and/or destruction of non-MHC antigens.

It appears that foreign thymic epithelium is involved in inducing tolerance of the host towards subsequent skin grafts of thymus MHC type (Katagiri & Tochinai 1987; Du Pasquier & Horton 1982; Nagata & Cohen 1984; Flajnik, Du Pasquier *et al.* 1985; Arnall & Horton 1986; Maeno, Nakamura *et al.* 1987). One of the above reports (Maeno, Nakamura *et al.* 1987) on allothymus-induced tolerance indicated that this tolerance is effected by selective deletion of anti-donor T lymphocytes in the thymus. However, it has generally been found that tolerance induced to skin of thymus donor type does *not* prevent a proliferative response when host splenic lymphocytes (and even host thymocytes in the foreign thymus implant) are cultured with donor stimulators in one-way MLC. Presumably, peripheral suppressive or anergic mechanisms are at play in these 'split-tolerance' situations.

The thymectomy/allothymus implantation approach failed to reveal a clearcut involvement of the thymus in MHC restriction: for example, T-B collaboration appears relatively normal in such *Xenopus* (Du Pasquier & Horton 1982). In contrast, studies on MHC-mismatched chimeric *Xenopus* have clearly shown that IgY responses are impaired, thereby suggesting that T cells educated by the foreign thymus epithelium cannot interact effectively with B cells of non-thymus MHC type. These experiments also suggested that the thymus is involved in selection of T cells that preferentially visualize minor histocompatibility antigens in the context of thymus MHC type (Flajnik, Du Pasquier *et al.* 1985).

Ontogeny of transplantation tolerance and metamorphosis

Studies on embryos and larvae

Although the thymus plays a crucial role in self-tolerance induction, it is clearly not the exclusive site where T cell tolerance occurs. Transplantation of tissues to embryonic, larval and metamorphic stages of amphibians is readily feasible and is providing valuable information on the cellular and genetic basis of such 'peripheral' tolerance. Most of this work has been done with allogeneic tissues, which by extrapolation provide information on tolerance induction to self. Some experiments, however, have directly probed *self* tolerance. For example, the demonstration that *Xenopus* grown without eyes or pituitaries are tolerant of subsequently-grafted, appropriate organs from either isogeneic or MHC-compatible *Xenopus* (Rollins-Smith & Cohen 1983; Maeno & Katagiri 1984) refuted the earlier, much quoted work of Triplett (1962), and illustrated that self-tolerance is not restricted to an early phase of ontogeny.

Immunocompetent *Xenopus* larvae, but not adults, can be rendered tolerant by transplantation of allogeneic (e.g. one MHC-haplotype disparate) skin from adult donors (see review by Cohen, DiMarzo *et al.* 1985). Some authors have found that allotolerance induction is particularly easy to achieve at metamorphosis (Du Pasquier & Chardonnens 1975). Others have found that the size of skin grafts applied to the larva and the degree of histoincompatibility appear to be critical to the outcome. For example, minor H antigen-disparate skin grafts are routinely tolerated by larval and perimetamorphic *Xenopus* (DiMarzo & Cohen 1982; Obara, Kawahara & Katagiri 1983), whereas a high degree of rejection occurs to two MHC-haplotype disparate skin grafts. We have recently found (J.D. Horton, Horton & Ritchie 1993) that the transplantation of *larval* (MHC class I^{-ve}) allogeneic skin or spleen (1 or 2 MHC-disparate) to larval *Xenopus* (LG clonal animals) can lead to donor-specific skin tolerance in the adult. Larval skin proves less effective in this respect than the spleen, which could well relate to the dearth of class II MHC-positive cells in larval skin (see Du Pasquier & Flajnik 1990) compared with the class II-rich cells found in larval spleen.

The tolerance induced in the foregoing experiments is judged on the basis of skin tolerance, specific to the donor, induced to secondary skin grafts applied to young adults. In general this tolerance is incomplete (or 'split' —see also p. 290) as it does not cause deletion of lymphocytes reactive with the tolerizing MHC (Cohen, DiMarzo *et al.* 1985). Such anti-donor MLC-reactive spleen cells can also be visualized *in vivo* (Arnall & Horton 1987), and tolerated skin grafts are subjected to some degree of T lymphocyte infiltration (Fig. 15.3) and increased MHC class II expression (J.D. Horton, Horton & Ritchie 1993). Failure to reject test skin grafts in these allotolerant animals appears to be maintained by active suppression (of cytotoxic T cells?—Arnall & Horton 1987), as evidenced by breakdown of tolerance following cyclophosphamide treatment (J.D. Horton, Horton, Varley & Ruben 1989), and the finding

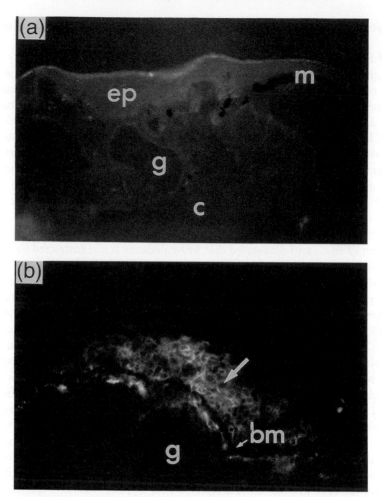

Fig. 15.3. Immunohistology of 'tolerated' skin allograft. Immunofluorescence of cryostat sections (6 μm) through (a) LG15 (MHC haplotype = ac) skin isograft and (b) LG5 (MHC haplotype = bc) skin allograft, three weeks after grafting to LG15, five-month-old *Xenopus*. Recipients had previously been implanted in larval life with a spleen from a larval LG5 donor, a procedure that is routinely effective at inducing permanent, donor-specific skin graft 'tolerance' in the adult. Both sections were stained with the mouse mAb AM22 (anti-CD8) and secondary, FITC-labelled anti-mouse Ig antibody. In (a) (×120) note the absence of staining within the isograft (ep, ventral skin graft epidermis; g, graft gland; c, collagen pad; m, melanin layer of adjoining dorsal host skin). In (b) (×240) patches of allograft epidermis are replete with AM22[+ve] cells (arrowed) (bm, basement membrane lying under the epidermis; g, graft gland). The infiltration by CD8[+ve] cells of the 'tolerated' foreign skin graft, which appeared in excellent health when examined externally, reveals some degree of persistent T cell reactivity towards the semi-allogeneic donor. (Data from J.D. Horton, Horton & Ritchie 1993.)

that tolerance can be transferred with splenic lymphoid cells (Nakamura *et al.* 1987).

It would seem that complete tolerance (*in vivo* and *in vitro*) can only be achieved when embryonic stem cells are transplanted, and when animals become chimeric, as in embryonic flank transplantation studies (Manning & Botham 1980). It has been suggested (Flajnik & Du Pasquier 1990a) that such tolerance may be achieved by haemopoietically-derived antigen-presenting cells. The latter may effect tolerance peripherally, but could conceivably gain access to the thymus and achieve 'central' deletion of appropriate T cell specificities. 'Split-tolerance', on the other hand, can be established by various class II[+ve] tissues that are introduced during early ontogeny, and is maintained through suppression or anergy (Houssaint & Flajnik 1990).

Metamorphosis

The question of how self-tolerance to adult-specific determinants is generated at metamorphosis continues to be explored in some depth in *Xenopus* (Flajnik, Hsu *et al.* 1987). The involvement of suppressor cell populations has been suggested, since lymphocytes (e.g. thymocytes) from transforming animals (but not adults) can suppress the response of young isogeneic froglets to minor histocompatibility-disparate grafts (Du Pasquier & Bernard 1980). This work was also important in showing that T memory cells were not as sensitive to suppression as cells responding in a first-set reaction: this would allow immune experiences of the larva to be remembered after metamorphosis. Thymectomy of five-week-old *Xenopus* larvae impairs the ease with which transplantation tolerance can be achieved at metamorphosis, possibly by lowering T suppressor cell numbers in the periphery (Barlow & Cohen 1983).

Hormonal control of altered immunoregulation at metamorphosis also appears likely (Ruben, Clothier *et al.* 1989). Thyroid hormones provide the driving force for both the regressive (removal of larval structures) and subsequent progressive (appearance of adult structures) developmental alterations at metamorphosis (Rollins-Smith, Parsons & Cohen 1988). High titres of thyroid hormones at metamorphosis are accompanied by a comparable dramatic transitory increase in both corticosteroid serum titres and corticosteroid receptors (especially on thymic and splenic lymphocytes). It has been suggested that corticosteroids directly inhibit T cell clonal expansion at metamorphosis through interference with regulatory cytokines (Ruben, Clothier *et al.* 1989). Despite corticosteroid inhibition of T cell proliferation, splenocytes of metamorphosing animals (but not adults) secrete *in vitro* 'thymus-replacing' factors (Ruben, Marshall *et al.* 1991). By acting directly on B cells, such factors would allow humoral responses to thymus-dependent pathogens to continue during metamorphosis, when T cell help is impaired.

Further studies on interplay between endocrine and immune systems (e.g. see Clothier, Ruben *et al.* 1992) in the regulation of cellular and humoral immunity in *Xenopus* are likely to shed more light on how amphibians escape the risk of dying from autoimmune disease at metamorphosis.

Acknowledgements

Thanks go to Paul Loftus (animal husbandry), David Hutchinson (photography) and to Ian Dimmick (Dryburn Hospital) and John Robinson (Newcastle University Medical School) for FACS facilities. The generosity of Max Cooper, Louis Du Pasquier, Martin Flajnik and Saburo Nagata in supplying mAbs is gratefully acknowledged. Grants from The Wellcome Trust, The Royal Society and the Nuffield Foundation are also gratefully acknowledged.

References

Arnall, J.C. & Horton, J.D. (1986). Impaired rejection of minor-histocompatibility-antigen-disparate skin grafts and acquisition of tolerance to thymus donor antigens in allothymus-implanted, thymectomized *Xenopus*. *Transplantation* **41**: 766–776.

Arnall, J.C. & Horton, J.D. (1987). *In vivo* studies on allotolerance perimetamorphically induced in control and thymectomized *Xenopus*. *Immunology* **62**: 315–319.

Barlow, E.H. & Cohen, N. (1983). The thymus dependency of transplantation allotolerance in the metamorphosing frog *Xenopus laevis*. *Transplantation* **35**: 612–619.

Bechtold, T.E., Smith, P.B. & Turpen, J.B. (1992). Differential stem cell contributions to thymocyte succession during development of *Xenopus laevis*. *J. Immunol.* **148**: 2975–2982.

Bell, E.B. (1989). Thymus-derived and non-thymus-derived T-like cells: the origin and function of cells bearing γδ receptors. *Thymus* **14**: 3–17.

Bernard, C.C.A., Bordmann, G., Blomberg, B. & Du Pasquier, L. (1979). Immunogenetic studies on the cell-mediated cytotoxicity in the clawed toad *Xenopus laevis*. *Immunogenetics* **9**: 443–454.

Bernard, C.C.A., Bordmann, G., Blomberg, B. & Du Pasquier, L. (1981). Genetic control of T helper cell function in the clawed toad *Xenopus laevis*. *Eur. J. Immunol.* **11**: 151–155.

Bleicher, P.A. & Cohen, N. (1981). Monoclonal anti-IgM can separate T cell from B cell proliferative responses in the frog, *Xenopus laevis*. *J. Immunol.* **127**: 1549–1555.

Clothier, R.H., Ali, I., Quaife, Y., Naha, B. & Balls, M. (1989). Skin xenograft rejection in *Xenopus laevis*, the South African toad. *Herpetopathologia* **1**: 19–28.

Clothier, R.H. & Balls, M. (1985). Structural changes in the thymus glands of *Xenopus laevis* during development. In *Metamorphosis*: 332–359. (Eds Balls, M. & Bownes, M.E.). Clarendon Press, Oxford.

Clothier, R.H., Balls, M. & Ruben, L.N. (1989). The immune system in *Xenopus laevis* (the South African clawed toad) after exposure to N-methyl-N-nitrosourea. *Herpetopathologia* **1**: 81–90.

Clothier, R.H., Last, Z., Somauroo, J.D., Ruben, L.N. & Balls, M. (1989). Differential cyclophosphamide sensitivity of suppressor function in *Xenopus*, the clawed toad. *Devl comp. Immunol.* **13**: 159–166.

Clothier, R.H., Ruben, L.N., Johnson, R.O., Parker, K., Sovak, M., Greenhalgh, L., Ooi, E.E. & Balls, M. (1992). Neuroendocrine regulation of immunity: the effects of noradrenaline in *Xenopus laevis*, the South African clawed toad. *Int. J. Neurosci.* 62: 123–140.

Cohen, N., DiMarzo, S., Rollins-Smith, L., Barlow, E. & Vanderschmidt-Parsons, S. (1985). The ontogeny of allo-tolerance and self-tolerance in larval *Xenopus laevis*. In *Metamorphosis*: 388–419. (Eds Balls, M. & Bownes, M.E.). Clarendon Press, Oxford.

Cohen, N., Effrige, N.J., Parsons, S.C.V., Rollins-Smith, L.A., Nagata, S. & Albright, D. (1984). Identification and treatment of a lethal nematode (*Capillaria xenopodis*) infestation of the South African frog, *Xenopus laevis*. *Devl comp. Immunol.* 8: 739–741.

Cohen, N. & Haynes, L. (1991). The phylogenetic conservation of cytokines. In *Phylogeny of immune functions*: 231–268. (Eds Warr, G. & Cohen, N.). CRC Press, Boca Raton.

Cohen, N. & Turpen, J.B. (1978). Early ontogeny of heterogeneous populations of lymphocytes in anuran amphibians. In *Animal models of comparative and developmental aspects of immunity and disease*: 37–47. (Eds Gershwin, M.E. & Cooper, E.L.). Pergamon Press, New York.

Cohen, N. & Turpen, J.B. (1980). Experimental analysis of lymphocyte ontogeny and differentiation in an amphibian model system. In *Biological basis of immunodeficiency*: 25–37. (Eds Gelfand, E.W. & Dosch, H.M.). Raven Press, New York.

DiMarzo, J.J. & Cohen, N. (1982). Immunogenetic aspects of *in vivo* allotolerance induction during the ontogeny of *Xenopus laevis*. *Immunogenetics* 16: 103–116.

Du Pasquier, L. (1982). Ontogeny of immunological functions in amphibians. In *The reticuloendothelial system: phylogeny and ontogeny*: 633–657. (Eds Cohen, N. & Sigel, M.M.). Plenum, New York.

Du Pasquier, L. & Bernard, C.C.A. (1980). Active suppression of the allogeneic histocompatibility reactions during the metamorphosis of the clawed toad *Xenopus*. *Differentiation* 16: 1–7.

Du Pasquier, L. & Chardonnens, X. (1975). Genetic aspects of the tolerance to allografts induced at metamorphosis in the toad *Xenopus laevis*. *Immunogenetics* 2: 431–440.

Du Pasquier, L. & Flajnik, M.F. (1990). Expression of MHC class II antigens during *Xenopus* development. *Dev. Immunol.* 1: 85–95.

Du Pasquier, L. & Horton, J.D. (1982). Restoration of antibody responsiveness in early thymectomized *Xenopus* by implantation of major histocompatibility complex-mismatched larval thymus. *Eur. J. Immunol.* 12: 546–551.

Du Pasquier, L., Schwager, J. & Flajnik, M.F. (1989). The immune system of *Xenopus*. *A. Rev. Immunol.* 7: 251–275.

Flajnik, M.F., Canel, C., Kramer, J. & Kasahara, M. (1991a). Evolution of the major histocompatibility complex: molecular cloning of MHC class I from the amphibian *Xenopus*. *Proc. natn. Acad. Sci. U.S.A.* 88: 537–541.

Flajnik, M.F., Canel, C., Kramer, J. & Kasahara, M. (1991b). Which came first, MHC class I or class II? *Immunogenetics* 33: 295–300.

Flajnik, M.F. & Du Pasquier, L. (1990a). The major histocompatibility complex of frogs. *Immunol. Rev.* No. 113: 47–63.

Flajnik, M.F. & Du Pasquier, L. (1990b). Changes in the expression of the major histocompatibility complex during the ontogeny of *Xenopus*. In *Developmental biology: UCLA symposium on molecular and cellular biology*: 215–224. (Eds Davidson, E., Rudeman, J. & Posakony, J.). Alan R. Liss, New York.

Flajnik, M.F., Du Pasquier, L. & Cohen, N. (1985). Immune responses of thymus/ lymphocyte embryonic chimeras: studies on tolerance and major histocompatibility complex restriction in *Xenopus*. *Eur. J. Immunol.* **15**: 540–547.

Flajnik, M.F., Ferrone, S., Cohen, N. & Du Pasquier, L. (1990). Evolution of the MHC: antigenicity and unusual tissue distribution of *Xenopus* (frog) class II molecules. *Molec. Immunol.* **27**: 451–462.

Flajnik, M.F., Hsu, E., Kaufman, J.F. & Du Pasquier, L. (1987). Changes in the immune system during metamorphosis of *Xenopus*. *Immunol. Today* **8**: 58–64.

Flajnik, M.F., Kasahara, M., Shum, B.P., Salter-Cid, L., Taylor, E. & Du Pasquier, L. (1993). A novel type of class I gene organization in vertebrates: a large family of non-MHC-linked class I genes is expressed at the RNA level in the amphibian *Xenopus*. *EMBO J.* **12**: 4385–4396.

Flajnik, M.F., Kaufman, J.F., Riegert, P. & Du Pasquier, L. (1984). Identification of class I major histocompatibility complex encoded molecules in the amphibian *Xenopus*. *Immunogenetics* **20**: 433–442.

Flajnik, M.F., Taylor, E., Canel, C., Grossberger, D. & Du Pasquier, L. (1991). Reagents specific for MHC class I antigens of *Xenopus*. *Am. Zool.* **31**: 580–591.

Harding, F.A. (1990). *Molecular and cellular aspects of the immune systems of lower vertebrates*. PhD thesis: University of Rochester, New York.

Haynes, L. & Cohen, N. (1993). Further characterization of an interleukin-2-like cytokine produced by *Xenopus laevis* T lymphocytes. *Devl Immunol.* **3**: 231–238.

Haynes, L., Harding, F.A., Koniski, A.D. & Cohen, N. (1992). Immune system activation associated with a naturally occurring infection in *Xenopus laevis*. *Devl comp. Immunol.* **16**: 453–462.

Haynes, L., Moynihan, J.A. & Cohen, N. (1991). A monoclonal antibody against the human IL-2 receptor binds to paraformaldehyde-fixed but not viable frog (*Xenopus*) splenocytes. *Immunol. Lett.* **26**: 227–232.

Ho, J. (1992). *Functional studies of cytokines produced by Xenopus lymphocytes following T cell mitogen stimulation*. MSc thesis: University of Durham, U.K.

Horton, J.D. (1994). Amphibian immunology. In *Immunology: a comparative approach*: 101–136. (Eds Turner, R.J. & Manning, M.J.). Wiley, Chichester.

Horton, J.D., Horton, T.L. & Ritchie, P. (1993). Incomplete tolerance induced in *Xenopus* by larval tissue allografting: evidence from immunohistology and mixed leucocyte culture. *Devl comp. Immunol.* **17**: 249–262.

Horton, J.D., Horton, T.L., Ritchie, P. & Varley, C.A. (1992). Skin xenograft rejection in *Xenopus*: immunohistology and effect of thymectomy. *Transplantation* **53**: 473–476.

Horton, J.D., Horton, T.L., Varley, C.A & Ruben, L.N. (1989). Attempts to break perimetamorphically induced skin graft tolerance by treatment of *Xenopus* with cyclophosphamide and interleukin-2. *Transplantation* **47**: 883–887.

Horton, J.D., Russ, J.H., Aitchison, P. & Horton, T.L. (1987). Thymocyte/stromal cell chimaerism in allothymus-grafted *Xenopus*: developmental studies using the X. *borealis* fluorescence marker. *Development* **100**: 107–117.

Horton, T.L., Horton, J.D. & Varley, C.A. (1989). *In vitro* cytotoxicity in adult *Xenopus* generated against larval targets and minor histocompatibility antigens. *Transplantation* **47**: 880–882.

Houssaint, E. & Flajnik, M.F. (1990). The role of thymic epithelium in the acquisition of tolerance. *Immunol. Today* 11: 357–360.

Hsu, E., Julius, M.H. & Du Pasquier, L. (1983). Effector and regulator functions of splenic and thymic lymphocytes in the clawed toad *Xenopus*. *Annls Inst. Pasteur, Immunol.* 134D: 277–292.

Ibrahim, B., Gartland, L.A., Kishimoto, T., Dzialo, R., Kubagawa, H., Bucy, R.P. & Cooper, M.D. (1991). Analysis of T cell development in *Xenopus*. *Fed. Proc.* 5: 7651.

Katagiri, C. & Tochinai, S. (1987). Ontogeny of thymus-dependent immune responses and lymphoid cell differentiation in *Xenopus laevis*. *Dev. Growth Differ.* 29: 297–305.

Kaufman, J., Flajnik, M.F. & Du Pasquier, L. (1991). The MHC molecules of ectothermic vertebrates. In *Phylogenesis of immune functions*: 125–149. (Eds Warr, G. & Cohen, N.). CRC Press, Boca Raton.

Kaufman, J.F., Flajnik, M.F., Du Pasquier, L. & Riegert, P. (1985). *Xenopus* MHC class II molecules. I. Identification and structural characterization. *J. Immunol.* 134: 3248–3257.

Langeberg, L., Ruben, L.N., Malley, A., Shiigi, S. & Beadling, C. (1987). Toad splenocytes bind human IL-2 and anti-human IL-2 receptor antibody specifically. *Immunol. Lett.* 14: 103–110.

Maeno, M. & Katagiri, C. (1984). Elicitation of weak immune response in larval and adult *Xenopus laevis* by allografted pituitary. *Transplantation* 38: 251–255.

Maeno, M., Nakamura, T., Tochinai, S. & Katagiri, C. (1987). Analysis of allotolerance in thymectomized *Xenopus* restored with semiallogeneic thymus grafts. *Transplantation* 44: 308–314.

Maeno, M., Tochinai, S. & Katagiri, C. (1985). Differential participation of ventral and dorsolateral mesoderms in the hemopoiesis of *Xenopus*, as revealed in diploid-triploid or interspecific chimeras. *Devl Biol.* 110: 503–508.

Manning, M.J. & Botham, P.A. (1980). The *in vitro* reactivity of lymphocytes in embryonically-induced transplantation tolerance. In *Development and differentiation of vertebrate lymphocytes*: 215–226. (Ed. Horton, J.D.). Elsevier, Amsterdam.

Manning, M.J. & Collie, M.H. (1977). The ontogeny of thymic dependence in the amphibian *Xenopus laevis*. In *Developmental immunobiology*: 291–298. (Eds Solomon, J.B. & Horton, J.D.). Elsevier, Amsterdam.

Manning, M.J. & Horton, J.D. (1982). RES structure and function of the Amphibia. In *The reticuloendothelial system: phylogeny and ontogeny*: 423–459. (Eds Cohen, N. & Sigel, M.M.). Plenum, New York.

Nagata, S. (1986). Development of T lymphocytes in *Xenopus laevis*: appearance of the antigen recognized by an anti-thymocyte mouse monoclonal antibody. *Devl Biol.* 114: 389–394.

Nagata, S. (1988). T cell-specific antigen in *Xenopus* identified with a mouse monoclonal antibody: biochemical characterization and species distribution. *Zool. Sci.* 5: 77–83.

Nagata, S. & Cohen, N. (1983). Specific *in vivo* and nonspecific *in vitro* alloreactivities of adult frogs (*Xenopus laevis*) that were thymectomized during early larval life. *Eur. J. Immunol.* 13: 541–545.

Nagata, S. & Cohen, N. (1984). Induction of T cell differentiation in early-thymectomized *Xenopus* by grafting adult thymuses from either MHC-matched or from partially or totally MHC-mismatched donors. *Thymus* 6: 89–104.

298   *John D. Horton, Trudy L. Horton and Pamela Ritchie*

Nakamura, T. (1985). Lethal graft-versus-host reaction induced by parental cells in the clawed frog, *Xenopus laevis. Transplantation* 40: 393–397.

Nakamura, T., Maeno, M., Tochinai, S. & Katagiri, C. (1987). Tolerance induced by grafting semi-allogeneic adult skin to larval *Xenopus laevis*: possible involvement of specific suppressor cell activity. *Differentiation* 35: 108–114.

Obara, N., Kawahara, H. & Katagiri, C. (1983). Response to skin grafts exchanged among siblings of larval and adult gynogenetic diploids in *Xenopus laevis. Transplantation* 36: 91–95.

Obara, N., Tochinai, S. & Katagiri, C. (1982). Splenic white pulp as a thymus-independent area in the African clawed toad, *Xenopus laevis. Cell Tissue Res.* 226: 327–335.

Ohinata, H., Tochinai, S. & Katagiri, C. (1989). Ontogeny and tissue distribution of leukocyte-common antigen bearing cells during early development of *Xenopus laevis. Development* 107: 445–452.

Ohinata, H., Tochinai, S. & Katagiri, C. (1990). Occurrence of nonlymphoid leukocytes that are not derived from blood islands in *Xenopus laevis* larvae. *Devl Biol.* 141: 123–129.

Rollins-Smith, L.A. & Blair, P.J. (1990). Expression of class II major histocompatibility complex antigens on adult T cells in *Xenopus* is metamorphosis-dependent. *Devl Immunol.* 1: 97–104.

Rollins-Smith, L.A., Blair, P.J. & Davis, A.T. (1992). Thymus ontogeny in frogs: T cell renewal at metamorphosis. *Devl Immunol.* 2: 207–213.

Rollins-Smith, L.A. & Cohen, N. (1983). The Triplett phenomenon revisited: self-tolerance is not confined to the early developmental period. *Transpl. Proc.* 15: 871–874.

Rollins-Smith, L.A., Parsons, S.C.V. & Cohen, N. (1988). Effects of thyroxine-driven precocious metamorphosis on maturation of adult-type allograft rejection responses in early thyroidectomized frogs. *Differentiation* 37: 180–185.

Ruben, L.N. (1986). Recombinant DNA produced human IL-2, injected *in vivo*, will substitute for carrier priming of helper function in the South African clawed toad, *Xenopus laevis. Immunol. Lett.* 13: 227–230.

Ruben, L.N., Clothier, R.H., Horton, J.D. & Balls, M. (1989). Amphibian metamorphosis: an immunologic opportunity. *Bioessays* 10: 8–12.

Ruben, L.N., Langeberg, L., Malley, A., Clothier, R.H., Beadling, C., Lee, R. & Shiigi, S. (1990). A monoclonal mouse anti-human IL-2 receptor antibody (anti-Tac) will recognise molecules on the surface of *Xenopus laevis* immunocytes which specifically bind rIL-2 and are only slightly larger than the human Tac protein. *Immunol. Lett.* 24: 117–126.

Ruben, L.N., Marshall, J.D., Langeberg, L., Johnson, R.O. & Clothier, R.H. (1991). Thymus-replacing activity from the metamorphic spleens of *Xenopus laevis. Cytokine* 3: 28–34.

Russ, J.H. & Horton, J.D. (1987). Cytoarchitecture of the *Xenopus* thymus following γ-irradiation. *Development* 100: 95–105.

Sato, K., Flajnik, M.F., Du Pasquier, L., Katagiri, M. & Kasahara, M. (1993). Evolution of the MHC: isolation of class II β-chain cDNA clones from the amphibian *Xenopus laevis. J. Immunol.* 150: 2831–2843.

Shum, B.P., Avila, D., Du Pasquier, L., Kasahara, M. & Flajnik, M.F. (1993). Isolation of a classical MHC class I cDNA from an amphibian. *J. Immunol.* 151: 5376–5386.

Smith, P.B., Flajnik, M.F. & Turpen, J.B. (1989). Experimental analysis of ventral blood island hematopoiesis in *Xenopus* embryonic chimeras. *Devl Biol.* **131**: 302–312.

Smith, P.B. & Turpen, J.B. (1991). Expression of a leukocyte-specific antigen during ontogeny in *Xenopus laevis*. *Devl Immunol.* **1**: 295–307.

Tochinai, S. (1978). Thymocyte stem cell inflow in *Xenopus laevis* after grafting diploid thymic rudiments into triploid tadpoles. *Devl comp. Immunol.* **2**: 627–635.

Triplett, E.L. (1962). On the mechanism of immunologic self recognition. *J. Immunol.* **89**: 505–510.

Turner, S.L., Horton, T.L., Ritchie, P. & Horton, J.D. (1991). Splenocyte response to T cell-derived cytokines in thymectomized *Xenopus*. *Devl comp. Immunol.* **15**: 319–328.

Turpen, J.B. & Cohen, N. (1976). Alternative sites of lymphopoiesis in the amphibian embryo. *Annls Immunol.* **127C**: 841–848.

Turpen, J.B., Cohen, N., Deparis, P., Jaylet, A., Tompkins, R. & Volpe, E.P. (1982). Ontogeny of amphibian haemopoietic cells. In *The reticuloendothelial system: phylogeny and ontogeny*: 569–588. (Eds Cohen, N. & Sigel, M.M.). Plenum Press, New York.

Turpen, J.B. & Smith, P.B. (1989). Precursor immigration and thymocyte succession during larval development and metamorphosis in *Xenopus*. *J. Immunol.* **142**: 41–47.

Varley, C.A. & Horton, J.D. (1991). Lack of expression of the T cell marker XTLA-1 in *Xenopus tropicalis*: exploitation in thymus restoration studies. *Devl comp. Immunol.* **15**: 307–317.

Watkins, D. (1985). *T cell function in* Xenopus: *studies on T cell ontogeny and cytotoxicity using an IL-2-like growth factor*. PhD thesis: University of Rochester, New York.

Watkins, D. & Cohen, N. (1987a). Mitogen-activated *Xenopus laevis* lymphocytes produce a T-cell growth factor. *Immunology* **62**: 119–125.

Watkins, D. & Cohen, N. (1987b). Description and partial characterization of a T-cell growth factor from the frog, *Xenopus laevis*. In *Immune regulation by characterized polypeptides*: 495–508. (Eds Goldstein, G., Bach, J.F. & Wizzell, H.). Alan R. Liss, New York.

Watkins, D., Harding, F. & Cohen, N. (1988). *In vitro* proliferative and cytotoxic responses against *Xenopus* minor histocompatibility antigens. *Transplantation* **45**: 499–501.

Watkins, D., Parsons, S.V. & Cohen, N. (1988). The ontogeny of interleukin production and responsivity in the frog, *Xenopus*. *Thymus* **11**: 113–122.

16 The immune system of *Xenopus*: with special reference to B cell development and immunoglobulin genes

LOUIS DU PASQUIER, MELANIE WILSON and JACQUES ROBERT

Synopsis

The immunoglobulin genes of *Xenopus* are organized in a manner similar to those of bony fish, birds and mammals, with multiple families of variable region elements, multiple Ds (for the heavy chain), multiple J segments, but a single gene for each constant region. B cells, in which these gene segments rearrange to build up functional immunoglobulin genes, develop early in larval life and express a repertoire of antibodies different from that of the adult and restricted by the small numbers of lymphocyte clones. The comparison of larval and adult cDNAs showed that the N diversification which contributes to the adult complexity of the third hypervariable region of heavy chains is missing in tadpoles. The adult repertoire is less heterogeneous than that of mammals, presumably because somatic hypermutations occurring in the V genes are not properly selected. Some polyploid species retained polysomic inheritance of immunoglobulin genes, whereas they functionally diploidized their major histocompatibility complex. A lymphoid cell line with a larval phenotype with two complete Ig heavy chain rearrangements has been isolated from a thymus tumour and grown *in vitro*: while expressing cell surface markers of a T cell, its immunoglobulin gene rearrangements are more typical of a B cell. This may contribute a new tool for *Xenopus* biologists in the future.

B cells and immunoglobulin genes in *Xenopus*

B cells and immunoglobulins

The classical T–B heterogeneity of lymphocytes can be detected in *Xenopus*. The immune system of *Xenopus*, as in mammals, can be divided into thymus-dependent and thymus-independent responses (Horton *et al.* 1979). The surface immunoglobulin (Ig) positive B cells can be found in all lymphoid organs. They represent a prominent fraction of gut lymphoid tissues in the adult, about 25% of spleen cells, variable fractions of kidney and liver

Table 16.1. Immunoglobulins of *Xenopus*

	Heavy[a] chains			Light[b] chains		
	IgM	IgX	IgY	1	2	3
MW + carbohydrate	73 000	80 000	69 000			
Without N linked carbohydrates	61 000	64 000	66 000	25 000	27 000	29 000

[a]Data from Hsu & Du Pasquier (1984c).
[b]Data from Hsu, Lefkovits *et al.* (1991).

lymphocytes, circulating lymphocytes and a small fraction of thymus cells. In their final differentiation stage, the plasma cells are found in large numbers in the gut and the liver, but also in smaller quantities in the other lymphoid organs like the thymus. *Xenopus* does not possess lymph nodes and its bone marrow is not lymphopoietic. So far, *Xenopus* B cells are characterized only by their surface Ig markers via monoclonal antibodies against heavy and light chain epitopes (Hsu & Du Pasquier 1984c; Hsu, Lefkovits *et al.* 1991). The properties of the immunoglobulin molecules synthesized by B cells and plasma cells are presented in Table 16.1 (reviewed in Du Pasquier, Schwager & Flajnik 1989).

Immunoglobulin genes in *Xenopus*

The heavy chain locus

The general architecture of the heavy chain locus (Fig. 16.1) is very similar to that of bony fish, birds and mammals and differs from that of chondrichthyans. Multiple V_H families (up to 11), the members of which show variable degrees of interspersion, represent roughly 80–100 V_H genes per haploid genome (Hsu, Schwager & Alt 1989; Schwager, Bürckert, Courtet & Du Pasquier 1989; Haire, Amemiya *et al.* 1990; Haire, Ohta *et al.* 1991). They are followed by at least 15 D segments of which only two have been identified in the germ line. The 13 others are derived from consensus obtained from cDNA sequences. Further, 3′, the J_H locus, contains eight J_H elements (of which J_H6 is a pseudo-gene) in *X. laevis* and nine J_H elements in *X. gilli* (see Schwager, Bürckert, Courtet & Du Pasquier 1991). Between J_H and the constant-region genes we find enhancer-like sequences and a large switch μ region of about 3.9 kb. For each isotype of a heavy chain constant region (IgM, IgX, IgY) a single gene with four C_H exons has been identified (μχυ) (Schwager, Mikoryak & Steiner 1988; Haire, Shamblott *et al.* 1989; Amemiya, Haire & Litman 1989). So far, the transmembrane exons have been identified for the μ and υ genes only. The structure of each V_H gene is similar to that of all vertebrate V_H genes with a promoter (octamer), a TATA box, a split leader and a coding region with two hypervariable regions (CDR1, CDR2). All the segments participating in

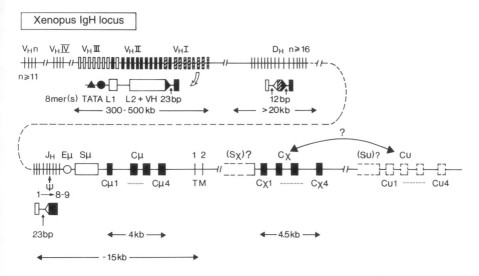

Fig. 16.1. Representation of the *Xenopus* immunoglobulin heavy chain locus.

the assembly of the functional Ig gene have the canonical recombination signal sequences.

The light chain loci

Three light chain constant region genes have been described so far, ρ, σ1 and σ2. ρ and σ are on separate chromosomes and have their own set of V_L and J_L segments. The homology (amino acid level) of Cσ to mouse κ and λ is 33/107 and 35/107 respectively, which does not allow predictions as to which *Xenopus* light chain is closer to κ or λ (Zezza *et al.* 1991; Schwager, Bürckert, Schwager & Wilson 1991). Yet from an organizational viewpoint, σ with its multiple isotypes resembles λ more than κ.

The mode of rearrangement

The mode of Ig gene rearrangement in *Xenopus* is similar to that found in all species with the same type of Ig locus architecture (Schwager, Grossberger & Du Pasquier 1988). D segments rearrange first to J_H, as evidenced by the presence of some D–J transcripts in cDNA libraries made from lymphocytes. Later, V_H genes arrange to DJ. The sequence of events might be more compressed in *Xenopus* because DJ rearrangements are very much rarer than they are in mammals. The rearrangement mechanisms involve the excision and circularization of the DNA fragment that is present between the rearranging elements. D segments are used in more flexible ways than in the mouse, with many examples of fusion, inversion and usages of all possible reading frames, a flexibility which somewhat resembles that in humans (Schwager, Bürckert, Courtet & Du Pasquier 1991). Allelic exclusion is also encountered in *Xenopus* (see Du Pasquier & Hsu 1983). B cell DNA shows on Southern blot the evidence of rearrangement on both chromosomes and sequences from B cell rearrangements at the genomic DNA level show about 50% of abortive VDJ rearrangements.

The resulting repertoire

As measured at the protein level, the antibody repertoire encoded by these elements is, in the case of the anti-DNP response or anti-sheep red blood cell response of isogenic *Xenopus*, surprisingly restricted and homogeneous when compared to mammals. Not only are antibody affinities similar between isogenic individuals, but idiotypes, isoelectrofocusing spectrotypes and affinity maturation profiles are often shared (Wabl & Du Pasquier 1976; Du Pasquier & Wabl 1978). Affinity maturation is poor in *Xenopus* (a three- to five-fold increase over four weeks and no further improvement during a year). All these features suggested that somatic mutations did not further diversify the repertoire of expressed germ-line genes (Du Pasquier & Schwager 1991). In the isogenic, Ig locus homozygous clone of *Xenopus* LG7, comparison of V_H1 cDNA obtained from immunized animals at the peak of their anti-DNP response to germ-line members of this family showed that somatic mutations had indeed occurred at a rate close to that estimated in mammals ($\approx 4 \times 10^4$ bp/generation) but that the resulting mutants were not efficiently selected in the lymphoid organ, perhaps because of the lack of germinal centres. The absence of germinal centres in poikilothermic vertebrates in general and in *Xenopus* in particular (Manning 1991) revealed an interesting lack of coevolution between the mechanism generating mutants and the organs necessary to select them (Wilson, Hsu *et al.* 1992).

A recently isolated lymphoid cell line derived from a spontaneous thymoma and showing characteristics of both B and T cells gives interesting information on allelic exclusion. This cell line, which has a T cell phenotype, has complete functional Ig heavy chain gene VDJ rearrangements on its two chromosomes (which is never encountered in normal B cells), but does not express any Ig on the surface or in the cytoplasm of the cell, nor does it produce any Ig mRNA (Du Pasquier & Robert 1992). This implies that, among models of allelic exclusion, those proposing that a feedback mechanism after the expression of a surface Ig molecule prevents the cell from rearranging further fit best with the type of rearrangement found in the cell line.

Larval and adult responses

The immune system during ontogeny (Fig. 16.2)

Lymphocytes are first produced by the thymus during the first week of life, whereas pre-B cells are present in the larval liver as early as 72 h after fertilization (Hadji-Azimi, Schwager & Thiébaud 1982). The spleen becomes colonized by T and B cells around 12 days after hatching. The number of lymphocytes steadily increases until stage 58, where they reach about 1–2×10^6 in the thymus, 5×10^5 in the spleen and an undetermined number in the

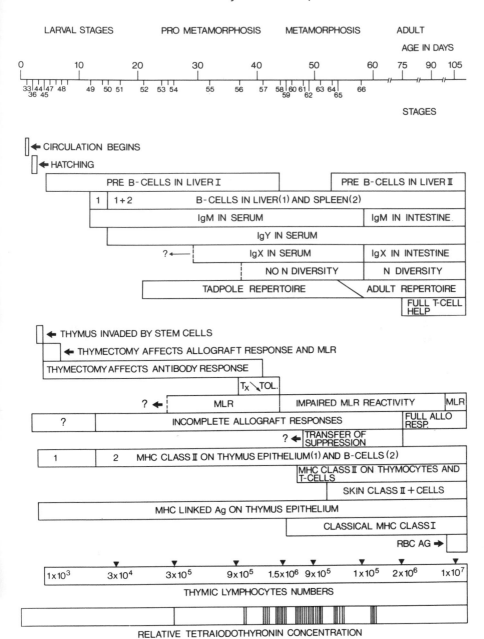

Fig. 16.2. The ontogeny of the immune system and of the immune responses in *Xenopus* (modified from Du Pasquier, Schwager & Flajnik 1989). Abbreviations: Tx—thymectomy; Tol—tolerance. Skin class II⁺ cells, appearance of class II positive cells in the skin. RBC AG, red blood cell antigen linked to MHC.

liver, kidneys and circulation. No lymphoid cell accumulations are visible in the skin or in the gut in larval *Xenopus*, unlike other anuran species. These organs will be colonized by lymphocytes after the metamorphosis, the skin mainly with class II positive T cells, the gut with T cells and IgM or IgX producing B cells and plasma cells. B cells represent about 45% of the population of splenocytes in tadpoles and 25% in adults. T cells are among the remainder; their number has not been measured in detail. It is known, however, that some of these T cells express some of the classical T cell markers such as CD8, X21.2 (Nagata 1985; Flajnik, Ferrone *et al.* 1990).

The expression of MHC molecules (Flajnik & Du Pasquier 1990; MHC molecules of *Xenopus* are presented in Table 16.2) on larval tissues is very different from that in adults. Class II molecules are expressed only by larval B cells and macrophages, and on some epithelia. All larval tissues, including haematopoietic tissues, are class I negative, as evidenced by immunoprecipitation data with monoclonal antibodies and alloantisera, as well as by immunofluorescence experiments (Flajnik, Kaufman, Hsu *et al.* 1986). At the time of metamorphosis, class II become expressed on all lymphocytes, including T cells, and MHC class I become ubiquitously expressed on cell membranes. The expression of these MHC molecules is under the control of different regulatory mechanisms, since animals blocked in their metamorphosis by chemical agents are unable to express class II on T cells whatever their age, whereas class I molecules do finally (i.e. nine months after blockage) become expressed in these experimental animals (Flajnik & Du Pasquier 1988; Du Pasquier & Flajnik 1990; Rollins-Smith & Blair 1990). The differentiation of class I positive cells has been followed in detail as far as the haematopoietic cells are concerned. Class I are expressed together with adult haemoglobin in the new erythroblasts that start appearing around stage 58 and differentiate into the red cells that replace the larval population during metamorphosis (Flajnik & Du Pasquier 1988). The differentiation is not due to a general signal giving 'the order' to all cells to express class I. These procedures remain to be investigated for other cell types. In adults, MHC molecules seem to achieve functions similar to those of their mammalian counterparts. T–B cell collaboration is indeed 'MHC restricted' in *Xenopus*.

All immunoglobulin isotypes are expressed in tadpoles. IgM, IgY and IgX can be detected at stage 52 in the serum or by immunofluorescence in lymphoid

Table 16.2. MHC molecules of *Xenopus*

	Class I[a]	β2-M[1]	Class IIα[b]	Class IIβ[b]	Class III (C₄)[c]
Nb of loci	>1	1?	>2	>2(5?)	1
MW + carbohydrates	40 000–44 000		33 000	33 000	
MW − carbohydrates	37 000–41 000	13 000	25 000	29 000	202 000

[a]Fourth component of complement. Its gene is linked to MHC (Nakamura *et al.* 1986).
[b]Data from Kaufman *et al.* (1985).
[c]Data from Flajnik, Kaufman, Riegert & Du Pasquier (1984).

organs. B cells expressing IgM can be seen in any lymphoid cell accumulation. IgY and IgX positive cells are more difficult to detect. Plasma cells producing these isotypes can, however, be seen in the liver of tadpoles. Thymectomy entirely suppresses the expression of IgY but not of IgM or IgX (Turner & Manning 1974; Collie, Turner & Manning 1975; Hsu, Flajnik & Du Pasquier 1985).

During metamorphosis, lymphoid organs are depleted of their lymphocytes and recover during a second histogenesis. The thymus loses 90% of its lymphocytes at metamorphosis. The spleen becomes haematopoietic after the metamorphosis and the cells of the haematopoietic layer of the liver remain active in *Xenopus* while they disappear in many other anuran species.

Onset of antibody responses

Xenopus larvae become immunologically competent during the first two weeks after fertilization. This can be deduced from assays probing T or B cell functions, whereas the first antibody synthesis can be measured around stage 51. However, these larval responses are different from those of the adult. On average, they show a poorer T cell help function than adult responses do, and helper function can be estimated by the degree of switch from IgM to IgY in antibody responses. Normal tadpoles require much stronger immunization protocols and higher temperatures to show a reasonable switch. If helped by adult T cells this step occurs more readily, showing that the lack of switch really was due to lack of T cell function and not to other larval-specific factors linked to the architecture of lymphoid organs or defects in antigen presentation (Hsu & Du Pasquier 1984a,b).

Larval responses

This is an area which has been investigated carefully at the serological, cellular and molecular level. At the early stages of development (up to stage 52), tadpole responses seem to be limited by B cells: fluctuations in anti-DNP response of stage 52 cannot be overcome with adult helper T cell injections, whereas they can after stage 52. This suggests that before stage 52 the specific B cells are at a limiting dilution in the various isogenic animals immunized simultaneously. Isoelectric focusing spectrotypes differ between isogenic larvae and adults whereas they are similar within isogenic animals of the same stage. The expression of the larval patterns of antibody seems to be more dependent on the T cell functions. If tadpoles which mount an almost exclusively IgM antibody response of low affinity are injected with primed adult T cells of the same genotype, the switch to IgY occurs, but the antibody affinity and the isoelectrofocusing pattern remain of the larval type (Hsu & Du Pasquier 1984a, b). These data can now be re-evaluated in the light of the recently published analysis done at the cDNA level.

The above-mentioned differences between the larval and the adult antibody repertoire seem to depend, among other things, on the way the CDR3 of the Ig molecule is made during larval and adult life. cDNA analyses of tadpoles have shown that larval CDR3, usually shorter than their adult counterpart, use fewer D elements. Larvae do not diversify the CDR3 by the random addition of N residues as do adults (Schwager, Bürckert, Courtet & Du Pasquier 1991). Whether larvae achieve somatic mutation is not yet known.

Metamorphosis

The immune responses of *Xenopus* during ontogeny correspond to two relatively stable patterns, a larval one that is functional and can be prolonged to a certain extent by blocking metamorphosis, which will cause it to retain its larval characteristics, and an adult one characterized by the ubiquitous expression of class I, a different class II distribution, and a more diverse antibody repertoire, with a more complete use of isotype switching and new areas (such as the gut) colonized by lymphoid cells.

From the viewpoint of antibody-production, the transition between the two repertoires is abrupt and strictly dependent on the morphological changes affecting the metamorphosis. Blocked tadpoles will retain their larval repertoire for months, even though the cell numbers increase in the spleen and the liver. Control toadlets left to metamorphose and immunized at the same time after fertilization show an adult response (Hsu & Du Pasquier 1992). The transfer of memory during the metamorphic changes remains an unresolved issue. Since cell populations with different characteristics seem to replace the larval ones at metamorphosis, it could be that what we take for a transfer of memory (in graft experiments or hapten or hapten carrier experiments) is indeed not the reflection of a transfer of memory cells to the adult but, rather, persistence of the antigen.

The metamorphosis period raises physiological questions, most of which remain unresolved. How does the animal escape autoimmunity? Can tadpoles make antibodies against adult structures? What is the meaning of the difference in MHC expression before and after metamorphosis? How is the new B cell repertoire generated? It is tempting to think that the lack of class I is linked to the lack of autoimmunity. In the absence of class I restricted T cells the new adult-specific antigens presented in the context of MHC I will not elicit a killer cell response, but rather a new selection of a T cell subset becoming tolerant to the adult self. With respect to B cells, the more heterogeneous repertoire of the adult is likely to be due to a new wave of rearrangements which occurs during the histogenesis of lymphoid organs and which allows addition of N regions in the CDR3, as well as more rearrangements. In support of this hypothesis, the study of pre-B cells during *Xenopus* ontogeny has revealed two phases during which pre-B cells could be detected in the liver and spleen—one larval, ending at stage 58, and the other starting at stage 60 (Hadji-Azimi, Coosemans & Canicatti 1990). The circular DNA from lymphocytes analysed during this period contains excision circles due to the rearrangements of immunoglobulin genes, thereby proving the ongoing buildup of Ig genes during at least 2–4 months after metamorphosis (Schwager, Bürckert, Courtet & Du Pasquier 1991).

Ig genes and antibody production in polyploid species

Xenopus offers other possibilities of addressing some pertinent questions about B cell clone development (Kobel & Du Pasquier 1986). Ig genes can be found in different numbers in various *Xenopus* species and related genera because of polyploidy. *Xenopus tropicalis* , also known as *Silurana*, with 20 chromosomes, has fewer Ig C and V genes than *Xenopus ruwenzoriensis* with 108 chromosomes. Potentially, *X. ruwenzoriensis* could make many more antibodies than *X. tropicalis* or *X. laevis*. All the *Xenopus* species studied show allelic exclusion, even though the highly polyploid ones are in principle more likely to produce several functional rearrangements (Du Pasquier & Hsu 1983). Indeed they do, since the number of isoelectrophoretic spectrotypes of anti-DNP antibodies found in these species is higher than in *X. tropicalis* (Du Pasquier & Blomberg 1982). However, it is likely that *X. ruwenzoriensis* does not fully exploit its potential. Indeed, the animal is smaller than many species with fewer chromosomes, therefore its B cell number is small. This is even further exaggerated because *X. ruwenzoriensis* cells are slightly bigger than those with simpler karyotypes. In other words, it is likely that cell clone size would be very limited in *X. ruwenzoriensis* during development and that many V_H have no chance to be expressed. Some evolutionary consequences of this could be that a vast amount of the Ig genome could disappear from *X. ruwenzoriensis* chromosomes without affecting its immune response, or could be turned into pseudogenes. If, during speciation events, such pseudogenes are inherited preferentially because of segregation, it could strongly affect the way repertoire would be built in the new descendant species by limiting it or by favouring new mechanisms to diversify it. One could speculate whether, for instance, the mechanism by which birds diversified their antibody repertoire from a large number of Ψ genes by conversion events may have developed from a situation of this type.

   In order to understand the situation better, Ig gene segregation has been analysed by Southern blotting in various families of *X. laevis* and *X. ruwenzoriensis*. As suggested by the phenotype of the antibody responses, Ig genes remained in large numbers in *X. ruwenzoriensis*, and remained under polysomic inheritance. There were differences in the fate of C and V_H genes. For the Ig constant region genes, they seem to be deleted on two chromosomes out of six (four remaining), whereas for Ig V_H genes, some families are present on all chromosomes. However, some families seem to have been deleted (perhaps together with the constant region) on one set of chromosomes. Moreover, deletion of entire families also suggests that even if the V_H genes show some degree of interspersion, they must be assembled in larger units grouping several families and not others on the chromosome. In other words, the V_H genes are not randomly distributed in the Ig locus (Wilson, Marcuz & Du Pasquier in prep.).

   It is interesting to mention that for other genes involved in the immune system, like the MHC, functional analysis suggests a diploidization of the locus in polyploid species, as if expression of too many MHC molecules at the surface of the cells would not be tolerated by the immune system (Du Pasquier, Miggiano *et al.* 1977). This issue can now be studied at the gene level, since a *Xenopus* class I gene has been isolated (Flajnik, Canel *et al.* 1991).

Conclusions

In order to answer the questions raised in the Introduction one can summarize as follows the peculiarities of the *Xenopus* immune system in general, and of its antibody responses in particular.

1. Under pressure to develop quickly a repertoire of recognition structure as diverse as possible (and thus with a small number of cells), the immune system in a first wave of rearrangements (not restricted to a small group of V region genes as in mammals) produces within 12–14 days after fertilization a repertoire which lasts without noticeable changes until metamorphosis. A new wave of rearrangements, with the introduction of more variability due to molecular events lacking in tadpoles, produces the adult repertoire, which differs significantly from the larval one. The contribution of somatic mutation to the functional repertoire is poor, even though hypermutation in the V genes takes place. The inability to select mutants may be linked to the simplicity of the lymphoid system, without lymph nodes and without germinal centres.

2. The *Xenopus* larval immune system, with an antibody repertoire different from the adult, functions without the classical MHC class I molecules expressed on cell surfaces. MHC serves as a marker of adulthood at metamorphosis in the haematopoietic lineage.

3. Gene duplication due to polyploidy reveals a trend toward the expression of a single MHC and a strong pressure for allelic exclusion at the Ig locus, even though polysomic inheritance of Ig genes is maintained in several polyploid species.

4. The ontogeny of the immune system and the study of the immune responses in larvae, blocked larvae, or adult *Xenopus* appears to be more and more a model for studying aspects of the differentiation of the immune system that cannot be studied in mammals. At least *Xenopus* offers different views on old questions. With the genetic refinements that the species offer (strains, clones, polyploid species) and with an increasing amount of knowledge about the molecular mechanism leading to the differentiation of the immune system, *Xenopus* increasingly provides results of general interest.

5. However, many things contribute to make *Xenopus* an imperfect model: lack of reagents, genetic complexity, lack of cell lines. In this respect, it is appropriate to conclude with the recent discovery of a lymphoid thymic

tumour in a MHC homozygous animal (mentioned above in the case of allelic exclusion), from which continuously growing cell lines have been derived *in vitro*. The cell line which represents a clone has a larval phenotype (no MHC class I or II molecules on the cell surface) and it combines features of both B and T cells: complete Ig gene rearrangements at the heavy chain locus on the two chromosomes but no expression of Ig; cell surface markers of T cells (CD8, X21.2) (Du Pasquier & Robert 1992). As a tumour it may become an interesting model for comparative studies since tumour cell lines are rare in lower vertebrates. In future, this cell line could be useful in many areas of *Xenopus* biology and immunology as a fusion partner, as a model for studying T and B cell differentiation by providing large amounts of larval mRNA for subtractive library screening, or as an abundant source of cell surface, cytoplasmic or nuclear material of specific interest for protein sequencing. In relation to this, a strain of clonable isogenic hybrid (LG7), homozygous at the immunoglobulin heavy chain locus, has been produced in order to simplify future immunological studies of antibody diversity (Wilson, Marcuz *et al.* 1992).

References

Amemiya, C.T., Haire, R.N. & Litman, G.W. (1989). Nucleotide sequence of a cDNA encoding a third distinct *Xenopus* immunoglobulin heavy chain isotype. *Nucleic Acids Res.* **17**: 5388.

Collie, M.H., Turner, R.J. & Manning, M.J. (1975). Antibody production to lipopolysaccharide in thymectomized *Xenopus. Eur. J. Immunol.* **5**: 426–427.

Du Pasquier, L. & Blomberg, B.B. (1982). The expression of antibody diversity in natural and laboratory-made polyploid individuals of the clawed toad *Xenopus. Immunogenetics* **15**: 251–260.

Du Pasquier, L. & Flajnik, M.F. (1990). Expression of MHC class II antigens during *Xenopus* development. *Devl Immunol.* **1**: 85–95.

Du Pasquier, L. & Hsu, E. (1983). Immunoglobulin expression in diploid and polyploid interspecies hybrids of *Xenopus*: evidence for allelic exclusion. *Eur. J. Immunol.* **13**: 585–590.

Du Pasquier, L., Miggiano, V.C., Kobel, H.R. & Fischberg, M. (1977). The genetic control of histocompatibility reactions in natural and laboratory-made polyploid individuals of the clawed toad *Xenopus. Immunogenetics* **5**: 129–141.

Du Pasquier, L. & Robert, J. (1992). In vitro growth of thymic tumor cell lines from *Xenopus. Devl Immunol.* **2**: 295–307.

Du Pasquier, L. & Schwager, J. (1991). Immunoglobulin genes and B cell development in amphibians. *Adv. exp. Med. Biol.* **292**: 1–9.

Du Pasquier, L., Schwager, J. & Flajnik, M.F. (1989). The immune system of *Xenopus. A. Rev. Immunol.* **7**: 251–275.

Du Pasquier, L. & Wabl, M.R. (1978). Antibody diversity in amphibians, inheritance of isoelectric focussing antibody patterns in isogenic frogs. *Eur. J. Immunol.* **8**: 428–433.

Flajnik, M.F. & Du Pasquier, L. (1988). MHC class I antigens as surface markers of adult erythrocytes during the metamorphosis of *Xenopus. Devl Biol.* **128**: 198–206.

Flajnik, M.F. & Du Pasquier, L. (1990). The major histocompatibility complex of frogs. *Immunol. Rev.* No. 113: 47–63.

Flajnik, M.F., Canel, C., Kramer, J. & Kasahara, M. (1991). Evolution of the major histocompatibility complex: molecular cloning of MHC class I from the amphibian *Xenopus. Proc. natn. Acad. Sci. USA* **88**: 537–541.

Flajnik, M.F., Ferrone, S., Cohen, N. & Du Pasquier, L. (1990). Evolution of the MHC: antigenicity and unusual tissue distribution of *Xenopus* (frog) class II molecules. *Molec. Immunol.* **27**: 451–462.

Flajnik, M.F., Kaufman, J.F., Hsu, E., Manes, M., Parisot, R. & Du Pasquier, L. (1986). Major histocompatibility complex-encoded class I molecules are absent in immunologically competent *Xenopus* before metamorphosis. *J. Immunol.* **137**: 3891–3899.

Flajnik, M.F., Kaufman, J.F., Riegert, P. & Du Pasquier, L. (1984). Identification of class I major histocompatibility complex encoded molecules in the amphibian *Xenopus. Immunogenetics* **20**: 433–442.

Hadji-Azimi, I., Coosemans, V. & Canicatti, C. (1990). B-lymphocyte populations in *Xenopus. Devl comp. Immunol.* **14**: 69–84.

Hadji-Azimi, I., Schwager, I. & Thiébaud, C. (1982). B-lymphocyte differentiation in *Xenopus laevis* larvae. *Devl Biol.* **90**: 253–258.

Haire, R.N., Amemiya, C.T., Suzuki, D. & Litman, G.W. (1990). Eleven distinct V_H gene families and additional patterns of sequence variation suggest a high degree of immunoglobulin gene complexity in a lower vertebrate, *Xenopus laevis. J. exp. Med.* **171**: 1721–1737.

Haire, R.N., Ohta, Y., Litman, R.T., Amemiya, C.T. & Litman, G.W. (1991). The genomic organization of immunoglobulin V_H genes in *Xenopus laevis* shows evidence for interspersion of families. *Nucleic Acids Res.* **19**: 3061–3066.

Haire, R.N., Shamblott, M.J., Amemiya, C.T. & Litman, G.W. (1989). A second *Xenopus* immunoglobulin heavy chain constant region isotype gene. *Nucleic Acids Res.* **17**: 1776.

Horton, J.D., Edwards, B.F., Ruben, L.N. & Mette, S. (1979). Use of different carriers to demonstrate thymic independent antitrinitrophenyl reactivity in the amphibian, *Xenopus laevis. Devl comp. Immunol.* **3**: 621–633.

Hsu, E. & Du Pasquier, L. (1984a). Ontogeny of the immune system in *Xenopus*. II. Antibody repertoire differences between larvae and adults. *Differentiation* **28**: 116–122.

Hsu, E. & Du Pasquier, L. (1984b). Ontogeny of the immune system in *Xenopus*. I. Larval immune response. *Differentiation* **28**: 109–115.

Hsu, E. & Du Pasquier, L. (1984c). Studies on *Xenopus* immunoglobulins using monoclonal antibodies.*Molec. Immunol.* **21**: 257–270.

Hsu, E. & Du Pasquier, L. (1992). Changes in the amphibian antibody repertoire are correlated with metamorphosis and not with age or size. *Devl Immunol.* **2**: 1–6.

Hsu, E., Flajnik, M.F. & Du Pasquier, L. (1985). A third immunoglobulin class in amphibians. *J. Immunol.* **135**: 1998–2004.

Hsu, E., Lefkovits, I., Flajnik, M. & Du Pasquier, L. (1991). Light chain heterogeneity in the amphibian *Xenopus. Molec. Immunol.* **28**: 985–994.

Hsu, E., Schwager, J. & Alt, F.W. (1989). Evolution of immunoglobulin genes: V_H families in the amphibian *Xenopus. Proc. natn. Acad. Sci. USA* **86**: 8010–8014.

Kaufman, J.F., Flajnik, M.F., Du Pasquier, L. & Riegert, P. (1985). *Xenopus* MHC class II molecules. I. Identification and structural characterization. *J. Immunol.* **134**: 3248–3257.

Kobel, H.R. & Du Pasquier, L. (1986). Genetics of polyploid *Xenopus. Trends Genet.* **2**: 310–315.

Manning, M.J. (1991). Histological organization of the spleen: implications for immune functions in amphibians. *Res. Immunol.* **142**: 355–359.

Nagata, S. (1985). A cell surface marker of thymus dependent lymphocytes in *Xenopus laevis* is identifiable by mouse monoclonal antibody. *Eur. J. Immunol.* **15**: 837–841.

Nakamura, T., Sekizawa, A., Fujii, T. & Katagiri, C. (1986). Cosegregation of the polymorphic C4 with the MHC in the frog *Xenopus laevis. Immunogenetics* **23**: 181–186.

Rollins-Smith, L. & Blair, P. (1990). Expression of class II major histocompatibility complex antigens on adult T cells in *Xenopus* is metamorphosis-dependent. *Devl Immunol.* **1**: 97–104.

Schwager, J., Bürckert, N., Courtet, M. & Du Pasquier, L. (1989). Genetic basis of the antibody repertoire in *Xenopus*: analysis of the V_H diversity. *EMBO J.* **8**: 2989–3001.

Schwager, J., Bürckert, N., Courtet, M. & Du Pasquier, L. (1991). The ontogeny of diversification at the immunoglobulin heavy chain locus in *Xenopus. EMBO J.* **10**: 2461–2470.

Schwager, J., Bürckert, N., Schwager, M. & Wilson, N.M. (1991). Evolution of immunoglobulin light chain genes: analysis of *Xenopus* Ig L isotypes and their contribution to antibody diversity. *EMBO J.* **3**: 505–511.

Schwager, J., Grossberger, D. & Du Pasquier, L. (1988). Organization and rearrangement of Immunoglobulin M genes in the amphibian *Xenopus. EMBO J.* **7**: 2409–2415.

Schwager, J., Mikoryak, C.A. & Steiner, L.A. (1988). Amino acid sequences of heavy chain from *Xenopus laevis* IgM deduced from cDNA sequence: implications for evolution of immunoglobulin domains. *Proc. natn. Acad. Sci. USA* **85**: 2245–2249.

Turner, R.J. & Manning, M.J., (1974). Thymic dependence of amphibian antibody responses. *Eur. J. Immunol.* **4**: 343–346.

Wabl, M.R. & Du Pasquier, L. (1976). Antibody patterns in genetically identical frogs. *Nature, Lond.* **264**: 642–644.

Wilson, M., Hsu, E., Marcuz, A., Courtet, M., Du Pasquier, L. & Steinberg, C. (1992). What limits affinity maturation of antibodies in *Xenopus*, the rate of somatic mutation or the ability to select mutants? *EMBO J.* **11**: 4337–4347.

Wilson, M., Marcuz, A., Courtet, M. & Du Pasquier, L. (1992). Sequences of C μ and the V_H1 family in LG7, a clonable strain of *Xenopus*, homozygous for the immunoglobulin loci. *Devl Immunol.* **3**: 13–24.

Zezza, D.J., Mikoryak, C.A., Schwager, J. & Steiner, L.A. (1991). Sequence of C region of L chains from *Xenopus laevis* Ig. *J. Immunol.* **146**: 4041–4047.

Phylogeny and speciation

17 An overview of the anuran fossil record

BORJA SANCHIZ and ZBYNĚK ROČEK

Synopsis

A general overview of the known anuran fossil record is presented, with an emphasis on diversity and extinct groups. The fossil record is analysed for all anurans at the family level, and palaeontological minimal ages are inferred. Most of the record can be referred to extant families, but a few exceptions remain: the South American Jurassic *Vieraella* and *Notobatrachus*, the Asiatic Cretaceous *Gobiates* and the holarctic palaeobatrachids are especially discussed in this regard. However, the real evolutionary pattern appears to include few examples of lost, extinct diversifications within the order Anura, unless this merely derives from a sampling bias of the known fossil record. Diversity in the past has not proven to be higher than today, and it seems to have been growing very slowly through time.

At least 10 Jurassic and Lower Cretaceous sites (dated >100 Ma) are known where multiple anuran remains have been recovered. In all these localities, one or a few anuran species are detected per site, but in no case have more than two very closely related genera been found. More diverse assemblages, including more than one family, are presently known only from the Upper Cretaceous and later. We consider the example of Europe, with a fairly rich fossil record, clearly documenting the role of addition, by means of transcontinental migration and minor speciation events, in the development of present anuran biodiversity.

Finally, consideration is given to the relationships of the Palaeobatrachidae. This extinct family, known from the Upper Cretaceous to the Plio–Pleistocene boundary (roughly 66 to 1.6 Ma) can be considered the sister-group and ecological equivalent to the living Pipidae.

Introduction

The fossil record of anurans has traditionally been considered rather poor and, in spite of its potential, has yielded very few informative clues to anuran evolutionary history. Since the last general overview by Estes & Reig (1973), only a few unexpected discoveries of fossil anurans have been made, but the fossil record has been substantially enlarged, and now includes representatives of many of the living lineages. It is becoming possible to analyse some aspects of anuran diversification, at least in a very general way. Such palaeontological study can provide not only knowledge about completely extinct groups, but also information complementary to the use of molecular techniques for elucidating evolutionary relationships and dating cladogenetic events.

Palaeontology is applied in this chapter in order to analyse briefly the extent of anuran global diversification with respect to (1) the relationships of extinct groups, an indication of the existence of taxonomic diversity different from the present one, and (2) the palaeofaunistic changes of diversity through time, as exemplified by the European fossil record. This general view will provide a basic setting to place pipoid frogs within anuran global diversity, and complements the chapter by Báez (this volume pp. 329–347) that presents a detailed review of the pipoid fossil record.

The fossil record of frogs

Phylogeny and fossil record

Estes & Reig (1973) provided the last general review of fossil anurans, and most of the morphological features analysed by these authors remain accurate. Nevertheless, some characters can be interpreted differently, owing to the use of other methodological approaches or to better knowledge of the variation within contemporary forms. In this section, we will comment upon the fossil record from the viewpoint of diversity and systematic assignments. Unless otherwise stated, the reader is refered to Estes & Reig (1973 and references therein) for older bibliographic references on these forms. A database for more than 860 fossil sites, taken from Sanchiz (in press), with updated faunistic lists at least at the generic level, has been used. The anuran fossil record, for family-level groups, is summarized in Fig. 17.1.

It is generally agreed (the senior author of this review being probably the only exception) that the Lower Triassic *Triadobatrachus massinoti*, discovered in Madagascar in the 1920s, remains the only representative of the ancestral grade that might have given rise to anurans. The morphology of this incomplete specimen has been extensively reviewed by Rage & Roček (1989), being noteworthy in that the cranial structure of this form shows a much closer relation to the unique anuran pattern than the locomotor structures do. The morphology of this specimen, even if interpreted as frog-like, does not give us clear hints on anuran origins, in the sense that it gives no indication on matters such as the mono- or polyphyletic origins of the Anura or the basic locomotor (saltatorial or swimming) adaptions of the group.

The Argentinian *Vieraella*, a single incomplete specimen from the Lower Jurassic, is the oldest unquestionable anuran known, showing many of the character states that incorporate the diagnosis of the order. As could be expected, *Vieraella* shows features (e.g. number of presacral vertebrae, carpals, ribs) that are considered primitive, and all the published interpretations of this

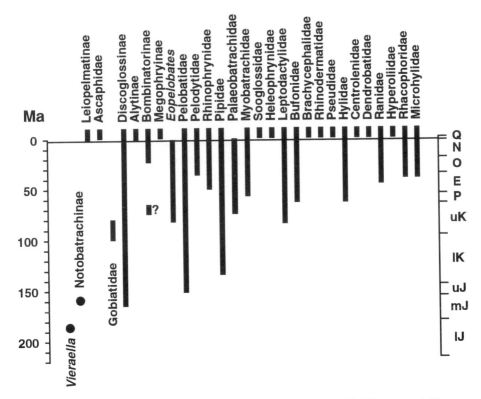

Fig. 17.1. Anuran fossil record at the family level. Q: Quaternary. N: Miocene and Pliocene. O: Oligocene. E: Eocene. P: Palaeocene. uK: upper Cretaceous. lK: lower Cretaceous. uJ: upper Jurassic. mJ: middle Jurassic. lJ: lower Jurassic.

unique specimen agree in considering it very close to the most primitive living frogs, Leiopelmatidae and Ascaphidae. In our opinion, the limited nature of the material prevents any detailed phylogenetic disquisition, and it should remain taxonomically as Anura *incertae sedis*, but it shows that the typical anuran structure has been present for at least 180 Ma.

The Upper Jurassic La Matilde Formation, from the Callovian–Oxfordian (154.7 to 161.3 Ma) of Argentina, has produced remains of more than 30 specimens of another primitive frog, *Notobatrachus degiustoi*. A reinterpretation of the features analysed by Estes & Reig (1973), such as the peculiar morphology of the pectoral girdle, suggests in our opinion that *Notobatrachus* should be included in Leiopelmatidae, although it may represent a distinct subfamily in this group. The living genus *Leiopelma* has an abundant fossil record in the New Zealand Pleistocene and Holocene (Worthy 1987). The family Ascaphidae, which can be considered the sister group of Leiopelmatidae following Clarke (1988) (but see Cannatella 1985 for a contrary view), has no fossil record.

The genus *Gobiates* was first described by Špinar & Tatarinov (1986) for Mongolian Upper Cretaceous articulated specimens previously assigned to the *Eopelobates* group, and it was proposed to place it in a new frog family. Nessov (1988, and references therein) described Cretaceous isolated bones, mainly from the Kizylkum desert in the Commonwealth of Independent States (C.I.S), that according to Roček & Nessov (1991) should also be refered to *Gobiates*. A detailed study of this material has recently been presented by Roček & Nessov (1993). These anurans are known in several localities from the uppermost Lower Cretaceous (Albian) to Upper Cretaceous (Campanian) age from the south-western part of the ancient Asian continent. It is noteworthy that these amphibians are members of assemblages which also contain marine vertebrates. In spite of being rather homogeneous, the group shows a certain taxonomic diversification, with several closely related species documented in its long stratigraphical range (Roček & Nessov 1993). In its morphology, the group combines characters of Leiopelmatidae, Ascaphidae and the discoglossoids, but it cannot be assigned clearly to any of those families and their phylogenetic relationships are still uncertain.

Discoglossoid frogs include the living genera *Bombina* and *Barbourula* (Bombinatoridae), *Discoglossus* (Discoglossidae *s.s.*), and the genus *Alytes*, which according to different opinions can be included (as a distinct subfamily) within the Bombinatoridae (e.g. Lanza, Cei & Crespo 1976) or within the Discoglossidae (e.g. Sanchiz 1984; Cannatella 1985), or can stand as a sister family to the Bombinatoridae plus Discoglossidae (e.g. Maxson & Szymura 1984; Clarke 1988). The three groups are here considered separately, and their relationships are not resolved. Discoglossids have a long fossil history, extending back to the middle Jurassic (161.3–166.1 Ma). *Eodiscoglossus oxoniensis* from Kirtlington in Great Britain is the oldest known anuran record after *Vieraella* (Evans, Milner & Mussett 1990). Four Lower Cretaceous Spanish sites have also produced abundant remains of the same genus (see review in Sanchiz in press), and it seems clear that an extreme morphological similarity exists between *Eodiscoglossus* and the living *Discoglossus*, to the extent that no clear apomorphies are found in *Eodiscoglossus* with relation to *Discoglossus* (Sanchiz 1984; Evans *et al.* 1990). This lineage seems to have been very conservative for more than 160 Ma. *Alytes* has no described fossil record earlier than the Pleistocene, and *Bombina* is known since the early European Miocene (Sanchiz & Schleich 1986). The latter group might be related to *Scotiophryne*, an extinct genus known from the uppermost Cretaceous and Palaeocene of North America (Estes & Sanchiz 1982), but there is no unequivocal evidence on the basis of the available information.

The family Megophryidae is not known in the fossil record (Sanchiz in press). Possibly related is the *Eopelobates* group, with an extended fossil record ranging from the Upper Cretaceous of North America to the Plio–Pleistocene boundary of Europe (Roček 1981). Pelobatid material has been reported from the Upper Jurassic of North America (Evans & Milner 1993). The genus *Pelobates* is known since the European basal Miocene (24 Ma;

Böhme, Roček & Špinar 1982), while *Scaphiopus* has its earliest record in the North American Orellan (32 Ma), and might be related to the Mongolian Oligocene *Macropelobates* (30–38 Ma; Roček 1984). Pelodytidae are known since the Upper Eocene of Europe.

The fossil record of pipids is reviewed by Báez (this volume pp. 329–347), and goes back to the Hauterivian–Barremian (124.5–135 Ma) of Israel. Rhinophrynidae are known since the Upper Palaeocene (late Tiffanian, 59–61 Ma) of North America. Palaeobatrachidae (see below) have been recorded in the Upper Cretaceous of North America and Europe, disappearing in the lowermost Pleistocene of Central Europe.

Myobatrachidae have a fossil record restricted to Australia, with an earliest datum in the Lower Eocene (Tyler & Godthelp 1993), while at present Sooglossidae and Heleophrynidae lack fossil representatives.

The family Leptodactylidae is known since the Campanian (74–83 Ma) and the Santonian–Maastrichtian of South America (Báez 1987), the latter being representatives of the Ceratophryinae. Bufonidae have been recorded in the Upper Palaeocene (Riochican) of Itaboraí (Brazil) (Estes & Báez 1985). Brachycephalidae and Rhinodermatidae have no fossil record.

The Hylidae is a family mentioned in faunistic lists from the late Cretaceous of India (e.g. Sahni, Rana & Prasad 1987) and the Upper Palaeocene of Brazil (Estes & Báez 1985), but we are not aware of any published description of these remains. Pseudidae and Centrolenidae have no fossil record known to us.

Within ranoids, the Dendrobatidae and Hyperoliidae have no fossil record, but the Ranidae has a very abundant one, albeit not very old. Besides an unconfirmed possibility of the presence of ranids in an upper Cretaceous site from Niger, the earliest recorded ranids are from the European Upper Eocene (Rage 1984a). The Rhacophoridae and the Microhylidae have both been detected in the European Upper Eocene (B. Sanchiz unpubl.).

If we consider the whole record, the most obvious pattern concerning diversity might be argued to be the scarcity of extinct clearly distinct lineages, an indication of the absence of adaptations within the order different from the present ones. Nevertheless, since there has been only limited research on this topic, some caution has to be applied to the conclusion, as it might merely derive from a sampling bias of the known fossil record.

Diversification of Mesozoic frogs

Only a few Jurassic sites are known bearing definitive anuran remains: Roca Blanca (Argentina; dated 178–194.5 Ma, *Vieraella herbsti*), Kirtlington (UK; 161.3–166.1 Ma, *Eodiscoglossus oxoniensis*), La Matilde (Argentina; 154.7–161.3 Ma, *Notobatrachus degiustoi*) and Quarry Nine (USA, two nominal taxa that are based on such fragmentary remains that it is impossible to make any definitive statement on their relationships; they might even be synonymous, *contra* Hecht & Estes 1960; see also Evans & Milner 1993).

At least in Kirtlington and La Matilde, where more than 30 individuals have been collected, only a single species seems to be present.

Lower Cretaceous sites where anurans have been reported at least at the family level include four localities in Spain: Uña, Galve, Santa María de Meyá, and Las Hoyas (Sanchiz in press), one still not published in detail from the Karakalpakia region of the C.I.S. (Chodzhakul: Nessov 1988), and two in Israel (Shomron and Makhtesh Ramon: Estes, Špinar & Nevo 1978). Both in Spain (Discoglossidae) and Israel (Pipidae), where the assemblages contain abundant remains and have been studied at the specific level, there are between one and three species per site, but only one subfamily is present at any site, represented by one or two closely related genera.

The Upper Cretaceous fossil record is much more abundant, with localities known in South America, North America, India, Asia, Europe and Africa. Los Alamitos and Alemanía (Argentina), Laguna Umayo (Peru), Tiupampa (Bolivia) and Peirópolis (Brazil) contain pipids and/or leptodactylids (Báez, this volume pp. 329–347). In the USA and Canada, Chris's Bonebed, the Hell Creek, Fruitland, Judith River and Lance Formations, as well as the El Gallo Formation in Mexico, contain bombinatorids, pelobatids, palaeobatrachids, discoglossids and *incertae sedis* (*Theatonius*), from one to five family-level units per site at the end of the Cretaceous (Estes & Sanchiz 1982). The Indian fauna has not been described in detail, but Asifabad, Pisdura and Gitti Khadan near Nagpur seem to contain discoglossoids, pelobatoids and hylids (Sahni *et al.* 1987). In Africa, the Nigerian In Beceten (Niger) fauna includes pipids and undetermined neobatrachians, and pipids have been described from Marydale in South Africa (Rage 1984a; Báez, this volume pp. 329–347). Central Asian faunas include several localities reviewed by Roček & Nessov (1993), with assemblages including gobiatids and discoglossoids.

Relationships of the Palaeobatrachidae

Palaeobatrachids are a homogeneous extinct family of anurans recorded from the late Cretaceous of North America (Estes & Sanchiz 1982) and throughout the Central European Tertiary. The group apparently became extinct in the lowermost Pleistocene. Wolterstorff (1885–1886) and Špinar (1972) remain the most detailed morphological studies available, although their taxonomic arrangements, especially their proliferation of species names, are open to question.

Palaeobatrachids have been reported twice from the lowermost Cretaceous of Santa María de Meyá (Spain). The first case was *Monsechobatrachus gaudryi*, originally assigned by Vidal (1902) to *Palaeobatrachus*. Estes & Reig (1973) indicated the possibility, pending confirmation through a study using modern techniques, that the original identification could be correct. However, our review of the specimen is in full agreement with that of Hecht (1963) and has convinced us that such a poorly preserved natural cast simply does not exhibit the minimum number of observable features to warrant any familial assignment. Approximate body proportions suggest that the specimen is probably only a late metamorphosing tadpole or froglet of the discoglossid *Eodiscoglossus santonjae*, common in the site.

Seiffert (1972) also described from Santa María de Meyá *Neusibatrachus wilferti*, a single specimen that he considered to be an ancestor to the Palaeobatrachidae and Ranidae. Estes & Reig (1973) considered it a definitive palaeobatrachid, with no clear indication of its relationship to other families. As will be discussed elsewhere, the review by one of us of this specimen, in the context of the ontogenetic development of living *Discoglossus galganoi*, suggests that several features of *Neusibatrachus* have been misinterpreted, and that this genus could also be considered a synonym of *Eodiscoglossus santonjae*, with no relation to palaeobatrachids or ranids (Sanchiz in press). No palaeobatrachid has so far been found in Europe before the Upper Cretaceous of Spain (B. Sanchiz pers. obs.).

Since the Palaeobatrachidae were established as a separate group at the family rank by Cope in 1865, it is generally agreed that their closest affinities are to the Pipidae. This view is based on the following essential similarities (which are not, however, of equal value). (1) Both groups are swimming anurans which is evidenced by considerably elongated metacarpals and metatarsals. (2) Their frontoparietal in adults is unpaired. In pipid tadpoles this arises from paired primordia which are, however, immediately afterwards obscured by overall centripetal ossification. In palaeobatrachid larvae, in the earliest preserved developmental stages, the frontoparietal is a single bone without any suture. From this it may be inferred that even development of the frontoparietal is similar in both groups and different from all other anurans. (3) Their parasphenoid lacks lateral alae. (4) Their larvae possess lateral barbels. (5) Palaeobatrachids have a parietal foramen. This is also present in *Xenopus* and one may suppose that it was originally present in all pipids. (6) In pipids the maxillary arch is incomplete (i.e. their quadratojugals are absent), similarly to most palaeobatrachids (except for *Palaeobatrachus diluvianus* in which this bone is vestigial). (7) The pectoral girdle is of arciferal or modified arciferal type.

Besides the similarities between palaeobatrachids and pipids mentioned above, which may be considered essential for assessing phylogenetic relationships, there is one striking difference: whereas pipids have opisthocoelous vertebrae, the shape of the vertebral centra in palaeobatrachids is uniformly procoelous. This was why Noble (e.g. 1922, 1931) classified palaeobatrachids in his group Procoela, together with 'modern' families such as Ranidae and Bufonidae.

It seems that, apart from the vertebrae, the osteological similarities between pipids and palaeobatrachids suggest their close relationship. There is also a classical agreement that considers both families as rather archaic frogs. This is evidenced by the presence of primitive characters such as free ribs in development, parasphenoid without lateral alae, well-developed mentomandibular (which is part of a branchial arch), and foramen parietale present at least in some species. Nevertheless, their consideration as primitive anurans is not supported by recent comparative anatomy studies (e.g. Cannatella 1985; Trueb, this volume pp. 349–377), as they show synapomorphies with the neobatrachians. Both palaeobatrachids and pipids are also well delimited, especially on the basis of developmental processes, from other groups of primitive frogs, namely from the Discoglossidae and Leiopelmatidae.

History of the European fauna

The present anuran fauna of Europe, excluding the C.I.S., is composed of eight genera, representing seven families, and includes approximately 28 species (e.g. Engelmann *et al.* 1986). The fossil faunas from a sample of 189 European Tertiary sites (excluding the C.I.S. countries) (taken from Sanchiz in press.) have been re-analysed, and even if the taxonomy at the specific level is in many instances dubious, the generic attributions can be established confidently. Figure 17.2 shows the fossil record for each continental stratigraphic stage. The number of fossil assemblages studied is not large, but the European record is the best available, and we think that it is probably adequate to show the general historical pattern.

The number of genera and the maximum number of species per site are indicated in Fig. 17.3. The 'Grande Coupure' or 'Stehlin faunistic turnover' (see for instance Rage 1984b, concerning herpetology), represents a major change for European faunas. This faunistic change reflects the Eocene–Oligocene boundary, a global event that palaeogeographically produced the final opening of the north Atlantic (now a barrier for anurans) and the disappearance of the Turgai Straits, no longer separating Europe and Asia (Prothero 1985). Rhacophorids, microhylids and leptodactylids are not recovered in younger sites, while *Bufo, Hyla, Bombina*, and perhaps *Pelobates*, should be considered immigrants from Asia. Within ranids a similar situation is observed at lower taxonomic levels, with the water frogs (the *Rana ridibunda* species group) immigrating in the basal Oligocene (Sanchiz, Esteban & Schleich 1993). This increase of diversity by migration is a transcontinental phenomenon: the assumed areas of origin are Gondwanan (Hylidae, Bufonidae) or eastern Asiatic (Bombinatoridae). Ranids are considered originally African (e.g. Duellman & Trueb 1986), but their immediate origin with relation to Europe is Asia, as indicated by immunologically-derived time estimations of the separation between western and eastern Palaearctic water frogs (Sanchiz *et al.* 1993). The pelodytids have a rather scarce fossil record, although they have been collected outside Europe in the North American Miocene, and their biogeographic status is unclear (Henrici 1994).

Eopelobates is recorded in Cretaceous North America, and it might have also been present in Europe, but there is no palaeontological evidence of this. Palaeobatrachids and discoglossids might be considered European natives, since they are already known by the end of the Mesozoic, both in Europe and

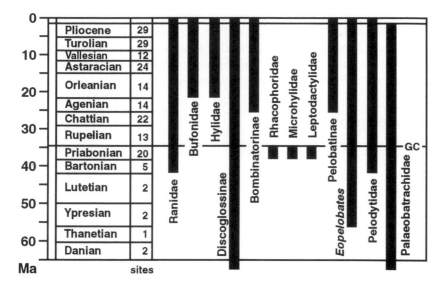

Fig. 17.2. Fossil record of European Tertiary anurans. Sites: number of sites in the corresponding stratigraphical level. GC: 'Grande Coupure'.

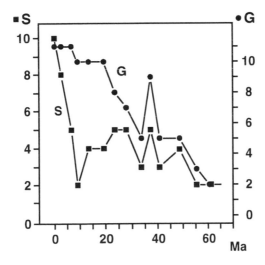

Fig. 17.3. Diversity increase in the anuran fossil record in Europe. S: maximum number of species per site for each stratigraphical level. G: total number of known genera in each stratigraphic level.

in North America. Both palaeobatrachids and especially discoglossids are the only groups that have several different genera in the European Cenozoic, and the only ones showing morphological evolutionary changes at the supraspecific level. The presumed immigrant groups, with stable integrated morphotypes, seem to show only minor speciation events within the European continent.

Acknowledgements

This work was supported in part by the Spanish grant DGICYT PB-910115.

References

Báez, A.M. (1987). The late Cretaceous fauna of Los Alamitos, Patagonia, Argentina. Part III. Anurans. *Revta Mus. argent. Cienc. Nat. Bernardino Rivadavia* 3(3): 121–130.

Böhme, W., Roček, Z. & Špinar, Z.V. (1982). On *Pelobates decheni* Troschel, 1861, and *Zaphrissa eurypelis* Cope, 1866 (Amphibia: Salientia: Pelobatidae) from the early Miocene of Rott near Bonn, West Germany. *J. vert. Paleont.* 2: 1–7.

Cannatella, D.C. (1985). *A phylogeny of primitive frogs (archaeobatrachians).* PhD thesis: University of Kansas.

Clarke, B.T. (1988). *Evolutionary relationships of the discoglossoid frogs: osteological evidence.* PhD thesis: City of London Polytechnic.

Duellman, W.E. & Trueb, L. (1986). *Biology of amphibians.* McGraw-Hill, New York.

Engelmann, W.E., Fritzsche, J., Günther, R. & Obst, F.J. (1986). *Lurche und Kriechtiere Europas.* Ferdinand Enke, Stuttgart.

Estes, R. & Báez, A. (1985). Herpetofaunas of North and South America during the late Cretaceous and Cenozoic: evidence of interchange? In *The great American interchange*: 139–197. (Eds Stehli, F.G. & Webb, S.D.). Plenum Press, New York and London. (*Topics Geobiol.* 4.)

Estes, R. & Reig, O.A. (1973). The early fossil record of frogs: a review of the evidence. In *Evolutionary biology of the anurans: contemporary research on major problems*: 11–63. (Ed. Vial, J.L.). University of Missouri Press, Columbia.

Estes, R. & Sanchiz, B. (1982). New discoglossid and palaeobatrachid frogs from the late Cretaceous of Wyoming and Montana, and a review of other frogs from the Lance and Hell Creek formations. *J. vert. Paleont.* 2: 9–20.

Estes, R., Špinar, Z.V. & Nevo, E. (1978). Early Cretaceous pipid tadpoles from Israel (Amphibia: Anura). *Herpetologica* 34: 374–393.

Evans, S.E. & Milner, A.R. (1993). Frogs and salamanders from the Upper Jurassic Morrison Formation (Quarry Nine, Come Bluff) of North America. *J. vert. Paleont.* 13: 24–30.

Evans, S.E., Milner, A.R. & Mussett, F. (1990). A discoglossid frog from the Middle Jurassic of England. *Palaeontology* 33: 298–311.

Hecht, M.K. (1963). A reevaluation of the early history of the frogs. Part II. *Syst. Zool.* 12: 20–35.

Hecht, M.K. & Estes, R. (1960). Fossil amphibians from Quarry Nine. *Postilla* No. 46: 1–19.

Henrici, A.C. (1994). *Tephrodytes brassicarvalis*, new genus and species (Anura:

Pelodytidae) from the Arikareean Cabbage Patch Beds of Montana, USA, and pelodytid–pelobatid relationships. *Ann. Carneg. Mus.* 63: 155–183.

Lanza, B., Cei, J.M. & Crespo, E.G. (1976). Further immunological evidence for the validity of the family Bombinidae (Amphibia Salientia). *Monit. zool. ital.* (N.S.) 10: 311–314.

Maxson, L.R. & Szymura, J.M. (1984). Relationships among discoglossid frogs: an albumin perspective. *Amphibia–Reptilia* 5: 245–252.

Nessov, L.A. (1988). Late Mesozoic amphibians and lizards of Soviet Middle Asia. *Acta zool. cracov.* 31: 475–486.

Noble, G.K. (1922). The phylogeny of the Salientia I. The osteology and the thigh musculature; their bearing on classification and phylogeny. *Bull. Am. Mus. nat. Hist.* 46: 1–87.

Noble, G.K. (1931). *The biology of the Amphibia.* Constable, London. (Reprinted 1954 by Dover Publ., New York.)

Prothero, D.R. (1985). North American mammalian diversity and Eocene–Oligocene extinctions. *Paleobiology* 11: 389–405.

Rage, J.C. (1984a). Are the Ranidae (Anura, Amphibia) known prior to the Oligocene? *Amphibia–Reptilia* 5: 281–288.

Rage, J.C. (1984b). La 'Grande Coupure' éocène/oligocène et les herpétofaunes (amphibiens et reptiles): problèmes du synchronisme des événements paléobiogéographiques. *Bull. Soc. géol. Fr.* 26: 1251–1257.

Rage, J.-C. & Roček, Z. (1989). Redescription of *Triadobatrachus massinoti* (Piveteau, 1936), an anuran amphibian from the Early Triassic. *Palaeontographica* (A) 206: 1–16.

Roček, Z. (1981). Cranial anatomy of frogs of the family Pelobatidae Stannius, 1856, with outlines of their phylogeny and systematics. *Acta Univ. Carol. Biol.* 1980 (1–2): 1–164.

Roček, Z. (1984). *Macropelobates osborni* Noble, 1924—redescription and reassignment. *Acta Univ. Carol. (Geol.)* 1982(4): 421–438.

Roček, Z. & Nessov, L.A. (1991). Cretaceous anurans from Central Asia. In *Czechoslovak paleontology 1990*: 24. (Ed. Roček, Z.). Carolinum Press, Charles University, Prague.

Roček, Z. & Nessov, L. (1993). Cretaceous anurans from Central Asia. *Palaeontographica (A)* 226: 1–54.

Sahni, A., Rana, R.S. & Prasad, G.V.R. (1987). New evidence for paleobiogeographic intercontinental Gondwana relationships based on late Cretaceous-earliest Paleocene coastal faunas from peninsular India. *Geophys. Monogr.* No. 41: 207–218.

Sanchiz, B. (1984). Análisis filogenético de la tribu Alytini (Anura, Discoglossidae) mediante el estudio de su morfoestructura ósea. In *Història biològica del Ferreret* (Baleaphryne muletensis): 61–108. (Eds Hemmer, H. & Alcover, J.A.). Moll, Palma de Mallorca.

Sanchiz, B. (In press). *Salientia. Handbuch der Palaeoherpetologie.* Gustav Fischer, Stuttgart.

Sanchiz, B., Esteban, M. & Schleich, H.H. (1993). Water frogs from the Lower Oligocene of Germany. *J. Herpet.* 27: 486–489.

Sanchiz, B. & Schleich, H.H. (1986). Erstnachweis der Gattung *Bombina* (Amphibia: Anura) im Untermiozän Deutschlands. *Mitt. bayer. St. Paläont. hist. Geol.* 26: 41–44.

Seiffert, J. (1972). Ein Vorläufer der Froschfamilien Palaeobatrachidae und Ranidae im Grenzbereich Jura-Kreide. *Neues Jb. Miner. Geol. Paläont Mh.* 1972(2): 120–131.

Špinar, Z.V. (1972). *Tertiary frogs from Central Europe.* Academia, Prague.

Špinar, Z.V. & Tatarinov, L.P. (1986). A new genus and species of discoglossid frog from the Upper Cretaceous of the Gobi desert. *J. vert. Paleont.* 6: 113–122.

Tyler, M.J. & Godthelp, H. (1993). A new species of *Lechriodus* Boulenger (Anura: Leptodactylidae) from the early Eocene of Queensland. *Trans. R. Soc. S. Aust.* 117: 187–189.

Vidal, L.M. (1902). Nota sobre la presencia del tramo kimeridgense en el Montsech (Lérida) y el hallazgo de un batracio en sus hiladas. *Mems R. Acad. Cienc. Artes Barcelona* (3) 4: 263–267.

Wolterstorff, W. (1885–1886). Über fossile Frösche insbesondere das Genus *Palaeobatrachus*. I und II. *Jber. naturw. Ver. Magdeburg* 1885: 1–82; 1886: 3–81.

Worthy, T.H. (1987). Osteology of *Leiopelma* (Amphibia, Leiopelmatidae) and description of three new subfossil *Leiopelma* species. *J. R. Soc. N. Z.* 17: 201–251.

18 The fossil record of the Pipidae

ANA MARÍA BÁEZ

Synopsis

Pipids have a relatively good fossil record, providing evidence of their evolutionary diversification. The oldest remains referred to this well-corroborated monophyletic group are from the Lower Cretaceous (around 120 Ma) of the Near East. Although their referral to the Pipidae may be questioned, they document the presence of primitive pipoids in the northern margin of Gondwana. All other fossils that clearly belong to this family have been found either in Africa or in South America, where they occur today. Moreover, their former geographical distribution on these continents was more extensive than is their current distribution.

Recent finds in Patagonia document the presence of pipids in South America at least from the middle Cretaceous (around 100–95 Ma). Most other known South American records are also from its southern part and from deposits of Cretaceous and Palaeogene age. Pipids are a component of the Patagonian palaeobatrachofauna and Cretaceous and Palaeogene representatives closely resemble the now strictly African *Xenopus*, a genus that, presumably, has diverged little from the ancestral pipid morphotype. However, possession of some characters unknown in any of the extant species of the genus can be demonstrated in some fossil taxa represented by reasonably complete remains.

In Africa, the oldest known pipids are early Senonian (around 90–80 Ma) in age. Among these is a peculiar pipine, the occurrence of which indicates that this derived group of pipids had already differentiated. The presence of fossil members of the clade [*Pipa* + [*Hymenochirus* + *Pseudhymenochirus*]] in South America, however, cannot be ascertained on the basis of the available material. Records of xenopines are known in Miocene and Pleistocene units of Africa.

The fossil record of pipids is interpreted in the context of the recent cladistic analysis of the interrelationships of extant pipid genera by Cannatella & Trueb (1988a). However, the incongruent combination of some character states suggests that some features used to define taxa might have been subject to convergence and/or that hypotheses of the directions of evolutionary change have to be re-examined.

Introduction

The palaeontological record of frogs is relatively poor in comparison with those of other vertebrates, although pipids are one of the best-represented anuran families. The record of pipids is discontinuous with respect to time and space and many significant episodes of their evolutionary history, which spans more than 100 million years, remain undocumented.

The diversity of preservation among pipid fossils constrains comparisons among them. Information on forms represented in microvertebrate assemblages obtained by washing and screening of sediments usually is restricted to a few isolated bones; however, the individual elements can be studied in detail. In contrast, gross (but not necessarily detailed) descriptive data may be obtained from the entire skeleton preserved as an articulated specimen flattened on a slab of rock.

A brief review of the fossil record of pipids is presented in this contribution. Described materials as well as unpublished finds are included. Unquestionable occurrences are restricted to Africa and South America, where the extant representatives occur today. These two landmasses were part of the western portion of the Gondwana supercontinent. However, it is important to remember that when the oldest recorded pipids existed, that supercontinent had already begun to split to form the South Atlantic Ocean (Emery & Uchupi 1984) (Fig. 18.1). Furthermore, as early as 110 Ma (by Albian time), only a tenuous continental link between north-eastern Brazil and western Africa existed and

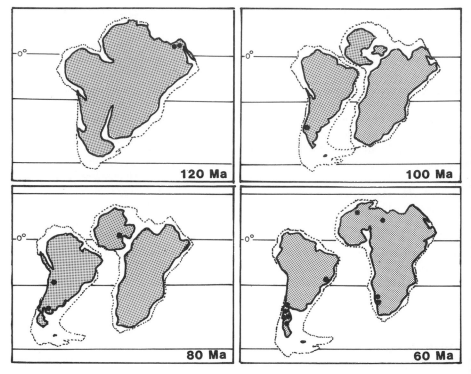

Fig. 18.1. Cretaceous and Palaeogene pipid fossil sites shown on palaeogeographical maps. Maps modified from Barron *et al.* (1981).

this physical connection disappeared entirely by Cenomanian–Turonian time (95 Ma) (Emery & Uchupi 1984: 796). Although a pattern of vicariance and subsequent endemism of pipids is to be expected, the characteristics of this continental decoupling process, along with palaeogeographical modifications related to global sea-level changes, might have resulted in multiepisodic biogeographical relationships between Africa and South America.

It is generally accepted that the extant Pipidae and Rhinophrynidae are sister taxa that comprise the superfamily Pipoidea. However, the extinct Palaeobatrachidae should also be included in this group and there is some evidence that they may be the sister group of the Pipidae (Estes & Reig 1973; Cannatella 1986). In a recent phylogenetic analysis of the extant pipids, Cannatella & Trueb (1988a) demonstrated their monophyly. They proposed that *Xenopus* is the sister taxon of all other living pipids, and that *X. tropicalis* and *X. epitropicalis* are more closely related to the monophyletic subgroup composed of *Pipa* and *Hymenochirus* than to other *Xenopus*; thus, the name *Silurana* was resurrected for these two species.

It has been suggested (Gauthier, Estes & De Queiroz 1988) that taxon names are stabilized by restricting them to clades in which at least two branches stemming directly from the basal node are represented by living organisms. Following this and according to the hypothesis of relationships of extant representatives, Pipidae could be defined as the clade that includes *Xenopus, Silurana* and the pipines and their most recent common ancestor. However, it is conceivable that some of the fossil taxa considered below might lie outside such a basal node. Because the relationships of several extinct taxa are still unclear, in order to avoid ambiguities a stem-based (*sensu* De Queiroz & Gauthier 1990) and consequently broader definition of Pipidae is adopted provisionally in this paper. Pipidae is thus defined as all pipoids sharing a more recent common ancestor with *Xenopus, Silurana* and the pipines than with palaeobatrachids or rhinophrynids.

Although some of the materials included in this survey currently are being studied and an integrative analysis of living and extant pipoids is in progress, some preliminary comments are offered on the possible phylogenetic significance of some features of the fossil forms and their relationships.

The Afro-Arabian record

The oldest remains that have been referred to the Pipidae are from Early Cretaceous units (around 120 Ma) in the Near East. The fossil sites are on the Arabian Peninsula, which is an extension of the African Plate (see Saint-Marc 1978) and lay close to the Tethys shoreline, within the tropical belt at that time (Parrish, Ziegler & Scotese 1982). About 100 tadpoles were recovered from probable Hauterivian beds near the base of the Tayasir Formation, exposed in the Shomron region, central Israel (Fig. 18.2, Wadi el Malih). Because their body shape resembles that of extant midwater filter-feeding larvae and they possess a narrow, lanceolate parasphenoid and an angular bone, they were considered pipoids (Estes, Špinar & Nevo 1978). The presence of an angular in the mandible, however, is probably a misinterpretation, as Trueb & Hanken (1992) have shown that this bone is absent in pipids. The wide mouth and the lack of cornified beaks and expanded lips bearing odontoids support their designation as Type 1 larvae, a highly derived type (Sokol 1975, 1977) that occurs in pipoids (including extinct palaeobatrachids). On the assumption that the shape of the larval parasphenoid reflects the morphology of the adult bone, these tadpoles were referred to a new genus and species, *Shomronella jordanica*, which possesses a wide diamond-shaped posterior portion of the parasphenoid (Estes *et al.* 1978). The shape of this bone in the most advanced ontogenetic stage represented in the fossil sample (Estes *et al.* 1978; Fig. 18.3a) resembles that of the adult *Eoxenopoides* from the Lower Tertiary of South Africa (see below); however, tadpoles of *Eoxenopoides* lack a posterior expansion of the parasphenoid.

Shomronella has a different ossification sequence from *Xenopus laevis*, a species extensively studied in this regard, but there is disagreement about the ossification sequence in the latter (Trueb & Hanken 1992). In *Shomronella*

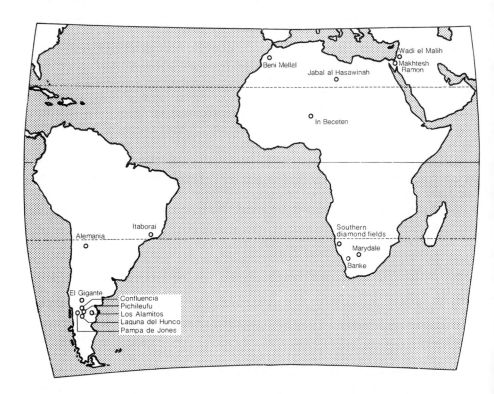

Fig. 18.2. Location of fossil sites mentioned in the text.

the vertebral column seems to ossify more precociously than it does in the living species, whereas osteocranial development is delayed. An exception among the skull bones is the sphenethmoid, which apparently begins to ossify relatively early as in *Rhinophrynus* (Trueb 1985). No evidence of barbels has been found in *Shomronella* even though external body outlines are clear in several examples (Estes *et al.* 1978). A barbel flanks each side of the mouth in tadpoles of *Xenopus* and *Silurana*, whereas more than one short 'antennae' were reported in palaeobatrachid larvae (Špinar 1972: 170, plates 52, 53). This evidence has been interpreted to indicate that barbels are a derived feature that was present in the tadpoles of the common ancestor of palaeobatrachids and pipids (Cannatella & Trueb 1988a), a clade also supported by several other synapomorphies (Cannatella 1986). In *Shomronella*, palaeobatrachids and pipids, the vertebral centra undergo epichordal development. However, the centra are structurally opisthocoelous in *Shomronella* and pipids and reportedly procoelous in palaeobatrachids; mainly for this reason *Shomronella* was included in the Pipidae by Estes *et al.* (1978).

An extensive series of postmetamorphic frogs also was found in the Lower Cretaceous of Israel and referred to the Pipidae (Nevo 1968). The specimens are from the Makhtesh Ramon region, Negev Desert (Fig. 18.2), from a slightly younger horizon than that yielding *Shomronella*. In spite of the large number of specimens recovered, many anatomical aspects are unclear because individuals are preserved mostly as flattened skeletons and coloured imprints; thus, restorations of taxa are highly diagrammatic. Two distinct types are clearly represented: *Thoraciliacus* and *Cordicephalus*, both of which share with other pipoids the presence of an azygous frontoparietal and a parasphenoid lacking subotic alae (Nevo 1968). Discussion of the evolutionary relationships of these forms involves the consideration of pipoid relationships. Some of the derived character states that unite palaeobatrachids and pipids also are present in these Early Cretaceous pipoids—e.g. the elongate and anteriorly directed columella, ossified pubis, elongate metapodials, perhaps monocuspid teeth and relatively large auditory capsules. However, the development of the vertebral centra has been described as perichordal, unlike the derived epichordal condition of palaeobatrachids and pipids. In these Early Cretaceous taxa the parasphenoid possesses a distinct posterior corpus similar to that of *Rhinophrynus*, although the cultriform process is long as in pipids and Caenozoic palaeobatrachids. Moreover, the sphenethmoid appears short, not extending far into the orbital region, and probably did not enclose the optic foramina; the cleithrum is neither expanded distally nor forked, and there is a well-developed preacetabular expansion on the body of the ilium. The sacral diapophyses have distinct, convex lateral margins, especially in *Cordicephalus*, as in *Rhinophrynus* and palaeobatrachids, although in most representatives of the latter several vertebrae tend to participate in the sacrum formation. Instead, in living and some extinct pipids the sacral margins are straight, a derived condition. Some of this evidence suggests that they may lie outside the dichotomy between northern palaeobatrachids and Gondwanan pipids. However, *Cordicephalus* and *Thoraciliacus* share with pipids at least two derived character states that are absent in palaeobatrachids—the lack of mentomeckelian bones and the presence of reduced vomers, the former also shared with rhinophrynids. The medial ramus of the pterygoid of *Thoraciliacus* may have concealed the Eustachian canal, at least partially, but in *Cordicephalus* the pterygoid does not seem to be expanded to form an otic plate. Certain resemblances to *Thoraciliacus* were noted on some poorly preserved specimens collected in Late Cretaceous crater-lake deposits in southern South Africa (Fig. 18.2, Marydale) (Van Dijk 1985), but the evidence supporting an actual relationship is weak.

The oldest reported remains of unquestionably pipid frogs from the present African continent were collected in the Upper Cretaceous (Lower Senonian, around 90–80 Ma) of In Beceten (=Ibeceten), Republic of Niger (Fig. 18.2). The presence of a form 'close to *Pipa* and *Hymenochirus*' and another one 'near *Xenopus*' was cited in a preliminary note on the fossil fauna from that locality (Broin *et al.* 1974). A subsequent study (Báez & Rage 1988 and in prep.) confirmed that a hyper-ossified form closely related to the clade *Pipa* + [*Hymenochirus* + *Pseudhymenochirus*] is represented. Moreover, this highly derived species possesses many synapomorphies of that clade; among those that could be verified are the following: wedge-shaped skull in lateral profile, absence of vomers, fusion of the first two presacral vertebrae and parasagittal spinous processes on presacral vertebrae. The configuration of the Eustachian canal, degree of fusion of endochondral bones in the orbital region and morphology of postzygapophyses suggest that it is related most closely to the hymenochirine pipids. This would give a minimum age for the separation of the lineages represented today by *Pipa* and [*Hymenochirus* + *Pseudhymenochirus*], accepting that they are sister groups (Báez 1981; Cannatella & Trueb 1988a, b).

A *Xenopus*-like pipid is also represented in the In Beceten fauna, documented by an incomplete, but well ossified, skull. However, the presence of any of the skeletal derived character states shared by the extant species of *Xenopus* cannot be assessed because the corresponding parts are not preserved, thus referral to that genus is not warranted. In this regard, one feature merits further comment—the cultriform process of the parasphenoid is slightly expanded laterally at the level of the antorbital plane, but there is no clear indication that a vomer is present or that it has fused to the overlying bones. The absence of vomers is one of the osteological synapomorphies uniting *Silurana* and the pipines, so this might suggest a sister-group relationship with this clade. However, with respect to the other derived character states of that node, there is no information on the first two presacral vertebrae and the sternum, and the frontoparietal is plesiomorphic in lacking anterolateral processes. A peculiar trait of this specimen is that the posterior terminus of the parasphenoid lies well anterior to the completely ossified margin of the foramen magnum.

All other known fossil pipids are of Tertiary age. A puzzling combination of character states is shown by *Eoxenopoides reuningi* Haughton represented by numerous specimens (Haughton 1931; Estes 1977) collected in lacustrine deposits of probable Late Eocene–Oligocene age (around 50–40 Ma) near Banke, Republic of South Africa (Fig. 18.2). *Eoxenopoides* (Fig. 18.3a) possesses some of the diagnostic synapomorphies of living Pipinae (e.g., straight anterior margin of the dorsal lamina of the atlas and reduced zygomatic process of the squamosal: Cannatella & Trueb 1988a). Also as in pipines and unlike *Xenopus* and *Silurana*, the sphenethmoid is flat ventrally and keeled laterally. Although the first two presacral vertebrae are fused, a derived character state that unites *Silurana* and the pipines, a single, rhomboid vomer occurs in *Eoxenopoides* (Estes 1977); this has been considered an autapomorphy of *Xenopus*. Parenthetically, it is noteworthy that in young specimens of *Xenopus* (e.g., *X. fraseri, X. borealis* pers. obs.) the vomers are paired, although it has been noted that in *X. laevis* no evidence for a paired origin is found even during metamorphosis (Paterson 1939; but see Trueb & Hanken 1992 for a different opinion). In *Eoxenopoides*, however, the vomer tends to fuse to the sphenethmoid and parasphenoid as in the Palaeocene *X. romeri* from Brazil (see below). Estes (1977) noted a number of resemblances between *Eoxenopoides* and *Xenopus* (including *X. tropicalis*), but the polarity of those character states was not determined. Apart from the azygous vomer, none of the osteological autapomorphies of *Xenopus* proposed by Cannatella & Trueb (1988a) is present in this taxon. Because of the presence of several presumably primitive character states absent in living pipids, such as the moderately expanded sacral diapophyses, lack of carpal torsion and relatively short and wide urostyle, it was considered (Estes 1977; Báez 1981) that the origin of *E. reuningi* preceded appearance of *Xenopus*.

During the early Palaeogene, Africa was moving northward; consequently, increasingly dry conditions developed in the northern part of the continent. By the mid-Miocene, a return to a monsoon-influenced global climate probably implies that a rainfall pattern similar to the present one prevailed at that time (Parrish *et al.* 1982). Pipids are recorded in the Lower Oligocene of Libya (Špinar 1980) and the Miocene of Morocco (Vergnaud-Grazzini 1966), north of their present range. The specimens from central Libya (Fig. 18.2, Jabal Al Hasawinah) are poorly preserved and most of them probably represent metamorphosing larvae or recently metamorphosed individuals. They were described as a new species of *Xenopus*, *X. hasaunus*, and allocated as the type of a new monotypic subgenus, *Libycus* (Špinar 1980). Although some bones might have been misidentified, some character states possessed by this frog (according to the original description) suggest that it is not referable to *Xenopus*. The characters that this taxon shares with *Xenopus* are also present in other pipids, but the wide cultriform process of the parasphenoid, fusion of the first two presacral vertebrae and apparent lack of vomers might indicate that it is related more closely to *Silurana* or the pipines. It is noteworthy that the anterior end of the parasphenoid has been described as blunt, recalling a derived condition present in *Hymenochirus*, but the shape and extent of the posterior part of this bone are not clear (Špinar 1980).

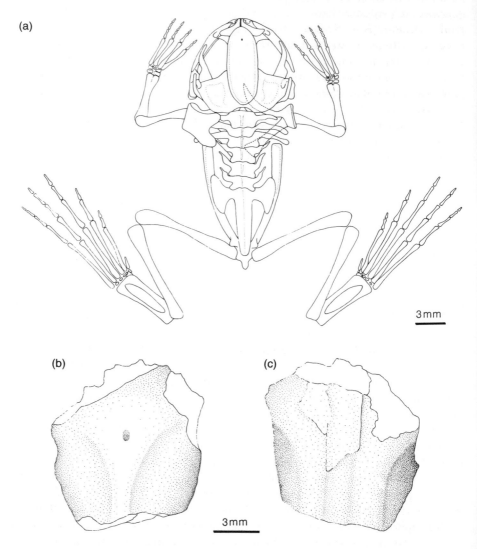

Fig. 18.3. (a) Restoration of *Eoxenopoides reuningi*, after Estes (1977). Courtesy of the *Annals of the South African Museum*. (b) cf. *Xenopus* sp., dorsal view of skull fragment. Museo Argentino de Ciencias Naturales (Argentina), RN-159. (c) Ventral view of same specimen.

The record from Morocco cited above consists of isolated remains, some of which might document the presence of *Silurana*. Minor differences from the living *Silurana* suggest that these specimens represent a different species (Vergnaud-Grazzini 1966).

Isolated bones from Early to Middle Miocene deposits of the present Namib Desert region also document the presence of pipids (Fig. 18.2, diamond fields south of Luderitz). They were referred to a new species, *Xenopus stromeri*, an action that was based on the large size of the bones (Ahl 1926). The figured skull fragment (Ahl 1926: figure 22) bears an overall resemblance to *Xenopus* species, particularly to *X. muelleri* in the shape of the dorsal skull table and general proportions (Estes 1975b). The ethmoidal region seems well ossified as it is in other fossil pipids, but not in the extant species of *Xenopus*. Other features are difficult to evaluate from the illustrations and, unfortunately, the material apparently is lost (Estes 1975b).

Considerable numbers of frog remains were recovered from Bed 1 at Olduvai Gorge, Tanzania, a fossiliferous level considered Lower Pleistocene in age. The presence of abundant remains that represent a species of *Xenopus* was reported (Leakey 1965: 70–72).

Remains of uncertain age collected in deposits from a former crater lake in Zaire are referable to Pipidae (R. Estes *in litt.* 1982). Three of the specimens are of metamorphosed, but young, individuals, whereas another retains remnants of a tail. Imprints of the skin and body outline were preserved in all four. This material is still undescribed.

The South American record

Recent finds extend the record of pipids in South America back to the Middle Cretaceous (around 100–95 Ma). Partially articulated remains, exposed in ventral view, were collected in the lower section of the Albian–Cenomanian Río Limay Formation in north-western Patagonia, Argentina (Fig. 18.2, El Gigante). Although the material requires further preparation, several derived character states are evident that, combined, support the referral of these fossils to the family Pipidae. Among these are opisthocoelous vertebrae that presumably are epichordal in development, prootics with transverse furrows to accommodate the Eustachian tubes, a fused sacrum and coxis, a well-developed, blade-like coronoid process of the angulosplenial, and a squamosal encircling an elongate columella (Báez & Calvo 1989 and in prep.). Although this taxon seems to be represented by a young individual, the first two vertebrae are fused, as in *Silurana* and the pipines. However, the nerve foramina between these vertebrae seem narrower than those of the extant species of *Silurana*, and no notch is visible between atlantal cotyles. The presence of other derived character states that would support a close relationship to the above-mentioned clade cannot be ascertained at present, but some other features should be mentioned. The clavicle is not fused to the capitate scapula, whereas it is fused in individuals of comparable stage of development in *Xenopus laevis*. In addition, the ilium lacks a crest on its shaft, unlike all living pipids.

All other Cretaceous fossils are from deposits dating from around 80 to 73 Ma. Numerous imprints of articulated skeletons of frogs were collected in beds ascribed to the Campanian Las Curtiembres Formation in north-western Argentina (Fig. 18.2, Alemania). The material includes many specimens corresponding to early post-metamorphic stages and tadpoles, representing the pipid taxon *Saltenia ibanezi* (Reig 1959; Báez 1981). Unlike the older pipid from the Río Limay Formation, there is no fusion of presacral vertebrae and the posterior presacral vertebrae bear styliform, anteriorly-oriented transverse processes. Moreover, the different configurations of both the pterygoids and the bony posteromedial processes of the hyolaryngeal apparatus indicate that the specimens from Río Limay and those from Las Curtiembres represent different taxa. It should be noted that although many specimens are available, the type of preservation and the grainy nature of the sediment in which they occur limit the interpretation of several character states present in *Saltenia*. Examination of a latex mould of a ventral imprint revealed the presence of an azygous vomer. This is the only derived character state among the few noted in the extant species of *Xenopus* (Cannatella & Trueb 1988a) present in *Saltenia* and one that also is shared with *Eoxenopoides* from South Africa (see above). The nasals are unfused and lack a distinct rostral projection, and the scapula, although cleft medially, is not synostotically united to the clavicle, unlike the conditions in *Xenopus*. In *Saltenia*, the dorsal margin of the atlas is straight or slightly convex anteriorly; this character was considered a synapomorphy of pipines (Cannatella & Trueb 1988a). However, this atlantal shape seems to be widespread among pipoids (e.g., the early Cretaceous *Shomronella, Thoraciliacus* and *Cordicephalus* from Israel, palaeobatrachids, *Eoxenopoides*); it might, therefore, be more parsimonious to consider this condition to be plesiomorphic at this level. In the possession of three ossified distal tarsals, *Saltenia*, like *Cordicephalus* and *Thoraciliacus* (Nevo 1968), might be considered more primitive than *Xenopus* and *Silurana*, which possess only two. This is suggested also by an extensive contact between the well-developed dorsal acetabular expansion of both ilia, a condition that occurs in the African *Eoxenopoides*.

The presence of pipids in slightly younger lacustrine deposits in north-eastern Patagonia is documented by a few isolated bones obtained from washing-and-screening operations in the Late Cretaceous Los Alamitos Formation (Fig. 18.2, Los Alamitos). The presence of a pipid comparable to the presently strictly African *Xenopus* is based on the features shown by an incomplete braincase (Fig. 18.3b,c), a basal portion of ilium, and several distal portions of humeri (Báez 1987). The presence of vomer, narrow cultriform process of parasphenoid, and closely spaced parasagittal crests on frontoparietal indicate that the taxon represented should not be included within the *Silurana* + Pipinae clade, but none of the proposed osteological derived character states for *Xenopus* (Cannatella & Trueb 1988a) can be demonstrated to be present because of the fragmentary condition of the material. At least some of the features in which this taxon more closely resembles *Xenopus* (e.g., the round section of the basal ilial shaft and the relatively wide dorsal acetabular expansion of the ilium) probably are primitive character states.

Even considering the changes in the shape of the frontoparietal crests during ontogeny (e.g., Reumer 1985), the very narrow dorsal table is unlike that of other pipid fossils; this suggests that these specimens from Patagonia represent a different species.

All other known records of pipids are from Tertiary deposits. Clastic fissure fillings in limestone near Rio de Janeiro, Brazil (Fig. 18.2, Itaborai), of Middle Palaeocene age (62–61 Ma) have yielded remains of a pipid frog, *Xenopus romeri* (Estes 1975a, b) (Fig. 18.4c). Referral to the genus was based on several character states (e.g., the distinctive morphology of the frontoparietal, parasphenoid with a narrow cultriform process, very short and cleft scapula, rod-like iliac shafts); however, the primitive or derived status of these characters was not evaluated and some of them might be primitive features for pipids in general. An actual relationship between *X. romeri* and *Silurana tropicalis* (auctorum) was proposed by Estes (1975b), a conclusion supported by the wide braincase, lateral position of retractor bulbi muscle scars, and fusion of the first two presacral vertebrae. The latter character state is one of the derived features used by Cannatella & Trueb (1988a) to diagnose the *Silurana* + Pipinae clade. However, vomers, which are absent in members of that clade, apparently are present in *X. romeri*. On the type specimen there is a large, rhomboidal, slightly concave area on the ventral surface of the sphenethmoid and cultriform process of the parasphenoid. In larger specimens, the same area is broader and thicker, suggesting the presence of a vomer; however, no clear sutures are visible. It seems logical to assume, as Estes did (1975b), that the vomer tends to fuse to the overlying parasphenoid and sphenethmoid in this Palaeocene species. It is noteworthy, however, that the cultriform process appears expanded at the level of the anterior margin of the sphenethmoid in extant *Silurana*, but no information on the development of this region is available. The nasals are preserved in some specimens of *X. romeri*; they are fused to each other and to the overlying frontoparietal.

Several finds document the presence of pipids in a lacustrine region that developed in association with a volcanic belt along the Pacific coast of northern Patagonia in early Tertiary times (around 60–55 Ma). A new pipid genus and species, *Shelania pascuali* (Fig. 18.5a), was erected for several articulated specimens (Casamiquela 1960, 1961, 1965) collected from a late Palaeocene sequence in the Laguna del Hunco region (Fig. 18.2). The specimens on which the taxon is based correspond to young post-metamorphic stages; the frontoparietal of the largest of those is shown in Fig. 18.4b. Because the skeletal elements of *Shelania* (as then known) were only slightly different from those of *Xenopus*, the former was included in the synonymy of the latter (Estes 1975b; Gasparini & Báez 1975). Subsequently, additional specimens were collected and a more comprehensive study is in preparation by the author. It is worth mentioning that even in young individuals, the nasals tend to fuse to each other anteromedially along the rostral process, although a weak suture is usually evident. The nasals, which rest on the well-ossified sphenethmoid, appear shallow in articulated specimens (Fig. 18.4a) but, in fact, a great part of their

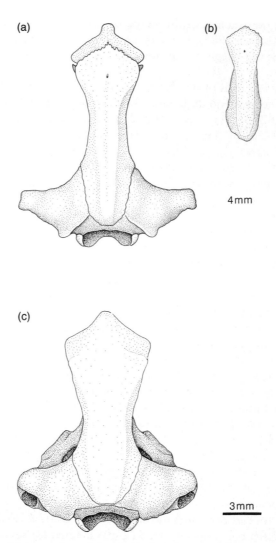

Fig. 18.4. (a) Partial restoration of the skull of *Shelania pascuali* in dorsal view. (b) Dorsal view of frontoparietal of *S. pascuali*, Museo de La Plata (Argentina) 62–XII–22–1. (c) Partial restoration of skull of *Xenopus romeri* based on Departamento Nacional da Producão Mineral, Paleontologia (Rio de Janeiro, Brazil) 569.

dorsal surface is covered posteriorly by the frontoparietal. This three-layered skull roof was also described in *Rhinophrynus* (Trueb & Cannatella 1982) and contrasts with the less ossified ethmoidal region of the skull in extant *Xenopus*. Unlike the living species of *Xenopus* investigated, the sphenethmoid bounds the anterior margin of the frontoparietal fontanelle. The configuration of the edentulous maxilla also differs from that of extant *Xenopus*, but no derived character state of the *Silurana* + Pipinae clade is present. The

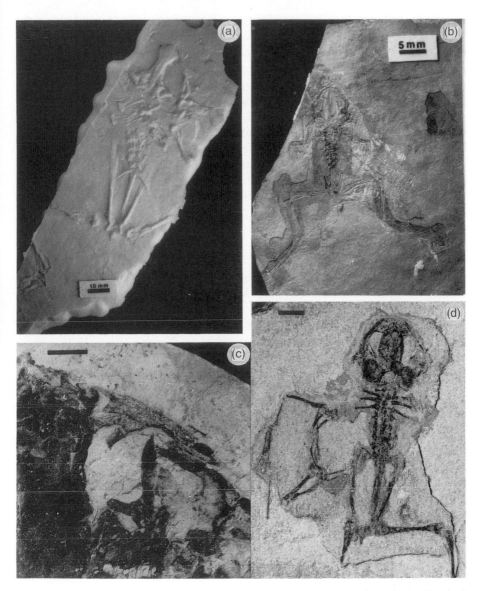

Fig. 18.5. (a) *Shelania pascuali*, latex mould of specimen 12219, Paleontología, Facultad de Ciencias Exactas, Universidad de Buenos Aires. (b) Specimen from Pampa de Jones, corresponding to an immature individual. Asociación Paleontológica Bariloche (Argentina) 3080–10. (c) Specimen from Confluencia. Asociación Paleontológica Bariloche 2361. (d) *Saltenia ibanezi* Museo de La Plata 62–XII–5–110. Black bar in (c) and (d) equals 3 mm. (c) and (d) courtesy of *Ameghiniana*.

available evidence justifies generic distinction of this taxon, and thus, the genus *Shelania* should be revalidated. In a nearby locality (Fig. 18.2, Río Pichileufu), coeval deposits have yielded delicate imprints of young individuals that possess general pipid features.

Another taxon is represented by imprints as well as sections of bones (Fig. 18.5c), collected in a series considered mid Palaeogene (around 50–35 Ma), in north-western Patagonia (Fig. 18.2, Confluencia). There, this species lived in bodies of water within forests dominated by conifers and southern beeches (Báez, Zamaloa & Romero 1990). The braincase is short and wide, instead of long and narrow as in *Shelania pascuali*. This species resembles *Xenopus romeri* in having the first two presacral vertebrae fused, a parabolic flat and unsculptured skull table extending to the posterior border of the frontoparietal and an anterior pineal foramen (Báez 1991b). Unfortunately, the presence or absence of vomers is not possible to determine on the available specimens. An isolated posterior portion of a hind limb preserved in a concretion indicates differences in arrangement, size, shape, and possibly number, of distal tarsal elements in contrast to the extant species of *Xenopus* and *Silurana* available for such comparisons (*X. laevis*, *X. muelleri*, *S. tropicalis*; Báez in prep.).

Eocene deposits exposed in two additional localities within the same general region have also yielded materials referable to Pipidae (Báez 1991a, pers. obs.). Imprints of articulated individuals with some fragments of bone left and corresponding to different ontogenetic stages were found in Early Eocene (Melendi, Scafati & Volkheimer in press) outcrops near the Lake Nahuel Huapi (Fig. 18.2, Pampa de Jones). Most remains, however, are those of early post-metamorphic stages or metamorphic larvae still showing three pairs of free ribs; one tadpole even has one barbel at each side of the mouth (Báez & Lavilla in prep.). This (Fig. 18.5b) material represents a taxon closely related to the one present in Confluencia. Very recently, the isolated prootics of a pipid frog were identified by the author among the material collected from the Late Eocene Vaca Mahuida Formation in the north-western fringe of Patagonia. Deep furrows to accommodate the Eustachian tubes are present and the posterior part of the parasphenoid, lacking alae, appears partially fused to the otic capsules. Presacral vertebrae with opisthocoelous and flattened centrum, among them one atlas fused to the second vertebra, are also present in the sample.

Fossil pipids and pipid evolutionary history

The distinctive morphology of pipids leaves little doubt about the familial allocation of most putative records, although in some cases these frogs are represented by fragmentary material that is not especially informative systematically. Thus, the assessment of many of the diagnostic pipid synapomorphies listed by Cannatella & Trueb (1988a) is restricted by the type and degree of preservation.

The large morphological hiatus distinguishing *Xenopus* (and *Silurana*) from the more derived pipine clade anticipates the existence of extinct forms with a combination of character states not found among extant representatives. However, some of the fossil taxa possess a combination of traits that obviates their inclusion in the lineages based on previous neontological studies without a considerable increase of homoplasy. Thus a re-analysis of pipid relationships considering the total evidence from both extinct and extant forms and assessments of evolutionary change is necessary. Even so, some aspects of the fossil record deserve to be commented on at this point.

As currently known, the fossil record is strongly biased geographically. Most fossil pipid remains from South America have been found in the southern part of the continent. This bias might reflect the long-term palaeontological efforts conducted in this region; nonetheless, it is evident that during the Late Cretaceous and Palaeogene, pipids were a common component of the freshwater fauna in areas where they do not occur today. Most of these fossils generally resemble *Xenopus* (or *Silurana*) in their possession of an unsculptured frontoparietal bearing parasagittal crests and a long narrow parasphenoid. Moreover, some of these specimens, collected in Brazil and southern Argentina, are considered to represent species of the now strictly African *Xenopus*, a genus that presumably has diverged little from the ancestral pipid morphotype. Potentially, this has significant macroevolutionary and biogeographical implications and, consequently, it seems pertinent to point out some interesting facts. The finding of additional specimens of '*Xenopus*' *pascuali* from the Palaeogene of Patagonia documents that the edentulous maxilla of this taxon bears a conspicuous process at the level of the planum antorbitale, forming a brace for the upper jaw (Fig. 18.4a). A detailed analysis of the feature is beyond the scope of this paper, but the presence of an osseous wall between the orbit and the nasal capsule involving the maxilla is unknown in extant *Xenopus* (and *Silurana*). This feature is also present in other fossil pipids from South America of different ages and representing more than one species (e.g. Fig. 18.5a–d). Such a maxillary process has not been described in any of the taxa recorded in Africa in which the maxilla is known. The identity of the 'frontal process' of the maxilla described by Nevo (1968) in *Thoraciliacus* is unclear at present. A blunt, low preorbital process is present in palaeobatrachids (Špinar 1972). The presence of a conspicuous maxillary process in the region of the planum antorbitale should be investigated further because if it is derived, it would suggest that these South American forms might comprise an endemic clade. These forms also seem to share the lack of teeth. This also poses a question: was such a process present in those taxa of which the maxilla is still unknown, such as '*Xenopus*' *romeri*? Some resemblance of this latter species to the Late Cretaceous '*Xenopus*' from In Beceten and to *Xenopus tropicalis* (= *Silurana*) was invoked to support an early diversification of *Xenopus* preceding the separation of Africa and South America (Estes

1975b), although an actual relationship based on shared derived character states has not been demonstrated.

Accepting the current hypothesis of relationships of extant pipid taxa, the occurrence of pipines in the lower Senonian of western central Africa indicates that the lineages represented today by *Xenopus* and *Silurana* should have existed already. Examination of the fossil record shows that although some Palaeogene taxa possess putative autapomorphies of *Xenopus* (i.e., azygous vomer, fused nasals), these characters occur in combination with other characters not present in the living representatives of the genus. Also, some presumably derived characters shared by all extant species of *Xenopus* (and *Silurana*) are lacking (i.e., medially expanded clavicles, complex zygapophyses).

The presence of some hymenochirine features in the Late Cretaceous pipine from Niger suggests that the vicariance between *Pipa* and *Hymenochirus* could have coincided with the final separation of north-eastern Brazil and equatorial West Africa. Thus, pipine ancestors were probably present in a still-connected Africa–South America block. There are fossil taxa from both sides of the Atlantic having some of the proposed synapomorphies for pipines. Fusion of the atlas and axis, shared with *Silurana* as well as with pipines, occurs in many forms on both continents. However, the fused element, at least in the South American forms (specimens from Los Gigantes, Itaborai, Confluencia and Puelén), is very short and has reduced nerve foramina unlike the fused atlas and axis of extant *Silurana*. Besides, it appears clear that at least some of these forms have discrete vomers.

The change of climatic and physiographical patterns, at global as well as local levels, surely were correlated, directly or indirectly, with the evolutionary history of pipid frogs. But at this point, it would be naive to propose a sequence of vicariant and/or dispersal events without having had resolved the cladistic relationship among extant and extinct representatives.

Acknowledgements

Many thanks are due to Professor Richard Tinsley (University of Bristol), Dr Hansruedi Kobel (Université de Genève) and the Zoological Society of London, who made my presence at this symposium possible. I also thank Drs Linda Trueb (University of Kansas), Richard Tinsley and Raymond Laurent (Instituto Lillo) for the gift and/or loan of specimens for comparisons. Thanks are extended to Dr Rosendo Pascual (Museo de La Plata), Dr José Bonaparte (Museo Argentino de Ciencias Naturales) and Ms Helga Smekal for the loan of fossil material under their care. I acknowledge the contribution of the Asociación Paleontológica Bariloche for assistance in the field work. I am especially indebted to Linda Trueb for her generosity during several stays at the Museum of Natural History, University of Kansas, and the critical reading of an early version of the manuscript. This research was funded in part by BID-CONICET grant 1435/91.

References

Ahl, E. (1926). *Xenopus stromeri* Ahl, n.sp. In *Die Diamantenwüste Südwest-Afrikas* 2: 141–142. (Ed. Kaiser, E.) Verlag von Dietrich Reimer (Ernst Vohsen), Berlin.

Báez, A.M. (1981). Redescription and relationships of *Saltenia ibanezi*, a Late Cretaceous pipid frog from northwestern Argentina. *Ameghiniana* 18: 127–154.

Báez, A.M. (1987). The Late Cretaceous fauna of Los Alamitos, Patagonia, Argentina. Part III. Anurans. *Revta Mus. argent. Cienc. nat. (Paleont.)* 3: 121–130.

Báez, A.M. (1991a). Anuros en el Eógeno de los alrededores del lago Nahuel Huapi, Neuquén Meridional. *Ameghiniana* 28: 403.

Báez, A.M. (1991b). New early Tertiary pipid frog from Patagonia and the evolution of pipids in South America. *Boletim de Resumos: XII Congresso Brasileiro de Paleontologia. São Paulo, Brasil*: 96.

Báez, A.M. & Calvo, J. (1989). Nuevo anuro pipoideo del Cretácico medio del noroeste de Patagonia, Argentina. *Ameghiniana* 26: 238.

Báez, A.M. & Rage, J.C. (1988). Evolutionary relationships of a new pipid frog from the Upper Cretaceous of Niger. *Program and Abstracts: Combined Meeting Herpetologists' League, American Elasmobranch Society, etc., The University of Michigan, Ann Arbor, Michigan*: 61.

Báez, A.M., Zamaloa, M.C. & Romero, E.J. (1990). Nuevos hallazgos de microfloras y anuros paleógenos en el noroeste de Patagonia: implicancias paleoambientales y paleobiogeográficas. *Ameghiniana* 27: 83–94.

Barron, E.J., Harrison, C.G.A., Sloan, J.L. & Hay, W. (1981). Paleogeography, 180 millions of years ago to the Present. *Eclog. geol. Helv.* 74: 443–470.

Broin, F. de, Buffetaut, E., Rage, J.C., Russell, D., Taquet, P., Vergnaud-Grazzini, C. & Wenz, S. (1974). La faune de vértebrés continentaux du gisement d' In Beceten (Senonien du Niger). *C.r. hebd. Séanc. Acad. Sci., Paris* 279: 2326–2329.

Cannatella, D.C. (1986). Phylogenetic position of the frog family Paleobatrachidae. *Am. Zool.* 26: 92A.

Cannatella, D.C. & Trueb, L. (1988a). Evolution of pipoid frogs: intergeneric relationships of the aquatic frog family Pipidae (Anura). *Zool. J. Linn. Soc.* 94: 1–38.

Cannatella, D.C. & Trueb, L. (1988b). Evolution of pipoid frogs: morphology and phylogenetic relationships of *Pseudhymenochirus*. *J. Herpet.* 22: 439–456.

Casamiquela, R. (1960) [1961]. Datos preliminares sobre un pipoideo fósil de Patagonia. *Actas Trab. Congr. sud-am. Zool.* 1 (4): 17–22.

Casamiquela, R. (1961). Un pipoideo fósil de Patagonia. *Revta Mus. La Plata (Sec. Paleont.)* (N.S.) 4: 71–123.

Casamiquela, R. (1965). Nuevos ejemplares de *Shelania pascuali* (Anura, Pipoidea) del Eoterciario de la Patagonia. *Ameghiniana* 4: 41–51.

De Queiroz, K. & Gauthier, J. (1990). Phylogeny as a central principle in taxonomy: phylogenetic definitions of taxon names. *Syst. Zool.* 39: 307–322.

Emery, K.O. & Uchupi, E. (1984). *The geology of the Atlantic Ocean*. Springer-Verlag, New York, Berlin, Heidelberg, Tokyo.

Estes, R. (1975a). *Xenopus* from the Palaeocene of Brazil and its zoogeographic importance. *Nature, Lond.* **254**: 48–50.

Estes, R. (1975b). Fossil *Xenopus* from the Paleocene of South America and the zoogeography of pipid frogs. *Herpetologica* **31**: 263–278.

Estes, R. (1977). Relationships of the South African fossil frog *Eoxenopoides reuningi* (Anura, Pipidae). *Ann. S. Afr. Mus.* **73**: 49–80.

Estes, R. & Reig, O.A. (1973). The early fossil record of frogs. A review of the evidence. In *Evolutionary biology of the anurans. Contemporary research on major problems*: 11–63. (Ed. Vial, J.L.). University of Missouri Press, Columbia.

Estes, R., Špinar, Z.V. & Nevo, E. (1978). Early Cretaceous pipid tadpoles from Israel (Amphibia: Anura). *Herpetologica* **34**: 374–393.

Gasparini, Z. & Báez, A.M. (1975). Aportes al conocimiento de la herpetofauna terciaria de la Argentina. *Actas Congr. argent. Paleont. Bioestratigr.* **1** (3): 377–415.

Gauthier, J., Estes, R. & De Queiroz, K. (1988). A phylogenetic analysis of Lepidosauromorpha. In *Phylogenetic relationships of the lizard families*: 15–98. (Eds Estes, R. & Pregill, G.K.). Stanford University Press, Stanford.

Haughton, S. (1931). On a collection of fossil frogs from the clays at Banke. *Trans. R. Soc. S. Afr.* **19**: 233–249.

Leakey, L.S.B. (1965). *Olduvai Gorge, 1951–1961. 1. A preliminary report on the geology and fauna.* Cambridge University Press, Cambridge.

Melendi, D.L., Scafati, L.H. & Volkheimer, W. (In press). Datación palinológica de un arco magmático del Paleógeno del noroeste de la Patagonia (Argentina). *Ameghiniana.*

Nevo, E. (1968). Pipid frogs from the Early Cretaceous of Israel and pipid evolution. *Bull. Mus. comp. Zool. Harv.* **136**: 255–318.

Parrish, J.T., Ziegler, A.M. & Scotese, C.R. (1982). Rainfall patterns and the distribution of coals and evaporites in the Mesozoic and Cenozoic. *Palaeogeogr. Palaeoclimat. Palaeoecol.* **40**: 67–101.

Paterson, N.F. (1939). The head of *Xenopus laevis. Q. J. microsc. Sci.* **81**: 161–234.

Reig, O.A. (1959). Primeros datos descriptivos sobre los anuros del Eocretáceo de la provincia de Salta (Rep. Argentina). *Ameghiniana* **1**: 3–8.

Reumer, J.W.F. (1985). Some aspects of the cranial osteology and phylogeny of *Xenopus* (Anura, Pipidae). *Revue suisse Zool.* **92**: 969–980.

Saint-Marc, P. (1978). Arabian Peninsula. In *The phanerozoic geology of the world. 2 The Mesozoic* A: 435–462. (Eds Moullade, M. & Nairn, A.E.M.). Elsevier Scientific Publishing Company, Amsterdam, Oxford, New York.

Sokol, O.M. (1975). The phylogeny of anuran larvae: a new look. *Copeia* **1975**: 1–23.

Sokol, O.M. (1977). The free swimming *Pipa* larvae, with a review of pipid larvae and pipid phylogeny (Anura: Pipidae). *J. Morph.* **154**: 357–426.

Špinar, Z.V. (1972). *Tertiary frogs from Central Europe.* Academia, Publishing House of the Czechoslovak Academy of Sciences, Prague.

Špinar, Z.V. (1980). The discovery of a new species of pipid frog (Anura, Pipidae) in the Oligocene of central Libya. In *The geology of Libya* **1**: 327–348. (Eds Salem, M.J. & Busrewil, M.T.). Academic Press, London.

Trueb, L. (1985). A summary of osteocranial development in anurans with notes on the sequence of cranial ossification in *Rhinophrynus dorsalis* (Anura: Pipoidea: Rhinophrynidae). *S. Afr. J. Sci.* **81**: 181–185.

Trueb, L. & Cannatella, D.C. (1982) The cranial osteology and hyolaryngeal apparatus of *Rhinophrynus dorsalis* (Anura: Rhinophrynidae) with comparisons to Recent pipid frogs. *J. Morph.* **171**: 1–40.

Trueb, L. & Hanken, J. (1992). Skeletal development in *Xenopus laevis* (Anura: Pipidae). *J. Morph.* **214**: 1–41.

Van Dijk, D.E. (1985). An addition to the fossil Anura of southern Africa. *S. Afr. J. Sci.* **81**: 207–208.

Vergnaud-Grazzini, C. (1966). Les amphibiens du Miocène de Beni-Mellal. *Notes mém. Serv. Mines Carte géol. Maroc* **27** (No. 198): 43–74.

19 Historical constraints and morphological novelties in the evolution of the skeletal system of pipid frogs (Anura: Pipidae)

LINDA TRUEB

Synopsis

Pipid frogs of the genera *Xenopus, Silurana, Hymenochirus, Pseudhymenochirus* and *Pipa* are dissimilar to all other living anurans and highly derived rather than primitive. Among living pipids, *Xenopus* is the most primitive—i.e., the most similar to the hypothesized ancestor of the Pipidae. Nonetheless, the adults of this taxon possess many morphological features that distinguish them from their closest living non-pipid relatives, *Rhinophrynus dorsalis* (Rhinophrynidae) and the pelobatoid frogs. Many of the morphological novelties seem to be associated with the evolution of an aquatic anuran from a terrestrial saltatorial ancestor. The sprawled, laterally positioned limbs and dorsoventral compression of the body in pipids that facilitate their swimming inhibit their hopping or jumping and are correlated with modifications of the thigh musculature and structures of the pectoral and pelvic girdles. Movement of the trunk is limited to anterior–posterior shifts of the pelvic girdle on the sacrum; dorsal–ventral rotation of the posterior trunk, which is critical to saltatorial locomotion, is not possible. Many cranial modifications in *Xenopus* seem to be associated with its mode of feeding without a tongue, but little is known about the feeding mechanism. The mandible is robust, whereas the upper jaw is weak and suspended from the skull by a modified suspensory apparatus. The hyolaryngeal apparatus is highly modified. The plectral apparatus is hypertrophied and the tympanic annulus is ossified and in synostosis with the dermal squamosal medially. The orbital region of the braincase is depressed and formed by membrane, rather than endochondral, bone.

Introduction

In comparison to the 4000 or so species of other anurans, pipid frogs have attracted a disproportionate amount of attention in the last century. They are biological curiosities—odd-looking frogs that are most certainly Gondwanan in origin and whose present distribution includes South America and Africa, and a family that has the most complete fossil record of any Recent amphibian group. Formerly thought to be one of the most primitive anuran groups, pipids are now considered to be among the most highly derived and advanced of the archaeobatrachian frogs (Fig. 19.1a; Cannatella 1985). Among living anurans, they are most closely related to the monotypic Rhinophrynidae, which together with the Pipidae compose the superfamily Pipoidea. Pipoids are related most closely to the Pelobatoidea (i.e., Megophryidae + Pelobatidae + Pelodytidae). The Pipoidea and Pelobatoidea comprise the clade Mesobatrachia, which is the sister group of the neobatrachians—i.e., all advanced anurans. Living pipids comprise a small group of 29 species placed in five genera (Fig. 19.1b; Cannatella & Trueb 1988b; Frost 1985), of which *Xenopus* is the largest and most primitive genus. The other genera are *Silurana*[1] (two species), *Pseudhymenochirus* (one species), *Hymenochirus* (four species), and *Pipa* (seven species).

Pipid frogs are aquatic and seemingly more specialized for an aquatic life style than any other group of anurans. This is suggested by their wide, flat habitus, dorsal eyes, possession of a lateral line system in the adult, extensive webbing, powerful hind limbs that cannot be folded under the body and specialized aquatic courtship rituals. Herein, the evolutionary trends that have produced the skeleton of pipid frogs are discussed and speculations are offered as to the possible functional significance of the peculiar skeletal features of pipids in comparison with their closest living relatives, *Rhinophrynus* and the pelobatoids, as well as other living anurans in general.

Preliminary considerations

In order to discuss evolutionary trends and novelties, one must identify the operant historical constraints—i.e., the ancestral shared, derived characters that diagnose a natural assemblage of organisms such as pipid frogs. Pipids are subject to the historical constraints of all anurans; for a recent account, see Trueb & Cloutier (1991). For example, pipids and all other anurans lack a tail and possess hind limbs that are elongated and longer than the forelimbs. The pelvic girdle is elongated and has a special articulation with a modified sacral vertebra. The presacral trunk is short and relatively inflexible; the postsacral vertebrae are fused into a bony rod. The skull has a reduced number of cranial elements and the pectoral girdle is hypertrophied. These are but a few of the skeletal characters that define the unique anuran bauplan and distinguish it from that of the sister group of salientians, the salamanders. Fundamentally, the evolution of the anuran bauplan or morphotype seems to have facilitated saltatorial locomotion in frogs and toads (Gans & Parsons 1966); thus, it can be assumed that the ancestor of pipids and their relatives, *Rhinophrynus* and the pelobatoids, was saltatorial.

The more proximal historical constraints of pipids are those diagnosing the clades Mesobatrachia and Pipoidea. There are five shared, derived

[1] Elsewhere in the volume, the conservative view is adopted that *Silurana* is a subgenus of *Xenopus*—ED.

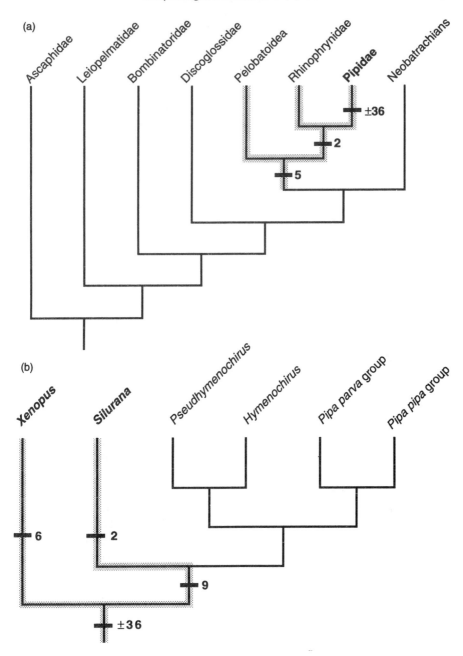

Fig. 19.1. Phylogenetic relationships of pipid frogs. (a) Phylogeny of archaeobatrachian anurans from Cannatella (1985). The Mesobatrachia is represented by the shaded clades. Synapomorphies are indicated by the numbers adjacent to heavy bars. (b) A phylogeny of the Pipidae based on Cannatella & Trueb (1988a,b).

characters for the Mesobatrachia (Fig. 19.1a; Cannatella 1985), three of which involve the skeleton—viz., the structure of the hyoid and the skull roof. The two synapomorphies of the pipoids are osteological characters of the skull. These constraints and others are described in more detail below.

Pipid frogs are obviously distinct from their relatives *Rhinophrynus* and the pelobatoids, which are terrestrial and include accomplished burrowers (e.g., *Scaphiopus* and *Rhinophrynus*). Thirty-six synapomorphies diagnose the Pipidae (Fig. 19.1; Cannatella & Trueb 1988a,b; this paper), of which 28 are skeletal characters and two are myological. These characters constitute the evolutionary novelties of the family Pipidae.

Herein, a broad definition of a morphological or evolutionary novelty is used in the sense of Futuyma (1986) and Cracraft (1990), wherein every shared, derived character of a taxon can be considered an evolutionary novelty. This application is less restricted than the interpretations of Wake & Roth (1989) and Müller & Wagner (1991) and includes the following classes of characters: (1) the loss of features that were present in the ancestor (e.g., some skull bones); (2) changes in shape of elements or architectural units (e.g., the lengthening of the proximal tarsal elements of anurans); (3) changes in numbers of homonomous characters (e.g., an increase or decrease in the number of phalangeal elements); (4) the appearance of a new structure (e.g., the presence of a novel bone in the dermal skull table); and (5) structures that have changed in their positional context (e.g., the stapes of lower tetrapods which is homologous with the hyomandibular bone of fishes).

This paper draws heavily on the data contained in several specific works—viz., Cannatella's (1985) study of the phylogeny of primitive frogs and four papers by Cannatella & Trueb (Trueb & Cannatella 1982, 1986; Cannatella & Trueb 1988a,b) dealing with the antomy and/or phylogenetic relationships of pipoids. To avoid redundancy, these sources are not repeatedly cited in the text. For elaboration of the historical relationships of pipids and their relatives and further explanation of anatomical features, the reader is referred to the latter works (and their reference lists), as well as the review of cranial osteology of *Xenopus* provided by Reumer (1985). General anatomical terminology follows Duellman & Trueb (1986).

Cranium

Historical constraints

From our understanding of the phylogenetic relationships among living mesobatrachians (Fig. 19.1a), it seems that the skull of *Xenopus* (Figs 19.2, 19.3) evolved from a type not dissimilar to that of some pelobatids. Plesiomorphic cranial features of relevance include the following that are

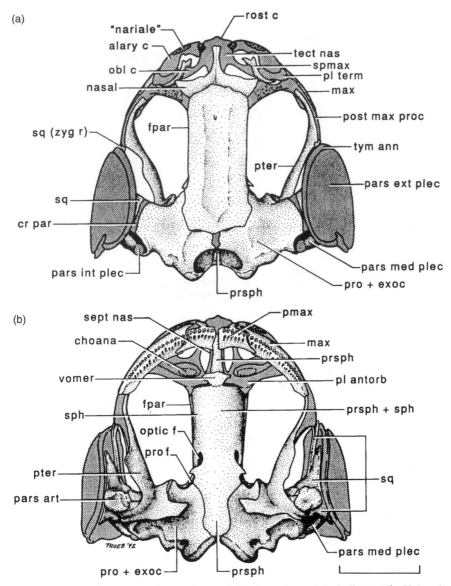

Fig. 19.2. *Xenopus laevis.* Dorsal (a) and ventral (b) views of an adult skull. KU (The University of Kansas Natural History Museum) 20957, male, 52.0 mm snout–vent length. Cartilage is shown in regular stipple pattern; irregular stippling (e.g., "nariale") indicates mineralized cartilage. alary c, alary cartilage; cr par, crista parotica; fpar, frontoparietal; max, maxilla; obl c, oblique cartilage; optic f, optic foramen; pars art, pars articularis of palatoquadrate; pars ext plec, pars externa plectri; pars int plec, pars interna plectri; pars med plec, pars media plectri; pl antorb, planum antorbitale; pl term, planum terminale; pmax, premaxilla; post max proc, posterior maxillary process; pro + exoc, fused prootic and exoccipital; pro f, prootic foramen; prsph + sph, fused parasphenoid and sphenethmoid; prsph, parasphenoid; pter, pterygoid; rost c, rostral cartilage; sept nas, septum nasi; spmax, septomaxilla; sph, sphenethmoid; sq (zyg r), zygomatic ramus of squamosal; sq, squamosal; tect nas, tectum nasi; tym ann, tympanic annulus. Adapted from Trueb & Hanken (1992). Scale = 5 mm.

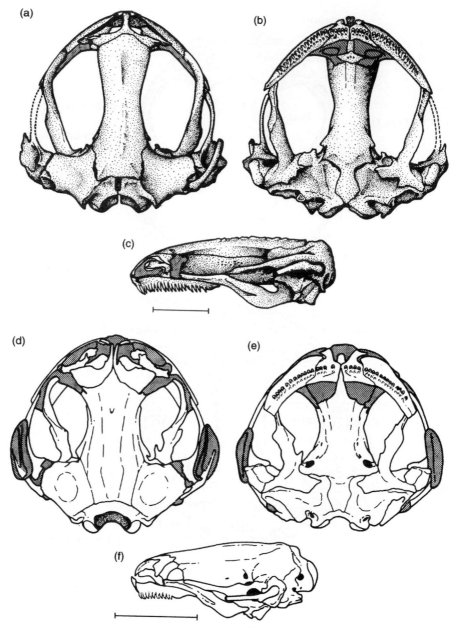

Fig. 19.3. Skulls of *Xenopus* and *Silurana*. (a–c) *Xenopus muelleri* (KU 196043, female, 74.9 mm snout–vent length) in dorsal (a), ventral (b) and lateral (c) aspects. (d–f) *Silurana epitropicalis* (KU 195660, female 61.1 mm) in dorsal (d), ventral (e) and lateral (f) aspects. Missing structures are indicated by dashed lines; cartilage is shown by stippled pattern. Adapted from Cannatella & Trueb (1988a). Scales = 5 mm.

depicted in Figs 19.4a–c and 19.5a: (1) head not notably flattened in lateral aspect; (2) paired frontoparietals and nasals that do not overlap one another; (3) nasal in contact or nearly in contact with the facial process of the maxilla, and lacking rostral processes; (4) parasphenoid terminating anteriorly near the anterior part of the orbit, not fused with other cranial bones, and bearing lateral alae posterolaterally; (5) vomers paired and bearing a postchoanal ramus; (6) maxilla terminating at about the midlevel of the orbit; (7) squamosal not co-ossified with the tympanic annulus if an annulus is present; (8) pterygoid lacking an expanded medial ramus and ventral flange, and having an anterior ramus that articulates with the lateral aspect of the upper jaw; (9) tympanic annulus lacking ossification; (10) stapes relatively short and straight, directed more or less laterally from fenestra ovalis; (11) teeth pedicellate, bicuspid, and not fused to the maxillae and premaxillae; (12) distinct frontoparietal fontanelle delimited anteriorly by the ethmoid cartilage or the sphenethmoid bone in adults; (13) sphenethmoid endochondral in origin, not fused to frontoparietals dorsally or parasphenoid ventrally, enclosing the orbitonasal foramen, and forming the posteromedial walls of the olfactory capsule; (14) septomaxillae small, complex elements; (15) mentomeckelian bones forming mandibular symphysis, which is not broadly overlapped by dentary bones anteromedially; and (16) coronoid flange of angulosplenial small or only moderately developed.

The pipid skull is further constrained by features that the Pipidae shares with its sister taxon, the Rhinophrynidae (Figs 19.4d–f and 19.5b). In both families, the frontoparietal is azygous (but arises from two centres of ossification as in other anurans: Trueb 1985; Trueb & Hanken 1992). Furthermore, the nasals bear distinct rostral processes and are overlapped by the frontoparietal posteriorly. The parasphenoid lacks posterolateral alae. The maxillary arcade is short and the articulation with the lower jaw lies anterior to the fenestra ovalis. Mentomeckelian bones are absent in the mandible and there is only a narrow gap between the dentary bones anteromedially. Meckel's cartilage is undivided in *Rhinophrynus* and some *Xenopus* (e.g., *X. boumbaensis*, *X. wittei*), but divided anteromedially in other *Xenopus* (e.g., *X. largeni*, *X. gilli*). The shared features involving the position and configuration of the upper and lower jaws probably relate to a modification in the feeding habits of the common ancestor. Primitive anurans are incapable of projecting the tongue from the mouth in the way that pelobatoid and neobatrachian frogs can (Regal & Gans 1976), and *Rhinophrynus* has a specialized mode of tongue protrusion that is correlated with the modifications of the mandible and hyoid (Trueb & Gans 1983). Given the constraints of a relatively inflexible mandible and unspecialized tongue, the absence of a tongue and elaboration of the coronoid flange in pipids may represent evolutionary novelties involved in feeding in the water; however, the feeding mechanism of pipids is undocumented to date.

356 *Linda Trueb*

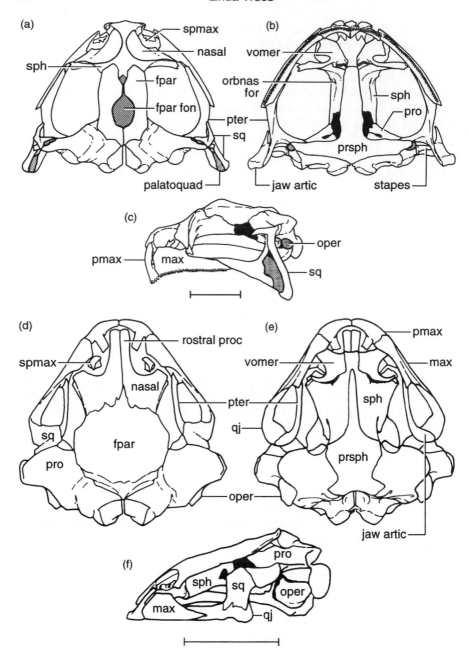

Fig. 19.4. Skulls of the pelobatid anuran *Spea*, and *Rhinophrynus*. (a–c) *Spea bombifrons* (KU 20991) in dorsal (a), ventral (b) and lateral (c) aspects. (d–f) *Rhinophrynus dorsalis* (KU 84886, adult female) in dorsal (d), ventral (e) and lateral (f) aspects. artic, articulation; fon, fontanelle; fpar, frontoparietal; max, maxilla; oper, operculum; orbnas for, orbitonasal foramen; palatoquad, palatoquadrate; pmax, premaxilla; pro, prootic; proc, process; prsph, parasphenoid; pter, pterygoid; qj, quadratojugal; sph, sphenethmoid; spmax, septomaxilla; sq, squamosal. Scales = 5 mm.

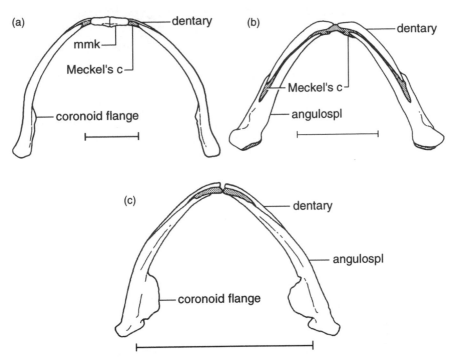

Fig. 19.5. Mandibles of mesobatrachian anurans in dorsal view. (a) The pelobatid *Spea bombifrons* (KU 20991). (b) *Rhinophrynus dorsalis* (TNHC [Texas Natural History Collection] 19327, female 57.8 mm snout–vent length). (c) *Xenopus vestitus* (KU 206873, female, 38.5 mm snout–vent length). angulospl, angulosplenial; c, cartilage; mmk, mentomeckelian bone. Cartilage is represented by stippled pattern. Scales = 5 mm.

Evolutionary novelties

Braincase and olfactory capsule

Relative to other anurans and especially their closest relatives, pipids have flat skulls (Figs 19.3c,f and 19.4c,f); whereas this is evident in *Xenopus* and *Silurana*, it is especially pronounced in the more derived hymenochirines and *Pipa*. In all other anurans for which we have data, the orbital walls of the braincase anterior to the optic foramen are formed by a cartilage replacement bone, the sphenethmoid, a bilateral ossification that usually forms laterally near the anterior part of the orbit. The sphenethmoid grows anterolaterally to enclose the orbitonasal foramen and posteriorly to form the lateral walls of the braincase. Usually, the developing bones meet ventromedially to form the floor of the braincase and unite dorsomedially to form the anterior margin of the frontoparietal fontanelle and the roof of the anterior part of the braincase. Depending on the amount of anterior elaboration, the sphenethmoid variably roofs the posteromedial part of the olfactory capsule and forms part of the medial wall separating the olfactory capsules. In pipids, however, the orbital cartilages disappear near the time of metamorphosis. The developing olfactory capsule is separated from the rest of the neurocranium, and the sphenethmoid is formed by membrane bone that appears near the optic foramen (Trueb & Hanken 1992). These lateral sheets of bone grow forward and dorsally to the frontoparietal; the sphenethmoid apparently does not participate in the formation of the braincase roof or the olfactory capsule. Owing to the erosion of chondrocranial cartilage in the orbital region, a circumscribed frontoparietal fontanelle is absent. As each sphenethmoid grows ventromedially, it contacts and fuses with the parasphenoid; thus, the floor of the braincase is formed by the combination of the sphenethmoid and parasphenoid. As the sphenethmoid grows posteriorly, it forms the entire margin of the optic foramen and the anterior margin of the prootic foramen and, therefore, the entire orbital wall of the braincase. In other anurans, the posterior limit of sphenethmoid ossification usually lies at, or anterior to, the optic foramen, and the posterior orbital wall normally is formed by cartilage or ossification of the prootic.

The flat, arcuate septomaxilla of pipids is unique among known anurans. The shape of the septomaxilla is correlated with its role in support in the olfactory capsule, which is discussed by Jurgens (1971).

Middle and external ear
Because of the anterolateral position of the jaw articulation relative to the fenestra ovalis, the stapes is long and curved anterolaterally, unlike its configuration in any other known living anuran. The tympanic apparatus is concealed in pipids; nonetheless, it is immense relative to the size of the head (Figs 19.2, 19.3). The peripheral rim of the tympanic annulus is cartilaginous, but the central, funnel-shaped portion is ossified and united to the squamosal bone medially to form a compound tympanosquamosal element, which is unknown in any other extant anuran. Similarly, the presence of an enlarged pars externa plectri that forms a circular plate closely bounded by the rim of the tympanic annulus is unknown in other living anurans.

Dermal investing bones
Because of the peculiar formation of the braincase, protection and support of the brain and olfactory capsules is limited to dermal elements. Thus, in all extant pipids, the frontoparietal overlaps the nasals and the nasals bear distinct medial rostral processes (fused in *Xenopus*) to form a moderately complete skull roof (Figs 19.2, 19.3). These features are also found in *Rhinophrynus* (Fig. 19.4d) and, hence, may not be evolutionary novelties of pipids. However, their evolutionary status cannot be determined until more is known about the formation of the anterior braincase and olfactory capsules in *Rhinophrynus*.

Ventral support and protection of the olfactory capsules is limited primarily to the parasphenoid, which is fused to the sphenethmoids and extends anterior, to the level of the premaxillae (Figs 19.2b, 19.3b,f). Presumably, this anterior extension of the parasphenoid supports the nasal capsule. Reduction and loss of vomers is a fairly common evolutionary event in anurans; however, some (but not all) pipids are the only known frogs that possess an azygous medial vomer that is fused to the composite sphenethmoid-parasphenoid bone.

Maxillary arcade and suspensorium

The pterygoid of pipids has several unusual features. The anterior ramus (incomplete in some *Pipa* and missing in hymenochirine pipids) has a broad, movable, dorsal articulation with the maxilla rather than a dorsolateral or lateral articulation. The medial ramus of the pterygoid is expanded into a broad otic plate that partially or totally covers the bony Eustachian canal in the floor of the otic capsule. Pipids bear a robust ventral flange on the anterior ramus of the pterygoid; although not unique to pipids, such a flange is rare among anurans and when present (e.g., the bombinatorid *Barbourula*), it is of a different shape.

The squamosal bone differs by its union with the ossified tympanic annulus described above. Among living pipids, only *Xenopus* and *Silurana* possess a long zygomatic ramus on the squamosal. The anterior end of the ramus articulates with the maxillary arcade via a dorsal process of the cartilaginous posterior maxillary process—an arrangement that may be unique among the living Anura.

Finally, the maxillary arcade of pipids lacks a pars dentalis and possesses nonpedicellate, monocuspid teeth that are fused to the maxillary and premaxillary bones to effect an acrodont condition.

Evolutionary and functional significance of cranial modifications

Fundamental differences between pipid skulls and those of other anurans are the structure and development of the braincase from the modified larval chondrocranium. Although I cannot explain the historical origin or adaptive significance of the peculiar braincase structure, the net result of the modifications is that the skull is depressed, the synchondrotic continuity between the rostrum and the braincase interrupted, and the braincase left open dorsally. While one could argue that a depressed, wedge-shaped skull might present less resistance to forward thrust in the water, I do not find this argument especially compelling.

Other modifications of pipid skulls seem to be logical consequences of the braincase structure. Hypertrophy of the nasals and frontoparietal provides protection for an otherwise open braincase. Hypertrophy of the parasphenoid structurally supports the rostrum anteriorly and, posteriorly, forms a floor to the braincase and transverse bridge between the auditory capsules. The elaboration of the pterygoid seems to provide added support for the poorly developed upper jaw and provides a site of attachment for the pterygomaxillary ligament. Shaw (1986) suggested that the maxilla and pterygoid slide past one another when the upper jaw is elevated and that the pterygomaxillary ligament would limit displacement of the maxillary arcade during feeding.

In most other anurans, the jaw is linked posteriorly to the palatoquadrate by means of a quadratojugal and medially to the palatoquadrate and auditory capsule by the pterygoid; anterior support of the jaw against the braincase is provided by the transverse planum antorbitale, which is synchondrotically united to the sphenethmoid. In addition to having a short maxilla and lacking a quadratojugal, pipids lack anterior support of the upper jaw, because the planum has no structural continuity with the braincase.

Given the absence of a tongue with which to manipulate prey against the palate, vomers seem unnecessary to support and protect the olfactory capsules or bear teeth against which prey might be impaled. The curious tooth implantation may have resulted from the reduction of the maxilla and premaxilla which includes loss of the pars dentalis, the tooth-bearing ventral ridge in other anurans. Perhaps because of the lack of a dental ridge for support, the teeth have become ankylosed to the remainder of the jaw. There is a tendency for teeth to be lost among pipids and the functional utility of teeth in those taxa that retain them (*Xenopus, Silurana, Pipa carvalhoi* and *P. arrabali*) seems to be limited to grabbing and trapping relatively long prey as the frog repeatedly opens and closes its jaws while pushing the prey into its mouth with the forelimbs (Shaw 1986).

Among the modifications of the ear, the length and curvature of the stapes surely are a result of the posterolateral orientation of the fenestra ovalis and the anterolateral position of the jaw articulation. However, the other evolutionary novelties involving the partial ossification of tympanic annulus and its fusion to the squamosal and the hypertrophy of the pars externa plectri may relate to hearing under water (see Elepfandt this volume pp. 177–193).

Hyolaryngeal apparatus

The hyolaryngeal apparatuses of *Rhinophrynus*, pelobatoids and most other anurans consist of a flat hyoid corpus that bears narrow posterolateral processes and stout posteromedial processes that flank the laryngeal apparatus—the cricoid ring that supports paired arytenoid cartilages, which house the vocal cords (Fig. 19.6a,b). Archaeobatrachians typically have a bone, the parahyoid, on the ventral surface of the hyoid corpus; this is present in *Rhinophrynus* (Fig 19.6b), but occurs in only one genus of pelobatoid, *Pelodytes*. Furthermore, the hyalia are divided in pelobatoids and *Rhinophrynus*. The proximal portion is attached to the hyoid corpus and medially expanded to enclose partially the hyoglossal sinus (through which the tongue muscles run from the ventral surface of the hyoid to the floor of the mouth); the distal portion of the hyale lies laterally adjacent to, but distinctly separated from, the main corpus of the hyoid.

The hyolaryngeal apparatus of pipids is markedly different from that of its closest living relatives and other anurans in general. The hyoid corpus is reduced to a small plate of cartilage that supports what may represent

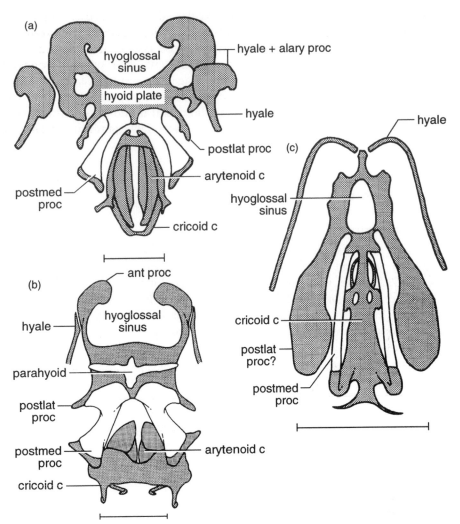

Fig. 19.6. Hyolaryngeal apparatuses of mesobatrachian anurans in ventral view. (a) The pelobatid *Scaphiopus intermontanus* (KU 79436, male, 51.3 mm snout–vent length). (b) *Rhinophrynus dorsalis* (TNHC 19327, female 57.8 mm snout–vent length). (c) *Xenopus wittei* (KU 195673, female, 55.0 mm snout–vent length). ant, anterior; c, cartilage; postlat, posterolateral; postmed, posteromedial; proc, process. Cartilage is represented by stippled pattern. Scales = 5 mm.

vastly enlarged posterolateral processes, posteromedial processes, and a pair of anterior processes that may be modified alary processes combined with the proximal hyalia (Fig. 19.6c). The hyoglossal sinus is small and completely enclosed by anteromedial fusion of these structures. The hyalia, which may be narrow (e.g., *Xenopus wittei*) or expanded (e.g., *X. muelleri*), diverge anterolaterally and dorsally from this medial bridge in *Xenopus* and *Silurana*;

the hyalia may be contiguous with the bridge (e.g., *X. laevis*, *X. muelleri*), narrowly separated from it (e.g., *X. wittei*, *X. boumbaensis*), or absent in hymenochirines and *Pipa*. The posteromedial processes are long, slender bones in *Xenopus* and *Silurana* that support the highly derived laryngeal apparatus of these frogs. The cricoid and thyrohyal cartilages are modified to form a long box that encloses modified arytenoid elements; vocal cords are absent. This suite of evolutionary novelties represents laryngeal modifications that enable pipids to vocalize under water (Rabb 1960; Yager 1992, and this volume pp. 121–141).

Axial skeleton

Historical constraints

As in most anurans, the vertebral columns of pelobatoids and *Rhinophrynus* have eight presacral vertebrae with the first (anterior) two presacrals unfused (Fig. 19.7a,b); the vertebral centra are procoelous or notochordal in the pelobatoids and notochordal in *Rhinophrynus*. Ribs are absent, but the transverse processes of presacral vertebrae II–IV are well developed and notably longer than the short transverse processes on the posterior vertebrae (V–VIII). As in many anurans, the neural arches of the anterior four vertebrae bear neural spines and are imbricate; the imbrication of the posterior vertebrae and the development of neural spines on these vertebrae is relatively unusual among anurans. The sacral diapophyses vary from broadly expanded to moderately dilated in the pelobatoids and are only moderately dilated in *Rhinophrynus*. Correlated with the degree of expansion of the sacral diapophyses is the occurrence of a Type II (of Emerson 1979) iliosacral articulation in *Rhinophrynus* and a Type I articulation in the pelobatoids. Among the taxa that have been surveyed, most have Type II articulations (Emerson 1979; pers. obs.) In most anurans having procoelous or notochordal vertebrae, the coccyx and sacrum have a synovial, bicondylar articulation; however, deviations from this pattern are not uncommon (e.g., synovial monocondylar or synchondrotic or synostotic fusion). *Rhinophrynus* has a synovial, bicondylar articulation, whereas the condition varies among pelobatoids from synovial, bicondylar to synovial monocondylar to synostotic or synchondrotic fusion. Insofar as is known, in all anurans except pipids, the articular surfaces between the pre- and postzygapophyses of adjacent vertebrae are simple.

Evolutionary novelties

The most apparent deviation of the pipid vertebral column from that of its relatives and most other anurans is its possession of opisthocoelous centra—a feature possessed only by bombinatorids and discoglossids—and ribs on presacral vertebrae II–IV (Fig. 19.7c,d). However, unlike the four

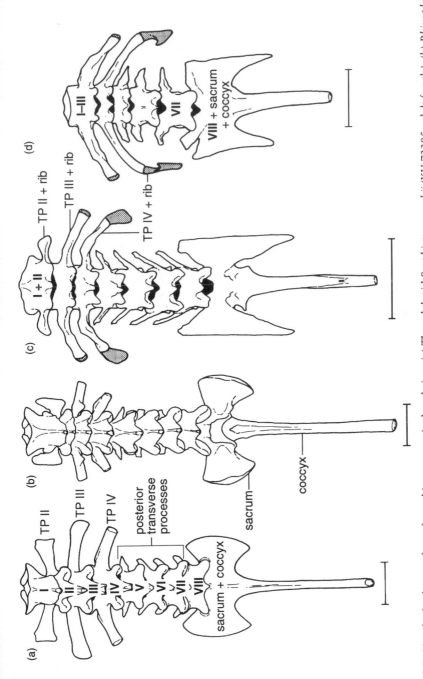

Fig. 19.7 Vertebral columns of mesobatrachian anurans in dorsal view. (a) The pelobatid *Scaphiopus couchii* (KU 73385, adult female). (b) *Rhinophrynus dorsalis* (KU 69084). (c) *Xenopus largeni* (KU 206863, adult male, 38.8 mm snout–vent length). (d) *Xenopus gilli* (KU 206865, adult female, 47.0 mm snout–vent length). The vertebral column is somewhat foreshortened as drawn owing to its curvature. I–VII, presacral vertebrae I–VII; TP II–IV, transverse processes of presacral vertebrae II–IV. Cartilage is represented by the stippled pattern. Scales = 5 mm.

primitive families of archaeobatrachians that possess ribs, the ribs of pipids are ankylosed to the transverse processes in the adults. In pipids, the number of functional presacral vertebrae is reduced through fusion of the first two (Fig. 19.7c) or three (Fig. 19.7d) presacrals and by incorporation of the last presacral into the sacrum (Fig. 19.7d); however, neither of these modes of shortening the vertebral column is unique to this family. In general, *Xenopus* tends to have eight functional presacrals (but see example of *X. gilli*, Fig. 19.7d), whereas in *Silurana* and *Pipa*, the first two are fused to yield seven functional presacrals, and in the hymenochirine pipids, there are only six. The vertebrae of *Xenopus* and *Silurana* are nonimbricate, unlike those of *Rhinophrynus*, the pelobatoids and most other pipids. In correlation with the poor development of the neural arches in the more primitive pipids, medial neural spines tend to be minimally developed. The sacrum is widely expanded in all pipids and fused to a relatively short coccyx. A unique evolutionary novelty of the vertebrae of pipids is the possession of complex articular surfaces, consisting of ridges and sulci on the pre- and postzygapophyses of adjacent elements (Cannatella 1985). Pipids are one of the few anuran families known to be uniformly characterized by a Type-I iliosacral articulation; the others are the pelobatoid anurans and the Bombinatoridae. The condition also occurs in *Discoglossus* and some neobatrachians (e.g., some hylids and microhylids and the Rhinodermatidae).

Evolutionary and functional significance of axial modifications

In all anurans, the vertebral columns tend be relatively short and inflexible. The presacral axial skeleton forms a semirigid framework for the attachment of epaxial musculature of two functional classes—those that both originate and insert on the vertebral column and those that originate on the pelvic girdle and insert on the axial column or vice versa. The former increase the rigidity and control the limited flexibility of the vertebral column so that the forward orientation of the anterior trunk and head is maintained as the body is thrust through the air (or water) in a straight line by the hind limbs (Gans & Parsons 1966). Presumably the shorter and less flexible the trunk, the more effective the anterior thrust, especially in the more dense aquatic medium. The reduction of the number of vertebrae along with the elaboration of the articulations between them, the ankylosis of the ribs to the transverse processes and the fusion of the sacrum and coccyx in pipids all would seem to strengthen the trunk.

The point of greatest movement in the trunk is at the iliosacral joint between the sacrum and the ilium of the pelvic girdle. The nature of the joint (i.e., Type I or II) determines the direction(s) of possible movement of the pelvis on the vertebral column (Emerson 1979; Emerson & De Jongh 1980). Depending on the configuration of the sacrum (i.e., dilated vs. narrow), Type-II anurans can rotate the pelvis laterally or dorsoventrally with respect to median axis of the vertebral column. However, in Type-I anurans, there is a sliding joint between the ilia and sacral diapophyses, which are united to one another by a broad, cufflike ligament. Owing to the great expansion of the sacral diapophyses and the nature of the ilia (see below) in pipids, little or no lateral or vertical rotation of the pelvis would be possible; movement is restricted to anterior and posterior sliding (Palmer 1960) which can result in a 15–20% change in body length during swimming, with the pelvis being in its most posterior position relative to the sacrum when the body attains its maximum extension (Emerson & De Jongh 1980).

Posterior appendicular skeleton

Historical constraints

Pelvis and the thigh muscles

In pelobatoids and *Rhinophrynus*, the pelvis consists of two ossified elements, the ilium and the ischium, and a cartilaginous pubis, which may be mineralized; the pelvic components are not fused to one another. The ilial shafts lack crests and, in dorsal aspect, configure a rounded (pelobatoids, Fig. 19.8b) or acute (*Rhinophrynus*, Fig. 19.8d) V-shape. If present, the dorsal ilial prominence is broad and low (Fig. 19.8a,c); three muscles originate from the area of the prominence—the m. glutaeus magnus, which moves the thigh dorsally; the m. iliofibularis, which flexes the knee joint and abducts the femur; and the m. iliofemoralis, which draws the femur dorsally.

The ilium forms the anterior half of the acetabulum, which is subcircular but nearly as broad as it is high, and a preacetabular zone that is expanded in the lateral plane and serves as the point of origin of a number of muscles that move the thigh. The mm. sartorius and semitendinosus abduct the femur, pull it ventrally, and flex the knee. In primitive anurans, including the pelobatoids, these two muscles are partially fused at their distal tendons, whereas in neobatrachians, the muscles are separate and have distinct insertions. In pelobatoids, the fused tendon of the mm. sartorius and semitendinosus passes ventral to that of the m. gracilis complex. In *Rhinophrynus*, the bellies are distinct, but the tendon of the m. sartorius inserts on that of the m. semitendinosus and then pierces that of the m. gracilis and passes dorsal to it. The m. gracilis complex consists of the mm. gracilis major and minor, which flex the knee joint and extend the hip joint by pulling the thigh backward. The origin of the m. gracilis minor is from the pelvis and the skin; the cutaneous origin is relatively restricted in most frogs including megophryine pelobatoids, but extensively cutaneous in other pelobatoids and *Rhinophrynus*. The mm. pectineus and adductor longus adduct the femur; the m. adductor longus (considered to represent a detached lateral margin of the m. pectineus) is absent in *Rhinophrynus* and pelobatoids of the genus *Scaphiopus*. The m. obturator internus pulls the femur dorsally and rotates it, and the m. cruralis moves the thigh ventrally.

Other muscles important in thigh movement include the m. pyriformis, which originates from the distal end of the coccyx and inserts on the

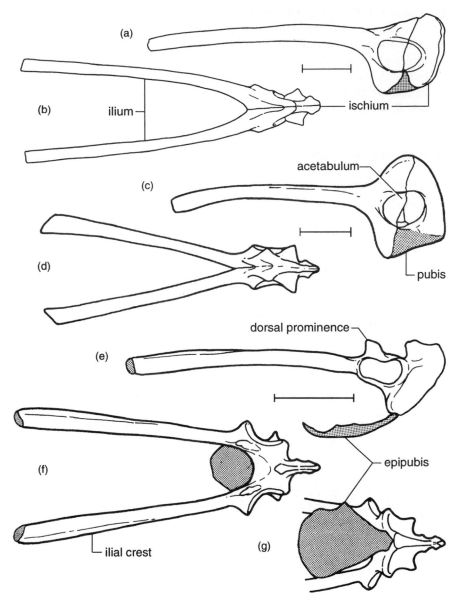

Fig. 19.8. Pelvic girdles of mesobatrachian anurans. (a–b) The pelobatid *Scaphiopus couchii* (KU 73385, adult female) in left lateral (a) and dorsal (b) views. (c–d) *Rhinophrynus dorsalis* (KU 186799) in left lateral (c) and dorsal (d) views. (e–g) *Xenopus largeni* (KU 206863, adult male, 38.8 mm snout–vent length) in left lateral (e), dorsal (f), and ventral (g) views. Cartilage is represented by the stippled pattern; cartilaginous tips of the ilia of *Scaphiopus* and *Rhinophrynus* were lost during preparation. Scales = 5 mm.

dorsal surface of the femur to abduct this element. This muscle is present in *Rhinophrynus*, as well as the other archaeobatrachians except the genus *Scaphiopus*. The m. tensor fasciae latae is a prefemoral thigh muscle that arises along the ventral margin of the posterior part of the ilial shaft and inserts on the fascia covering the m. cruralis; together these two muscles extend the knee joint and flex the hip joint.

Foot

The long bones of the hind limb are not included in this discussion, but some brief comments on the feet are warranted. The feet of *Rhinophrynus* and many of the pelobatoids deviate considerably from those of most other anurans. The prehallux is enlarged as a support for a digging 'spade' on the foot (Fig. 19.9a–d). In *Rhinophrynus*, the first digit is modified for digging and the tibiale and fibulare frequently are short, robust and frequently fused in large individuals. The distal tarsal elements consist of a centrale and two tarsals in pelobatoids (Fig. 19.9a,b). In *Rhinophrynus*, the centrale is absent; one tarsal may be present (Fig. 19.9c,d), but in larger individuals it may be indistinguishable owing to fusion with the tibiale and fibulare (pers. obs.).

Evolutionary novelties

Pelvis and the thigh muscles

The design of the pelvis in pipids differs substantially from that of other anurans. The ilia and ischia are fused to one another and medial sutures between these paired bones are absent (Fig. 19.8e–g) in many *Xenopus* and hymenochirines. The ilium bears a well-developed dorsolateral crest and, together, the ilia configure a broad U-shape in dorsal aspect (Fig. 19.8f). The dorsal ilial prominence is large and bladelike in *Xenopus* and *Silurana*, but acuminate or reduced (rarely) in other pipids. The acetabulum is markedly longer than high and slightly constricted dorsally and ventrally to form a dumbbell shape in most taxa (Fig. 19.8e). The preacetabular zone is transverse rather than lateral. The pubis develops from a pair of thick, platelike, transversely oriented endochondral ossifications beneath the acetabulum; in the adult, these are indistinguishably fused to one another, the ilium anteriorly and the ischium posteriorly. Some pipids (*Xenopus*, *Silurana* and *Pseudhymenochirus*) have a broad, transverse plate of cartilage, the epipubis, that extends anteriorly beneath the ilial shafts from the pubis (Fig. 19.8g). Elsewhere among living anurans, an epipubis resembling that of the pipids is present in *Ascaphus, Leiopelma*; the epipubis of discoglossids is a narrow strip of cartilage (Cannatella 1985).

There are several interesting differences in the muscles that typically originate from the preacetabular ilium. In *Xenopus* and *Silurana*, the mm. sartorius and semitendinosus are fused proximally instead of distally as in other archaeobatrachians; fusion is more extensive in *Silurana*, in which the proximal three quarters of the length of the compound muscle is fused. In *Pipa*

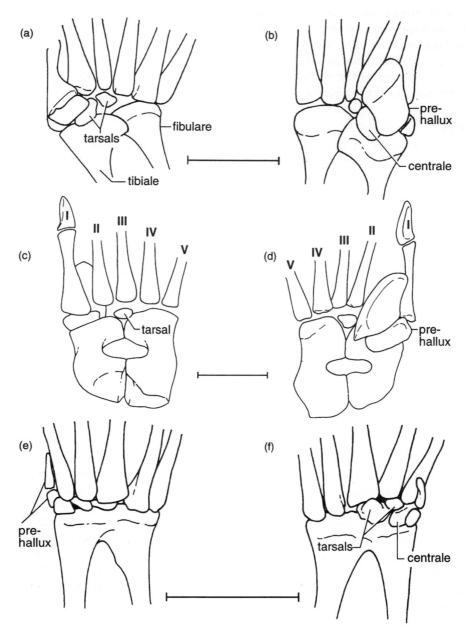

Fig. 19.9. Right feet of mesobatrachian anurans. (a–b) The pelobatid *Scaphiopus intermontanus* (KU 79436, male, 51.3 mm snout–vent length) in dorsal (a) and plantar (b) aspects. (c–d) *Rhinophrynus dorsalis* (TNHC 19327, female 57.8 mm snout–vent length) in dorsal (c) and plantar (d) aspects. (e–f) *Xenopus vestitus* (KU 206873, adult female, 38.5 mm snout–vent length) in dorsal (e) and plantar (f) aspects. Scales = 5 mm.

and *Hymenochirus*, most of the proximal parts of the muscles are fused and distally, fusion is complete. The relationship of the tendon of the mm. sartorius and semitendinosus to that of the m. gracilis in *Pipa* and *Hymenochirus* is the same as that in *Rhinophrynus*. However, in *Xenopus* and *Silurana*, in which the distal parts of the compound muscle are discrete, the tendon of the m. sartorius inserts on the tendon of the m. gracilis. In *Silurana*, the tendon of the m. semitendinosus pierces that of the m. gracilis while passing dorsal to it as it does in *Rhinophrynus, Pipa* and *Hymenochirus*; however, *Xenopus* is uniquely characterized by having the tendon of the m. semitendinosus pass dorsal to that of the m. gracilis complex and remain completely free of it. The origin of the m. gracilis minor in pipids differs from that of other anurans because the muscle originates only from the pelvis, rather than from the pelvis and the skin; moreover, *Hymenochirus* lacks the muscle. Like *Rhinophrynus* and pelobatoids of the genus *Scaphiopus*, pipids lack the m. adductor longus.

The m. pyriformis is present in *Xenopus*, but greatly reduced in thickness relative to its state in other anurans; the muscle is absent in other pipids. Whereas the m. tensor fasciae latae has a fleshy origin from the ilium in other known anurans, the origin is tendinous in pipids.

Foot

The foot in pipids is less derived than that of either *Rhinophrynus* or the pelobatoids. Neither the prehallux nor the first digit is markedly enlarged and there is the usual complement of a centrale and two distal carpal elements (Fig. 19.9e,f). The foot is distinguished by the length and slenderness of the metatarsal and phalangeal elements; these effectively increase the length of the foot relative to that of most other living anurans.

Evolutionary and functional significance of pelvic modifications

Owing to the lack of functional locomotor studies on pipids, there is little definitive that can be said about the significance of the configuration of the pelvic girdle and the changes that have occurred in the pelvic-femoral musculature. The dumbbell shape of the acetabulum suggests that rotational movement of the hip joint is restricted in comparison to other known anurans and that the principal plane of movement of the femur is fore-to-aft with limited dorsal and ventral excursions. This is consistent with the lateral, sprawled position of the limbs typical of pipids and their apparent inability to adduct and fold the limb at the side beneath the body.

The elaboration of a dorsolateral crest and dorsal ilial prominence suggests expanded areas of origin for the m. iliacus externus, which flexes the hip joint and draws the femur forward, and the mm. iliofibularis, iliofemoralis and glutaeus magnus, which abduct and pull the femur dorsally and flex the knee joint. The development of tendinous origin for the m. tensor fasciae latae may indicate strengthening of the action of this prefemoral muscle that extends the knee joint and moves the femur forward and up. Strengthening of these adductors might be expected in anurans that pull their legs forward through water and may account for the elaboration of the bony pubes in the transversely oriented preacetabular zone from which many of the prefemoral muscles originate.

In contrast to the prefemoral muscles, there seems to be simplification in postfemoral muscles given that the m. pyriformis is absent and the m. gracilis complex is reduced to a single point of origin, with at least one taxon having lost the m. gracilis minor. Seemingly contradictory trends in the relationship of the mm. sartorius and semitendinosus are apparent in pipids. In *Xenopus* and *Silurana*, this abductor complex that pulls the leg ventrally and flexes the knee has lost its distal fusion and changed slightly the nature of the tendinous insertions of the muscles, whereas in other pipids, the muscles tend to be fused throughout their lengths. The functional effects of these changes in configuration are unknown.

The lengthening of the digits is consistent with the development of complete digital webbing to provide an extensive surface against which to develop propulsive thrust when moving through water.

Anterior appendicular skeleton

Historical constraints

The pectoral girdles of *Rhinophrynus* and the pelobatoids are arciferal. They are characterized by broad epicoracoid cartilages that freely overlap one another (Fig. 19.10a,b). The clavicles are deeply concave (anteriorly) and the medial ends are distinctly narrower than the lateral ends; the clavicles are not fused to the scapulae. The robust scapulae are long, and the moderately short coracoids are slightly wider at their glenoid ends than at their sternal ends. Some pelobatoids and *Rhinophrynus* have narrow cleithra and suprascapular cartilages, whereas most anurans have more expanded, fan-shaped suprascapulae. A prezonal element is variable in its occurrence in both archaeobatrachians and neobatrachians, but it occurs in all pelobatoids; *Rhinophrynus* lacks a prezonal element and is one of the only anurans that lacks a sternum (Fig. 19.10b).

The hands of *Rhinophrynus* and pelobatoids are composed of the usual anuran complement of four large proximal carpal elements—the ulnare, radiale and pre- and postaxial centralia—along with three distal carpals and a bipartite prepollex.

Evolutionary novelties

The pectoral girdles of pipid frogs have a suite of characters that render them unique among known living anurans. The zonal portion of the girdle is greatly expanded in the longitudinal plane as a result of the long coracoid bones, the sternal ends of which lie much farther posterior than in other anurans,

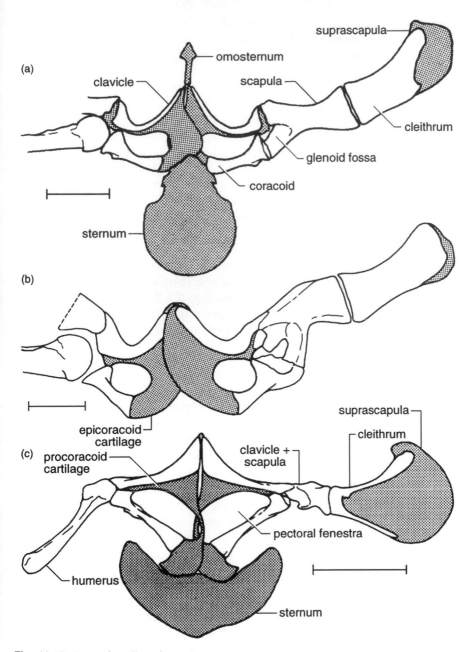

Fig. 19.10. Pectoral girdles of mesobatrachian anurans in ventral view. The left scapula and suprascapular blade have been deflected into the ventral plane for purposes of illustration. (a) The pelobatid *Spea bombifrons* (KU 205038, 59.0 mm snout–vent length). (b) *Rhinophrynus dorsalis* (TNHC 19327, female 57.8 mm snout–vent length). (c) *Xenopus boumbaensis* (KU 206928, male, 34.6 mm snout–vent length). Cartilage is represented by the stippled pattern. Scales = 5 mm.

the epicoracoid cartilages that are expanded posteriorly and posterolaterally around the sternal ends of the coracoids, and the orientation of the clavicles along the anterior margin of the zonal girdle (Fig. 19.10c). The sternum always is well developed and fused to the elaborated posterior epicoracoid cartilages in all pipids except *Xenopus* and *Pipa*. The clavicles are robust and fused to the scapulae except in *Pipa*, in which each clavicle broadly overlaps the pars acromialis of the scapula (similar to the relationship of these bones in *Rhinophrynus*). Typically, the clavicle is only slightly concave. It is narrowest at its glenoid end and markedly expanded medially, with the anteromedial ends lying far anterior to the level of the glenoid fossae (Fig. 19.10c); this is true of all pipids except the hymenochirines, in which the clavicle has a nearly transverse orientation just anterior to the level of the glenoid fossae. The scapulae are short and stocky and the suprascapulae extraordinarily expanded into a broad fan shape in all pipids. Unlike the condition in other arciferal anurans (Fig. 19.10a,b), the epicoracoid cartilages do not broadly overlap one another medially. In pipids, the anteromedial margins of the epicoracoids tend to abut one another and the posteromedial margins narrowly overlap one another (Fig. 19.10c); the posteromedial margins are fused in *Silurana*, and the epicoracoids are fused throughout their lengths in hymenochirines.

The digits of the hand, like those of the foot, are long and slender. The dorsal side of the hand in *Xenopus* and *Silurana* frequently bears a sesamoid bone dorsal to and between the radiale and ulnare (Fig. 19.11e). The palmar surfaces of the postaxial centrale and carpals 2 and 3 tend to bear well-developed knobs for the attachment of ligaments and muscles (Fig. 19.11f). There tends to be a reduction in distal carpal elements; thus, in *Pipa*, carpal 1 presumably is incorporated into the preaxial centrale, and in the hymenochirines, both carpals 1 and 2 are absent. There is a subtle difference in the palmar plane of the hand to the axis of the radioulna in pipids. Whereas in other anurans the palmar plane of the resting hand is parallel with the substrate, in pipids the hands are turned in such a way that the palmar plane is more vertical than horizontal; thus, at rest, the dorsal surface of the hand nearly parallels the longitudinal axis of the body and lies adjacent to the body with the palmar surface facing away from the body.

Evolutionary and functional significance of pectoral modifications

Much less is known about the functional morphology of the pectoral skeleton in anurans than the pelvic. In her functional analysis of anuran pectoral girdles, Emerson (1983) pointed out that there are two points of flexion in the girdle—one in the cartilage at the ventral midline, and the other at the junction between the scapula and suprascapula—and that in most anurans in which the scapula is long, the greatest amount of girdle flexion would occur at the joint between the scapula and suprascapula. Emerson (1983) also hypothesized that compressive loading of the pectoral girdle during movement of frogs in a terrestrial environment would produce major differences in stress distribution

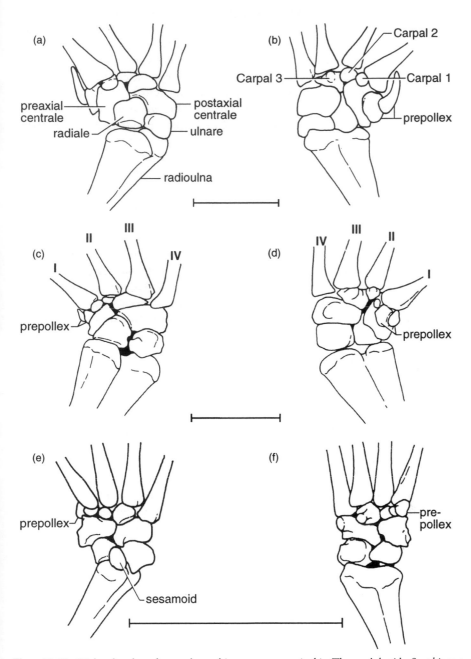

Fig. 19.11. Right hands of mesobatrachian anurans. (a–b) The pelobatid *Scaphiopus intermontanus* (KU 79436, male, 51.3 mm snout–vent length) in dorsal (a) and palmar (b) aspects. (c–d) *Rhinophrynus dorsalis* (TNHC 19327, female 57.8 mm snout–vent length) in dorsal (c) and palmar (d) aspects. (e–f) *Xenopus vestitus* (KU 206873, adult female, 38.5 mm snout–vent length) in dorsal (e) and palmar (f) aspects. Scales = 5 mm.

in girdles with freely overlapping medial halves (arciferal) as opposed to those in which the girdle elements are fused medially (firmisternal). She suggested that in arciferal girdles, the coracoids would absorb force by compression and the epicoracoid cartilages by tension, and that owing to the biomechanics of the mobile arciferal girdle, the clavicles and anterior epicoracoid cartilage might be necessary to retain structural integrity during tensile stress distribution. In the more solidly constructed firmisternal girdle, force would be absorbed by compression of the coracoids and fused epicoracoid cartilage between the bones.

In terms of historical origin, the pipid pectoral girdle is arciferal, but modifications of its structure in this family seem to have rendered it functionally firmisternal; this is arguably true of *Xenopus*, and certainly true of all more advanced taxa. Moreover, the only significant point of flexion in the pipid girdle is the joint between the short scapula and the suprascapula. If Emerson (1983) is correct in her assessment of the evolutionary importance of the clavicles, then it might be assumed that the cartilaginous zonal pectoral girdle in pipids is subject to a great deal of tensile stress during swimming and that clavicles have been elaborated to maintain the configuration of the girdle. The reduced length of the scapulae simply may reflect the dorsoventral compression of the body. The marked expansion of the suprascapula may provide an enlarged area of origin for the m. dorsalis scapulae, which, together with the m. latissimus dorsi, inserts on the deltoid crest of the humerus and circumducts the humerus dorsally and backward.

The change in the orientation of the resting position of the hand in pipids (and possibly the elaboration of the carpals and postaxial centrale in many) may relate to two behavioural changes. First, because they do not normally move in the terrestrial environment, there is no need for pipids to position the hands so that they broadly contact the substrate and absorb the force of the frog's jump; when not in water, pipids literally 'swim' in a skittering manner across the substrate and are incapable of leaping or hopping. Second, freed from the functional constraints associated with terrestrial movement, the hands become useful adjuncts to feeding in pipid frogs, which routinely use the backs of their hands to shove prey into their mouths.

Discussion

Although not all of the novel features of the skeletal system of pipids (e.g., the structure of the braincase and correlated cranial changes) can be attributed to the evolution of an aquatic organism from a semiaquatic or terrestrial, saltatorial ancestor, most seem to be associated with three fundamental activities in the aquatic medium—viz., locomotion, feeding and communication. Of these, the evolution of a suite of novelties that enable pipids to swim effectively seems to have had the most profound effect on the specialization and divergence of the pipid bauplan from the more generalized anuran morphotype. The morphological novelties include changes in the head, axial column, suspension and nature of the posterior appendicular skeleton, and the anterior appendicular skeleton. The head, like the rest of the body, is relatively flat. This feature is more pronounced in hymenochirines and *Pipa* (which has a wedge-shaped head in lateral profile because of the extraordinarily flat rostral portion of the skull) than it is in the more generalized *Xenopus* and *Silurana*, which have proportionally deeper rostral areas. The vertebral column is short, compact and less flexible than that of most other anurans as a result of complex articulations between adjacent vertebrae and multiple fusions (e.g., sacrum and coccyx, presacral vertebrae). The trend toward loss of axial flexibility culminates in the vertebral columns of hymenochirines and *Pipa*, which have completely imbricate vertebrae with well-developed neural spines and accessory parasagittal processes on the neural arches. Dorsoventral and lateral rotation of the pelvic girdle on the sacrum has been eliminated by expansion of the sacrum and development of a cuff-like binding ligament that permits only fore and aft motion in the horizontal plane of the pelvis with respect to the vertebral column. Dorsal and ventral excursions of the hind limb are limited by the dumbbell configuration of the acetabulum; thus, rotation of the limbs is limited and the primary movement would be in the longitudinal axis and the horizontal plane of the body. Adduction of the hind limb and extension of the knee seem to be facilitated by elaboration of the prefemoral musculature; in contrast, postfemoral muscles involved in abduction are characterized by loss and simplification of muscle complexes. The evolution of a large, fully webbed hind foot provides an extensive surface against which to develop propulsive thrust. The large, platelike, dorsoventrally compressed pectoral girdle apparently is functionally firmisternal, with the only significant point of flexion being between the scapula and suprascapula. The functional significance of the expanded zonal and postzonal girdle is unclear.

Several cranial novelties of pipids seem to be associated with their feeding behaviour, which, in the absence of a tongue, probably combines lunging and forcing food into the mouth with the modified forelimbs. The upper jaw is relatively mobile and is buttressed against the neurocranium by a movable, sliding joint with the pterygoid; some stability may be provided by the pterygomaxillary ligament. Although the zygomatic ramus extends to the posterior end of the maxilla in primitive pipids (e.g., *Xenopus* and *Silurana*), it articulates loosely with a dorsal process of the cartilaginous posterior maxillary process, rather than with the maxilla. In contrast to the upper jaw, the mandible is robust. The dentaries extend around the anterior margin of the jaw and a symphysial joint is absent; thus, the lower jaw seems rigid in comparison to that of most other anurans that project their tongue to capture prey. The lack of a tongue apparently obviates the need for vomers and extensive bony support of the palate.

Because pipids lack a tongue, the hyoid is relieved of one of its primary functions in other anurans—viz., serving as a basal plate of origin for the extensive tongue muscles (although traces of the mm. hyoglossus and genioglossus are fused to the floor of the mouth: Horton 1982). In pipids the primary function of the hyoid seems to be to provide support for the elaborate and enlarged laryngeal apparatus that is modified to produce clicking sounds. Similarly, elaboration of the tympanic annulus and plectral apparatus must be related to hearing under water.

Acknowledgements

I thank The Zoological Society of London and Richard Tinsley for their invitation to participate in this timely symposium and for their provision of local hospitality. I am particularly indebted to several of my colleagues who shared results of their own studies with me and discussed some of the ideas presented herein. In particular, I thank David Cannatella, Rafael de Sá, A.M. Báez, W.E. Duellman and David Kizirian. Research and travel was supported by a grant from the National Science Foundation (BSR 89–18161).

References

Cannatella, D.C. (1985). *A phylogeny of primitive frogs (archaeobatrachians)*. PhD diss.: The University of Kansas.
Cannatella, D.C. & Trueb, L. (1988a). Evolution of pipoid frogs: intergeneric relationships of the aquatic frog family Pipidae (Anura). *Zool. J. Linn. Soc.* **94**: 1–38.
Cannatella, D.C. & Trueb, L. (1988b). Evolution of pipoid frogs: morphology and phylogenetic relationships of *Pseudhymenochirus*. *J. Herpet.* **22**: 439–456.
Cracraft, J. (1990). The origin of evolutionary novelties: pattern and process at different hierarchical levels. In *Evolutionary innovations*: 21–44. (Ed. Nitecki, M.H.). The University of Chicago Press, Chicago.
Duellman, W.E. & Trueb, L. (1986). *Biology of amphibians*. McGraw-Hill Book Company, New York.
Emerson, S.B. (1979). The ilio-sacral articulation in frogs: form and function. *Biol. J. Linn. Soc.* **11**: 153–168.
Emerson, S.B. (1983). Functional analysis of frog pectoral girdles. The epicoracoid cartilages. *J. Zool., Lond.* **201**: 293–308.
Emerson, S.B. & De Jongh, H.J. (1980). Muscle activity at the ilio-sacral articulation of frogs. *J. Morph.* **166**: 129–144.
Frost, D.R. (Ed.) (1985). *Amphibian species of the world: a taxonomic and geographical reference*. Allen Press and the Association of Systematics Collections, Lawrence, Kansas.
Futuyma, D.J. (1986). *Evolutionary biology*. Sinauer Associates, Sunderland, Mass.
Gans, C. & Parsons, T.S. (1966). On the origin of the jumping mechanism in frogs. *Evolution*, **20**: 92–99.
Horton, P. (1982). Diversity and systematic significance of anuran tongue musculature. *Copeia* 1982: 595–602.

Jurgens, J.D. (1971). The morphology of the nasal region of Amphibia and its bearing on the phylogeny of the group. *Ann. Univ. Stellenbosch (A)* **46**: 1–146.

Müller, G.B. & Wagner, G.P. (1991). Novelty in evolution: restructuring the concept. *A. Rev. Ecol. Syst.* **22**: 229–256.

Palmer, M. (1960). Expanded ilio-sacral joint in the toad *Xenopus laevis*. *Nature, Lond.* **187**: 797–798.

Rabb, G.B. (1960). On the unique sound production of the Surinam toad, *Pipa pipa*. *Copeia* **1960**: 368–369.

Regal, P.J. & Gans, C. (1976). Functional aspects of the evolution of frog tongues. *Evolution* **30**: 718–734.

Reumer, J.W.F. (1985). Some aspects of the cranial osteology and phylogeny of *Xenopus* (Anura, Pipidae). *Rev. suisse Zool.* **92**: 969–980.

Shaw, J.P. (1986). The functional significance of the pterygomaxillary ligament in *Xenopus laevis* (Amphibia: Anura). *J. Zool., Lond. (A)* **208**: 469–473.

Trueb, L. (1985). A summary of osteocranial development in anurans with notes on the sequence of cranial ossification in *Rhinophrynus dorsalis* (Anura: Pipoidea: Rhinophrynidae). *S. Afr. J. Sci.* **81**: 181–185.

Trueb, L. & Cannatella, D.C. (1982). The cranial osteology and hyolaryngeal apparatus of *Rhinophrynus dorsalis* (Anura: Rhinophrynidae) with comparisons to Recent pipoid frogs. *J. Morph.* **171**: 11–40.

Trueb, L. & Cannatella, D.C. (1986). Systematics, morphology, and phylogeny of the genus *Pipa* (Anura: Pipidae). *Herpetologica* **42**: 412–449.

Trueb, L. & Cloutier, R. (1991). A phylogenetic investigation of the inter- and intrarelationships of the Lissamphibia (Amphibia: Temnospondyli). In *Origins of the higher groups of tetrapods: controversy and consensus*: 223–313. (Eds Schultze, H.-P. & Trueb, L.). Cornell University Press, Ithaca & London.

Trueb, L. & Gans, C. (1983). Feeding specializations of the Mexican burrowing toad, *Rhinophrynus dorsalis* (Anura: Rhinophrynidae). *J. Zool., Lond.* **199**: 189–208.

Trueb, L. & Hanken, J. (1992). Skeletal development in *Xenopus laevis* (Anura: Pipidae). *J. Morph.* **214**: 1–41.

Wake, D.B. & Roth, G. (1989). The linkage between ontogeny and phylogeny in the evolution of complex systems. In *Complex organismal functions: integration and evolution in vertebrates*: 361–377. (Eds Wake, D.B. & Roth, G.). John Wiley & Sons, Chichester & New York. (*Life Sci. Res. Rep.* **45**.)

Yager, D.D. (1992). A unique sound production mechanism in the pipid anuran *Xenopus borealis*. *Zool. J. Linn. Soc.* **104**: 351–375.

20 Molecular approaches to the phylogeny of *Xenopus*

JEAN-DANIEL GRAF

Synopsis

Comparative molecular studies of *Xenopus* species have addressed four main questions: (1) the monophyly or polyphyly of the genus *Xenopus*, (2) the subdivisions of the genus and the relationships among species groups, (3) the age of major splitting events within the genus, (4) the age of polyploidization events. The first question has been examined in two studies, based on immunological distances and ribosomal DNA sequences, which strongly suggest that the genus *Xenopus*, including *X. tropicalis*, is monophyletic with respect to *Hymenochirus*. The second question has been considered in several studies using electrophoresis of proteins, peptide mapping of albumin or restriction mapping of mtDNA. Different criteria produce different classifications, although some consensus is obtained on the existence of a *tropicalis* group, a *fraseri* group and a *laevis* group. A partial answer to the third question has been provided by a comparison of nucleotide sequences of globin genes: the divergence between the two most distant species groups, i.e. the *tropicalis* group and the rest of the genus, is estimated to be about 120 million years old. The same study has produced a tentative answer to the fourth question: the divergence time between the homoeologous globin genes of the tetraploid-derived *X. laevis* is estimated at about 40 to 60 million years ago. However, if tetraploidy resulted from allopolyploidization, then this estimate corresponds to the age of the diploid parental species, and not to the date of polyploidization, which remains unknown.

Introduction

One of the most outstanding characteristics of the genus *Xenopus* is the fact that it is mostly composed of polyploid species, with degrees of ploidy ranging from 2C to 12C. For several reasons, discussed by Kobel (this volume pp. 391–401), the most likely explanation for the origin of polyploid *Xenopus* is allopolyploidy, i.e. interspecific hybridization followed by genome duplication. If this hypothesis is correct, then the phylogeny of *Xenopus* should not be described simply in terms of genetic divergence, but should include figures illustrating the rejoining of separate lineages into new, hybrid-derived polyploid species. This condition is very similar to an evolutionary pattern known to occur in certain groups of plants, and named 'reticulate evolution' by Wagner (1954).

The application of molecular techniques to evolutionary problems has considerably improved our understanding of phylogenetic relationships in a great variety of taxonomic groups. Therefore, one might expect that the analysis of molecular genetic markers in *Xenopus* species should provide answers to current questions about *Xenopus* evolution, i.e. the monophyly or polyphyly of the genus, the phylogenetic relationships among species and the age of the various lineages.

In fact, there have been several attempts to infer evolutionary trees from comparative analyses of molecular characters in *Xenopus*. However, these attempts have been either extensive in terms of number of informative characters but focused on few species (e.g. Knöchel *et al.* 1986), or comprehensive in terms of species sampled but limited to a small number of characters (e.g. Graf & Fischberg 1986). Although some questions have been unequivocally answered, the information drawn from molecular comparisons is still too fragmentary to allow a detailed phylogeny of *Xenopus* to be inferred. In particular, owing to the suspected prevalence of interspecific hybridization as a mechanism of speciation, inferences about the network of phylogenetic relationships among polyploid species require that all homoeologous loci of a given nuclear gene family be separately analysed in all species that are compared. Therefore, usual techniques of molecular taxonomy generating a single distance value for each pair of species (e.g. immunotaxonomy, DNA/DNA hybridization) or relying on uniparentally transmitted extranuclear genome (e.g. mitochondrial DNA analysis) are not suitable for detecting the hybrid origin of polyploid species. Several studies using these techniques will be discussed in this review, however, since they can provide average estimates of genetic divergence between distantly related species or species groups.

Classifications based on quantitative analysis of molecular characters

Immunological comparison of albumins

A classification of *Xenopus* species based on immunological comparison of their albumins was proposed by Bisbee *et al.* (1977). For that study, antisera were raised against purified albumins from six *Xenopus* species. Each antiserum was then used to evaluate the degree of cross-reactivity of albumins from different species, including representatives of the genera *Hymenochirus* and *Pipa*. The results of these comparisons are expressed as units of immunological distance, which are thought to correlate with the number of amino acid substitutions between the compared albumins.

The distance matrix for the albumins of the *Xenopus* species tested and

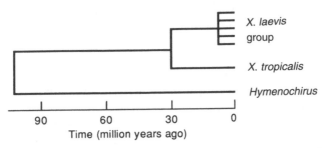

Fig. 20.1. Dendrogram illustrating the relationships of similarity among serum albumins from *Hymenochirus* and *Xenopus* species. The time scale is based on the assumption of a linear relationship between evolutionary time and albumin immunological distance in vertebrates. (After Bisbee *et al.* 1977.)

a representative of the pipid genus *Hymenochirus* was used to generate a dendrogram summarizing the relationships of similarity among them (Fig. 20.1). From these results, the following phylogenetic inferences could be drawn: (1) the genus *Xenopus* is monophyletic with respect to the related genera *Hymenochirus* and *Pipa*; (2) the genus *Xenopus* consists of two widely divergent lineages, one leading to *X. tropicalis*, the other leading to the rest of the genus, referred to as the *X. laevis* species group (Bisbee *et al.* 1977); (3) the splitting that gave rise to the *X. tropicalis* and *X. laevis* lineages can be dated to about 30 million years ago.

The immunological comparison of albumins did not allow elucidation of the relationships of similarity within the *X. laevis* group. The main reason might be that all extant species of this group are polyploid-derived and express two or more albumin loci (Graf & Fischberg 1986). In addition, evidence from artificial crosses and cytogenetics strongly suggests that these species originated by interspecific hybridization and subsequent polyploidization (Kobel & Du Pasquier 1986). Therefore, inferences about phylogenetic relationships between *Xenopus* species require that the products of all albumin loci of the various species be separately compared with one another. In the study of Bisbee *et al.* (1977), this condition was not fulfilled.

Sequence analysis of globin genes

Knöchel *et al.* (1986) compared the nucleotide sequences of adult globin genes of *X. tropicalis*, *X. laevis* and *X. borealis*. While the tetraploid-derived *X. laevis* and *X. borealis* have two nonallelic sets of α- and β-globin genes, the 'true' diploid *X. tropicalis* has only one set (Jeffreys *et al.* 1980). Consequently, the nucleotide sequences of five α- and five β-globin genes had to be determined. Pairwise comparisons within each group of five genes produced a matrix of divergence values calculated from the percentage of nucleotide substitutions leading to amino acid changes (Knöchel *et al.* 1986).

382   *Jean-Daniel Graf*

The highest divergence values (12.2 to 16.3%) were observed in comparisons between the unique globin cluster of *X. tropicalis* and clusters I and II of both *X. laevis* and *X. borealis*. Conversely, the lowest values (1.5 to 5.6%) were found in comparisons between *X. laevis* and *X. borealis* (cluster I[laevis] versus I[borealis], cluster II[laevis] versus II[borealis]), whereas intermediate values (3.1 to 9.1%) were observed between the homoeologous (i.e. nonallelic) clusters within each tetraploid species.

Compared to the immunological study of albumins by Bisbee *et al.* (1977), the sequence analysis of globin genes presents a major advantage in allowing the homoeologous genes of any polyploid species to be compared with one another as well as with those of other species. The results of these comparisons for *X. laevis*, *X. borealis* and *X. tropicalis* (Fig. 20.2) support the early divergence of the *X. tropicalis* lineage from the common stem of the *X. laevis* and *X. borealis* lineages. In addition, they suggest that tetraploidization was anterior to the partition of the lineages leading to *X. laevis* and *X. borealis*, as the divergence values between homoeologous genes within each species are larger than they are between the two species.

Using a calibrated time-divergence curve of globin evolution, Knöchel *et al.* (1986) estimated the *tropicalis/laevis* separation to be 110 to 120 million years old, whereas the *borealis/laevis* separation was dated to about 15 to 20 million years ago. Similarly, the divergence between the two homoeologous genomes within each of the two tetraploid species examined (*X. laevis* and *X. borealis*) was estimated to be 40 to 60 million years old. It is worth noting that the age of tetraploidy in the *X. laevis* group cannot be determined on the sole basis of nucleotide sequence comparisons. Indeed, if tetraploidy resulted from hybridization of different species (evidence reviewed in Kobel & Du Pasquier 1986), then the divergence value between clusters I and II of *X. laevis* (or *X. borealis*) reflects the age of the split between the two parental diploid species, and does not allow dating of the genome duplication event.

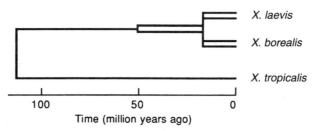

Fig. 20.2. Dendrogram showing the relationship of similarity among globin genes from one diploid (*X. tropicalis*) and two tetraploid *Xenopus* species. Double lines indicate the presence of two homoeologous (i.e. nonallelic) genomes in the tetraploid species. The time scale is based on the assumption that globin genes accumulate substitutions at a constant rate in vertebrates. (After Knöchel *et al.* 1986.)

Classifications based on cladistic analysis of molecular characters

Comparison of restriction maps of mitochondrial DNA

Carr, Brothers & Wilson (1987) isolated mtDNA from nine *Xenopus* taxa. For each species or subspecies, they established a map showing the sites of cleavage of the mtDNA by 11 restriction endonucleases. Each cleavage site was considered a cladistic character. Comparisons of the nine restriction maps using a procedure that minimizes the number of nucleotide substitutions required to account for the observed diversity (i.e., maximum parsimony analysis) produced different possible trees with similar degrees of parsimony. Only two clusters are conserved among the four most parsimonious trees. The first is composed of *X. borealis* and *X. clivii*, and the second includes *X. laevis laevis* and *X. laevis victorianus*. The dodecaploid *X. ruwenzoriensis* clusters either with *X. muelleri* or with *X. laevis*.

When interpreting the results presented by Carr *et al.* (1987), one should keep in mind that transmission of mtDNA is clonal and matrilinear. Consequently, polyploid species of hybrid origin should have the mtDNA of one parental species only. For instance, if one assumes that the dodecaploid *X. ruwenzoriensis* is the outcome of several rounds of allopolyploidization, then the sequence of its mtDNA provides information on only one of the six ancestral lineages.

Comparison of ribosomal DNA sequences

To test the monophyly of the genus *Xenopus*, De Sà & Hillis (1990) conducted a cladistic analysis of nuclear ribosomal DNA sequences from *X. laevis*, *X. tropicalis* and the related pipid species *Hymenochirus cultripes*. The maximum parsimony procedure produced a single most parsimonious tree that places *X. tropicalis* as the sister taxon of *X. laevis*. Trees showing other arrangements could also be obtained, but they required substantially larger numbers of steps. The rDNA sequence comparison therefore supports the monophyly of the genus *Xenopus* with respect to *Hymenochirus*.

Classifications based on qualitative comparison of molecular characters

Electrophoretic comparison of sperm nuclear proteins

Mann *et al.* (1982) compared the sperm-specific basic nuclear proteins from 17 species and subspecies of *Xenopus* on the basis of their electrophoretic profiles in polyacrylamide gels containing acetic acid, urea and a nonionic detergent. This electrophoretic technique allows the detection of small structural differences in otherwise similar polypeptides. The basic pattern of sperm nuclear proteins comprises six zones within which the number and precise localization of bands vary in a species-specific manner. The great complexity of the pattern in some species, as well as the absence of certain bands in others, prevents any attempt to quantify the level of biochemical–genetic divergence between the species compared. Nevertheless, the authors discussed their results in relation to taxonomy, and interpreted the overall similarity of the electrophoretic patterns of some species as an argument in favour of their phylogenetic propinquity.

For comparative purposes, I summarize the results and interpretation of Mann *et al.* (1982) in a scheme illustrating the clustering of species resulting from their analysis (Fig. 20.3a). A first group comprises *X. tropicalis* and *X. epitropicalis*, provisionally named *Xenopus* sp. n. (Zaire) in the original article. These two species show electrophoretic patterns very distinct from all other *Xenopus* species. A second group comprises the subspecies of *X. laevis*, with protein profiles very similar to one another. In addition, six species, namely *X. fraseri, X. amieti, X. ruwenzoriensis, X. wittei, X. vestitus* and *X. largeni*, show electrophoretic patterns quite similar to that of *X. laevis*, whereas *X. borealis* and *X. muelleri* are very distinct from *X. laevis* with respect to this criterion.

Electrophoretic comparison of adult globins

Using an electrophoretic technique similar to the one mentioned in the preceding section, Bürki & Fischberg (1985) compared the profiles of globin polypeptides from 18 species and subspecies of *Xenopus*. The comparison revealed extensive interspecific variation in the number and migration distance of globin bands, as well as evident similarity between the profiles of certain species. On this basis, the 18 *Xenopus* taxa were arranged in four clusters (Fig. 20.3b).

The two species *X. tropicalis* and *X. epitropicalis* share the same basic pattern, and constitute a first cluster. Similarly, the six subspecies of *X. laevis* have almost identical globin profiles, which in turn show some resemblance with those of *X. largeni , X. borealis, X. muelleri* and *X. clivii* (second cluster). The three species *X. fraseri, X. amieti* and *X. ruwenzoriensis* form a third cluster, whereas *X. vestitus* and *X. wittei* constitute a fourth one. Interestingly enough, the dodecaploid species *X. ruwenzoriensis* has a globin profile similar to the superposition of the profiles of the tetraploid *X. fraseri* and the octoploid *X. amieti*, supporting the view that *X. ruwenzoriensis* has a hybrid origin and possibly results from a past hybridization between species of the *fraseri* and *amieti* lineages.

Comparison of peptide patterns of serum albumin

All but one *Xenopus* species have two or more serum albumin isoproteins. The only exception is the diploid *X. tropicalis* (see Graf & Fischberg 1986). In a comparative study of 21 *Xenopus* taxa, the albumins of each

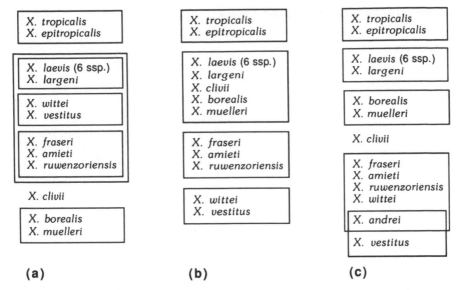

Fig. 20.3. Clustering of *Xenopus* species as inferred from three comparative studies of molecular characters: (a) electrophoresis of sperm nuclear proteins (Mann *et al.* 1982); (b) electrophoresis of adult globins (Bürki & Fischberg 1985); (c) electrophoresis and peptide mapping of serum albumins (Graf & Fischberg 1986). Enclosed in the same rectangle are species showing higher degrees of similarity with respect to the molecular character examined.

species or subspecies were separately characterized for molecular weight and peptide pattern after partial cleavage by a specific protease (Graf & Fischberg 1986). Divergent peptide patterns indicate differences in amino acid sequence resulting in gain or loss of cleavage sites. Conversely, the occurrence of identical peptide patterns in two albumins indicates that these polypeptides share the same cleavage sites, presumably conserved from the time preceding their evolutionary separation.

Variation in the molecular weight of albumins allowed subdivision of the genus into three groups (Fig. 20.3c). The first comprises *X. tropicalis* and *X. epitropicalis*. The second group is constituted by the six subspecies of *X. laevis* and *X. largeni*, which share the peculiarity of having two serum albumins with different molecular weights (i.e., 70 kDa and 74 kDa). The third group comprises the rest of the genus. On the basis of shared peptide patterns, this group can be further subdivided. A first subgroup is composed of *X. fraseri*, the three octoploid species *X. amieti*, *X. wittei* and *X. andrei*, and the dodecaploid *X. ruwenzoriensis*. The octoploid species *X. vestitus* has an isolated position with respect to this subgroup. Interestingly, the octoploid *X. andrei* possesses two *fraseri*-type and one *vestitus*-type peptide patterns, suggesting that *X. andrei* arose by hybridization between taxa of the *X. fraseri* and *X. vestitus* lineages.

A second subgroup comprises *X. borealis* and *X. muelleri*, whereas *X. clivii* has an isolated position.

Synthesis: phylogeny of *Xenopus*

Monophyly of the genus *Xenopus*

On the basis of a comparative morphological study of the Pipidae, Cannatella & Trueb (1988) challenged the view that the genus *Xenopus* is monophyletic. Their cladistic analysis suggested that *X. tropicalis* and *X. epitropicalis* are more closely related to the genus *Hymenochirus* than they are to other *Xenopus* species. Consequently, these authors resurrected the genus *Silurana*, with the species *Silurana tropicalis* and *S. epitropicalis*, as a new taxonomic name for the *X. tropicalis* group.

This new phylogenetic configuration is in complete contradiction to the molecular phylogeny based on albumin immunological distances (Fig. 20.1). Indeed, the immunological data presented by Bisbee *et al.* (1977) unequivocally associate *X. tropicalis* with the other *Xenopus* species, whereas *Hymenochirus* is only distantly related to the *Xenopus* cluster.

Another molecular study (De Sà & Hillis 1990) undertaken to assess the phylogenetic status of the genus *Xenopus* clearly supports the phylogeny proposed by Bisbee *et al.* (1977). In addition, De Sà & Hillis (1990) conducted a combined cladistic analysis of their data and those of Cannatella & Trueb (1988), which resulted in a tree associating *X. tropicalis* with *X. laevis* and isolating *Hymenochirus* in an early diverging lineage.

In conclusion, the available molecular evidence unequivocally supports the monophyly of the genus *Xenopus*.

Relationships among species

The question of interspecific relationships within the entire genus *Xenopus* has been treated in three comprehensive electrophoretic studies (Mann *et al.* 1982; Bürki & Fischberg 1985; Graf & Fischberg 1986). However, since all three studies were based on a small number of characters, quantitative analysis could not be applied to the data. Nevertheless, the observed biochemical phenotypes revealed patterns of similarity that could be used to subdivide the genus *Xenopus* into several groups of species. The three resulting classifications (Fig. 20.3) show overall concordance on the composition of species groups, with few exceptions.

Perfect concordance was realized as to the association of *X. tropicalis* and *X. epitropicalis* in a cluster well-differentiated from all other species, in agreement with karyology (Tymowska & Fischberg 1982) and morphology (Reumer 1985). Similarly, the three classifications agree on the inclusion of the tetraploid *X. fraseri*, the octoploid *X. amieti* and the dodecaploid *X. ruwenzoriensis* in a single cluster. Graf & Fischberg (1986) associate the two octoploid species *X. wittei* and *X. andrei* with this cluster (Fig. 20.3c), and Mann *et al.* (1982) add the octoploid *X. vestitus* to the same cluster (Fig. 20.3a). As already pointed out by Reumer & Graf (1986), the composition of this 'X. *fraseri* cluster' precisely corresponds to the *X. fraseri* group as defined by Reumer (1985) on the basis of morphological criteria. It is worth mentioning that Mann *et al.* (1982) noted a striking similarity between the electrophoretic phenotypes of *X. laevis* and those of the *X. fraseri* group. This association was not observed in other comparative studies.

The subspecies of *X. laevis* showed essentially similar phenotypes, and associate with one another in the same cluster. In two of the three classifications presented (i.e., Fig. 20.3a & c), *X. largeni* is included in the same cluster as *X. laevis* because of the similarity of their biochemical phenotypes.

Finally, there is some evidence in favour of a close association of *X. borealis* and *X. muelleri* (Fig. 20.3), whereas the relationships between this cluster, *X. clivii* and *X. laevis* are still obscure. On the basis of cranial osteology, Reumer (1985) defined a *X. laevis* group that included *X. laevis*, *X. clivii*, *X. borealis* and *X. muelleri*. He noted, however, that the group so defined is not as coherent as the *X. tropicalis* or the *X. fraseri* groups.

Age of major splitting events

Using a time scale based on the assumption of a linear relationship between immunological distance and divergence time, Bisbee *et al.* (1977) estimated that the separation of the *tropicalis* lineage from the stem leading to the other *Xenopus* species occurred some 30 million years ago. On the basis of a comparable assumption applied to the divergence of globin gene sequences, Knöchel *et al.* (1986) estimated that the lineage leading to *X. tropicalis* separated from the one leading to *X. laevis* and *X. borealis* approximately 110 million years ago, whereas the split giving rise to separate *laevis* and *borealis* lineages occurred some 15 to 20 million years ago.

In both studies, the time scale was calibrated by using observed correlations between time and molecular divergence in taxonomic groups where fossil evidence allows the dating of splitting events. It is worth noting, however, that the time/divergence equation used by Knöchel *et al.* (1986) included a correction for multiple substitutions at one site, whereas the one used by Bisbee *et al.* (1977) did not.

The large discrepancy between the two dates proposed for the separation of the *tropicalis* and *laevis* lineages (30 million years and 110 million years) suggests that either the hypothesis of a constant rate of amino acid substitution in a given protein is incorrect or the methods used in one or both studies are not adequate for inferring a molecular clock. Since immunological comparison is only an indirect method for estimating the number of amino acid differences between proteins, an approach using direct nucleotide sequence comparison is more accurate. Consequently, dating based on corrected percentages of nucleotide substitutions should provisionally be considered more reliable.

Age of polyploidization events

Theoretically, one might expect that a comparison of nucleotide sequences of homoeologous genes in a polyploid species would allow the dating of the genome duplication event by providing an estimate of the time elapsed since the two genes started evolving separately. However, as pointed out by Kobel & Du Pasquier (1986) and Knöchel et al. (1986), this reasoning does not hold for species of allopolyploid origin. In this case, molecular analysis of homoeologous genes would provide information on the time elapsed since the two parental species separated from one another, and not on the date of hybridization and genome duplication. Consequently, the range of 40 to 60 million years inferred from globin gene sequences by Knöchel et al. (1986) constitutes only an upper estimate of the age of tetraploidy in the lineage leading to X. laevis and X. borealis.

References

Bisbee, C.A., Baker, M.A., Wilson, A.C., Hadji-Azimi, I. & Fischberg, M. (1977). Albumin phylogeny for clawed frogs (Xenopus). Science 195: 785–787.

Bürki, E. & Fischberg, M. (1985). Evolution of globin expression in the genus Xenopus (Anura: Pipidae). Molec. Biol. Evol. 2: 270–277.

Cannatella, D.C. & Trueb, L. (1988). Evolution of pipoid frogs: intergeneric relationships of the aquatic frog family Pipidae (Anura). Zool. J. Linn. Soc. 94: 1–38.

Carr, S.M., Brothers, A.J. & Wilson, A.C. (1987). Evolutionary inferences from restriction maps of mitochondrial DNA from nine taxa of Xenopus frogs. Evolution 41: 176–188.

De Sà, R.O. & Hillis, D.M. (1990). Phylogenetic relationships of the pipid frogs Xenopus and Silurana: an integration of ribosomal DNA and morphology. Molec. Biol. Evol. 7: 365–376.

Graf, J.-D. & Fischberg, M. (1986). Albumin evolution in polyploid species of the genus Xenopus. Biochem. Genet. 24: 821–837.

Jeffreys, A.J., Wilson, V., Wood, D., Simons, J.P., Kay, R.M. & Williams, J.G. (1980). Linkage of adult α- and β-globin genes in X. laevis and gene duplication by tetraploidization. Cell 21: 555–564.

Knöchel, W., Korge, E., Basner, A. & Meyerhof, W. (1986). Globin evolution in the genus Xenopus: comparative analysis of cDNAs coding for adult globin polypeptides of Xenopus borealis and Xenopus tropicalis. J. molec. Evol. 23: 211–223.

Kobel, H.R. & Du Pasquier, L. (1986). Genetics of polyploid Xenopus. Trends Genet. 2: 310–315.

Mann, M., Risley, M.S., Eckhardt, R.A. & Kasinsky, H.E. (1982). Characterization of spermatid/sperm basic chromosomal proteins in the genus Xenopus (Anura: Pipidae). J. exp. Zool. 222: 173–186.

Reumer, J.W.F. (1985). Some aspects of the cranial osteology and phylogeny of Xenopus (Anura, Pipidae). Rev. suisse Zool. 92: 969–980.

Reumer, J.W.F. & Graf, J.-D. (1986). Contribution to the phylogeny of Xenopus (Anura: Pipidae). In Studies in herpetology: 107–110. (Ed. Roček, Z.). Charles University, Prague.

Tymowska, J. & Fischberg, M. (1982). A comparison of the karyotype, constitutive heterochromatin, and nucleolar organizer regions of the new tetraploid species *Xenopus epitropicalis* Fischberg and Picard with those of *Xenopus tropicalis* Gray (Anura, Pipidae). *Cytogenet. Cell Genet.* **34**: 149–157.

Wagner, W.H. (1954). Reticulate evolution in the Appalachian aspleniums. *Evolution* **8**: 103–118.

21 Allopolyploid speciation

HANS RUDOLF KOBEL

Synopsis

Since all but one of the *Xenopus* species are polyploid in various degrees, polyploidy should have resulted from relatively few and uncomplicated mechanisms that operated repeatedly in this genus. This contrasts with other vertebrates where polyploidy is rare and most often accompanied by an aberrant mode of reproduction. In both animals and plants, polyploid species are often of allopolyploid origin and there is evidence that this also applies to *Xenopus*. Conditions promoting bisexual allopolyploidy include (1) mixed populations that hybridize, (2) polyploidization in the germ line of hybrids, (3) compatibility of polyploidy with sex determination, and (4) reproductive success of polyploids equal to that of their parental species. Several of these requirements have been investigated in the laboratory. While male hybrids are sterile owing to univalent meiosis, female hybrids frequently produce unreduced eggs through premeiotic endoreduplication. When fertilized by spermatozoa of either parental species these develop into 3n females which again may produce unreduced eggs, giving rise to tetraploid backcross offspring of both sexes; the sex ratio depends on the species combination and on the temperature during sexual differentiation.

A hybridizing mixed population of two appropriate species could thus automatically generate tetraploid offspring by means of backcrosses. The fate of such a population (disappearance of one or both parental species or elimination of hybrids) depends mainly on assortative mating and fitness of the various genotypes. While nothing is known about the former, one may argue that reproductive success of allopolyploids could equal or exceed that of parental species, these being on the edge of their respective ecological ranges. The hybrids could be favoured by their allopolyploid constitution which assembles genetic adaptations of both parental species. Allopolyploid speciation in *Xenopus* appears to be preprogrammed by the capacity of hybrids to produce unreduced eggs and to circumvent genetic sex determination.

Introduction

The extraordinary range from diploid to dodecaploid bisexual species in *Xenopus* contrasts with the situation in other vertebrates where polyploidy is rare and most often accompanied by unisexuality and an aberrant mode of reproduction, i.e. partheno-, gyno- and hybridogenesis (Schultz 1980; Bogart

1980). The case of *Xenopus* suggests that the mechanisms causing repeated polyploidization in this genus should be relatively simple and/or that its genetic composition especially favours polyploid speciation.

Two forms of polyploidy have been distinguished, autopolyploidy and allopolyploidy; the former is defined as doubling of chromosome sets within a species, whereas allopolyploidy results from genome doubling provoked by hybridization (Kihara & Ono 1926). The effective causes that primarily lead to unreduced gametes are not known for either type. Autopolyploidy suffers from meiotic multivalent synapsis resulting in random reduction of chromosome numbers and, as a consequence, impaired fertility or sterility. Allopolyploidy on the other hand, by assembling genomes sufficiently diverged to prevent homoeologous chromosome pairing, may restore the fertility of otherwise sterile diploid hybrids. Examples are numerous in plant breeding and it also holds true for experimental *Rana* hybrids (Kawamura & Nishioka 1983). Polyploid species are generally of allopolyploid origin in both animals and plants and there is evidence that this also applies to the polyploid bisexual *Xenopus*.

Conditions promoting bisexual allopolyploid speciation

There are at least four requirements that must be fulfilled in order to permit bisexual allopolyploid speciation:

- mixed interbreeding populations;

- polyploidization in the germ lines of hybrids;

- compatibility of polyploidy with sex determination;

- reproductive success of polyploids comparable to that of the parental species.

Some of these points have been investigated experimentally in the laboratory; for others some field observations are available, but these are scarce and much remains speculative.

Occurrence of hybrids in mixed populations

Although many *Xenopus* species have a parapatric distribution, field samples from contact zones have regularly recorded two or more species sharing the same confined habitat. However, interspecific hybrids are documented in only three cases, i.e. between *X. gilli* and *X. l. laevis* (Rau 1978; Kobel, Du Pasquier & Tinsley 1981; Picker 1985; Picker, Harrison & Wallace this volume pp. 61–71), between *X. borealis* and *X. l. victorianus* (Yager this volume pp. 121–141) and between *X. muelleri* and *X. l. laevis* (Poynton & Broadley

1985; A. Elepfandt pers. comm.). The presumed hybrids between X. *fraseri* and X. *tropicalis* (see Knoepffler 1967) have to be reconsidered because X. *fraseri* includes several cryptic species not recognized at that time (see Kobel, Loumont & Tinsley this volume pp. 9–33).

In fact, assortative mating normally should prevent the blending of species, and natural hybrids demonstrate a failure of pregametic isolation mechanisms that may be due either to a constitutive leakiness in mate choice or to particular circumstances. In the absence of pertinent investigations, one may speculate that numerical imbalance between cohabiting species allows little chance for minority females to encounter a conspecific male, eventually limiting the choice to alien males. On the other hand, it is also conceivable that a recent expansion in geographical range may bring together formerly isolated species which lack mutual discrimination.

Polyploidization in the germ line of female hybrids

Many hybrid crosses have been obtained and analysed for fertility at the Station de Zoologie Expérimentale of the University of Geneva (see Kobel, Loumont *et al.*, this volume pp. 9–33). In the majority of species combinations, experimentally produced *Xenopus* hybrids are viable and possess well developed gonads, though certain combinations show rudimentary gonads that can be almost empty of germ cells.

Male hybrids are sterile and even a concentrated cell suspension made from hybrid testes is only occasionally able to fertilize an egg. The causes of male sterility lie in the failure of meiotic chromosome pairing; subsequent univalent segregation leads to aneuploid spermatids and results in cell lethality or abnormal differentiation of spermatozoa.

Fertility characteristics of hybrid females are more complex. Since meiosis takes place only after the complete differentiation of oocytes, hybrid eggs are functional gametes with regard to fertilization and early development. However, owing to univalent meiosis (Müller 1977), zygotes have an aneuploid constitution and die at a relatively advanced stage of embryogenesis (Kobel, Egens de Sasso & Zlotowski 1979). Survival of backcross offspring is therefore low or nil.

However, hybrid females may produce a second class of eggs which result from endoreduplicated polyploidized oocytes. These have about double the volume of the others and exclusively contain bivalents in numbers corresponding to the somatic chromosome number of the mother (Müller 1977; Müller & Kobel 1977). The interpretation that these bivalents represent autobivalents formed through an extra replication of all chromosomes has been proven correct by the isogenic constitution of gynogenetic offspring of 'big' eggs (Kobel & Du Pasquier 1975). Meiotic chromosome segregation in such polyploidized oocytes proceeds normally and contributes to the egg nucleus a complete genome of both parental species. After fertilization, they develop into triploid females which again may produce the two egg classes, small and big eggs, the latter then having three chromosome sets and giving rise to tetraploid offspring. The frequency with which polyploid eggs are produced depends on the species combination, on the particular parents, and on individual hybrid females. In certain species combinations, almost all hybrids lay big eggs, with frequencies within a clutch of 90% and more; in others the percentage of females may be low and variation between females very high (Table 21.1). The same variability also holds true for polyploid backcross females. That endoreduplication frequencies are conditioned by genetic factors, despite the above-mentioned rather unpredictable variability, is suggested by gynogenetic isogenetic offspring which tend to produce big eggs in amounts comparable to those produced by their mothers.

In general, the production of 'unreduced' gametes is a well-known and widespread phenomenon in animal and plant hybrids. In females of a large variety of *Xenopus* hybrids, it occurs quasi-automatically and results from polyploidization through premeiotic endoreduplication.

Sex determination in polyploid hybrids

In *Xenopus*, sex determination operates with a WZ/ZZ mechanism: females are heterozygous for dominant female-determining W factors and for recessive male-determining Z factors, while males are homozygous for the latter. This system applies to all *Xenopus* species so far studied by means of hormonal sex reversal (Chang & Witschi 1956), germ cell transfer (Blackler & Gecking 1972), or gynogenesis and sex ratio in F_1 hybrids between and within the various ploidy levels (Kobel 1985). So far, no morphologically distinct sex chromosomes corresponding to W and Z have been identified (Tymowska & Kobel 1972; Schmid & Steinlein 1991), although the character 'sex' can be mapped genetically with respect to syntenic marker genes and centromere position (Graf 1989a,b; Colombelli, Thiébaud & Müller 1984; Reinschmidt *et al.* 1985). W and Z thus can be regarded as alleles of a single locus (Engel & Kobel 1983).

The question then arises as to how the female-dominant genetic sex determination could have been conserved through the passage of polyploidy. Endoreduplicated big eggs of hybrids offer the opportunity to address this question experimentally. Since such eggs have the same heterozygous constitution as their mothers and spermatozoa contribute a Z-genome, the genetic constitution with respect to sex-determining factors of polyploid offspring is defined: WZZ in the triploid F_2 and WZZZ in the next tetraploid generation (Fig. 21.1). All these animals should differentiate into females because of the dominance of the female-determining W. Surprisingly, such is not the case: the dominance of W diminishes with rising amounts of Z-factors and both sexes may develop from zygotes of identical heterozygous constitution. Moreover, by combining a W-genome of *X. l. laevis* with different numbers of Z-genomes from *X. borealis*, *X. gilli*, *X. l. laevis* and *X. muelleri*, different species have been shown to possess species-specific strengths of Z-, and by consequence

Table 21.1. Incidence (%) of endoreduplicated big eggs in hybrid females from various interspecific crosses

Hybrid	Chromosome number	% big eggs	Females producing big eggs (n)	Total females (n)
♀l. laevis × ♂ gilli	18 + 18	20 – 95	8	8
l. laevis × fraseri	18 + 18	0 – 100	5	6
l. laevis × muelleri	18 + 18	0 – 45	12	20
borealis × l. poweri	18 + 18	0 – 3	1	6
borealis × muelleri	18 + 18	0 – 1	3	4
fraseri × l. laevis	18 + 18	40 – 86	4	4
vestitus × l. poweri	36 + 18	0 – 60	2	4
vestitus × wittei	36 + 36	0 – 10	3	4
(laevis × gilli) × muelleri	18 + 18 + 18	10 – 40	10	10

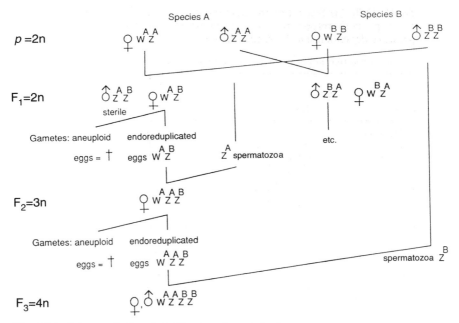

Fig. 21.1. Scheme of allopolyploidization through backcrosses in an interbreeding population of two *Xenopus* species. W = female-, Z = male-determining factors.

W-factors, probably in the above order, with X. *muelleri* having the weakest (Table 21.2, and additional unpublished data).

What determines sex in animals where the species combination produces a ploidy level at which female- and male-determining factors equal each other? It seems possible that environmental conditions interfere. This was indeed the case when tadpoles of a balanced constitution were reared at different temperatures: 16 °C produced more male, 26 °C more female adults (Kobel 1985; A. Rohrbach & H.R. Kobel unpubl. data). Table 21.2 summarizes results (including unpublished data) of temperature-shifting experiments conducted just after a developmental stage known to be sensitive to hormonal sex reversal (Chang & Witschi 1956). Inspection of adults showed no signs of hermaphrodism in their anatomy: females have well developed ovaries and oviducts, while in males oviducts are reduced and testes histologically normal.

Obviously, polyploidization disturbs the genetic sex determination system and switches it to an environmental mode. Eventually, this condition may facilitate if not be essential for bisexual polyploid speciation because, sex becoming independent from a particular genotype, it allows both sexes to exist at the WZZZ condition that is inevitably reached first. Moreover, it is only with such males that the new ploidy level can be perpetuated into the next generation.

The fertility of such males has not been tested although various *Rana* species hybrids with an allotetraploid (amphidiploid) constitution produced almost

Table 21.2. (a) Sex ratios in experimental polyploids of known genotype with respect to sex-determining factors; and (b) temperature dependence at W:Z equilibrium

(a) Sex condition in polyploid species combinations

		Percentage of		
Cross	Genotype	♀	♂	Number
♀ LG × ⚥ UV	$2n = W^L Z^G$	100	0	300
♀ LG × B	$3n = W^L Z^G Z^B$	2	98	288
× G	$= W^L Z^G Z^G$	12	88	51
× L	$= W^L Z^G Z^L$	100	0	100
× M	$= W^L Z^G Z^M$	100	0	58
♀ LGL × G	$4n = W^L Z^L Z^G Z^G$	0	100	13
× L	$= W^L Z^L Z^L Z^G$	43	57	112

(b) Temperature dependence of sex ratio at W:Z equilibrium

		Percentage of		
Genotype	Temperature (°C)	♀	♂	Number
$3n = W^L Z^G Z^B$	16	0	100	
	20	5	95	
	26	45	55	300
Change at stage 53	16 to 26	1	99	
	26 to 16	52	48	163
$4n = W^L Z^L Z^L Z^G$	16	19	81	
	20	43	57	
	26	78	22	112

W = dominant female, Z = recessive male sex-determining factors. B = *X. borealis*, G = *X. gilli*, L = *X.l. laevis*, M = *X. muelleri*; all with 2n=36.

exclusively diploid gametes and were fertile (Kawamura & Nishioka 1983; Nishioka & Okumoto 1983).

Population genetical considerations

The two phenomena discussed above, spontaneous polyploidization in the female germ line and abolition of genetic sex determination in new polyploids, suggest the way in which bisexual polyploid species could have arisen in *Xenopus*. Assuming that hybrid females (F_1, F_2) produce a high percentage (e.g. 50%) of endoreduplicated eggs, and provided that the allopolyploid genome combination permits an equal sex ratio under the local environmental conditions, an interbreeding mixed population of two *Xenopus* species should evolve, at least theoretically, into a single polyploid species. Figure 21.2 illustrates the fertile genotypes and the possible contribution of females to the various genotypes. Additional genotypes also generated are genetically unbalanced and lethal or sterile; they do not contribute to the next generation. The production of the various genotypes is governed by four variables.

1. The initial admixture of species in the population, $p + q = 1$;

2. Adult survival: a reasonable figure could be that half of the parents survive and, together with their offspring, form the next breeding generation, $S = 0.5$;

3. Assortative mating: $C = 0$ to 1 (0 = random mating, 1 = no interbreeding);

4. Reproductive success of the various genotypes: parental species may be attributed an equal fitness, $V_1 = 1$; diploid and triploid hybrids, because they produce two egg classes at 50% of which only one is viable, may have a $V_2 = 0.5$; tetraploid females can also produce two egg types, although so far observed at low frequency, possibly because the very large size may impair the growth of endoreduplicated oocytes, V_3 likely to be higher than V_2 since most of the eggs produced are diploid. One may further speculate that the fitness of allopolyploids eventually equals or is even superior to that of parental species because they assemble possible genetic adaptations of both species, whereas parental species actually may be on the edge of their respective ecological ranges and therefore relatively maladapted.

Such a population is highly unstable and there are only two possible outcomes: either the allopolyploid genotype eliminates both parental species, or the species with the higher initial frequency eliminates the other as well as the temporarily-existing hybrids. Figure 21.3 gives four examples of numerical calculations, with selected values of variables, showing the influence of mate

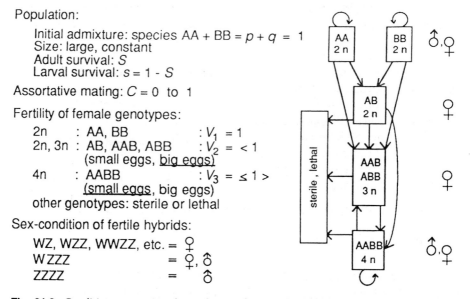

Population:

Initial admixture: species AA + BB = $p + q = 1$
Size: large, constant
Adult survival: S
Larval survival: $s = 1 - S$

Assortative mating: $C = 0$ to 1

Fertility of female genotypes:

2n	: AA, BB	: $V_1 = 1$
2n, 3n	: AB, AAB, ABB	: $V_2 = < 1$
	(small eggs, big eggs)	
4n	: AABB	: $V_3 = \leq 1 >$
	(small eggs, big eggs)	

other genotypes: sterile or lethal

Sex-condition of fertile hybrids:

WZ, WZZ, WWZZ, etc.	= ♀
W ZZZ	= ♀, ♂
ZZZZ	= ♂

Fig. 21.2. Conditions governing the evolution of an interbreeding population of two *Xenopus* species and the contributions of females to the different genotypes.

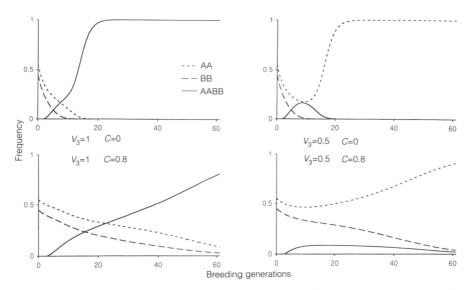

Fig. 21.3. Examples of numerical calculations through 60 breeding seasons of an interbreeding population of two *Xenopus* species that generates allopolyploid offspring (see text). Basic conditions: $p = 0.55$; $S = 0.5$; $V_1 = 1$; $V_2 = 0.5$; V_3 and C as given in the graphics. Other segregating genotypes are omitted from the graphics.

choice and of fitness of allopolyploids. The initial species proportion is also decisive: too low a frequency of one of the species eliminates both the minor species and the hybrids because it favours unilateral backcrosses that are sterile.

If such calculations allow assessment of the relative importance of variables, they should not be mistaken for a realistic model. Hybridization is more likely to take place in numerically-limited populations with animals frequently migrating between habitats. An initial F_1 could equally result from single females which become isolated in 'foreign territory', such as another pond in the neighbourhood, and only the return of F_2 backcrosses would be necessary in order to produce allopolyploid offspring. Little is known about the mate-choice behaviour of *Xenopus*, nor about such factors as survival rate, population density, or attractiveness of allopolyploid males. Nevertheless, some of the parameters that have permitted the extraordinary polyploid speciation of *Xenopus* have now become identifiable; among the most relevant are:

- spontaneous polyploidization in the female germ line of hybrids;

- temporary abolition of genetic sex determination in favour of an environmental mode of sex determination;

- Production of allopolyploids through backcrosses;

- Fitness of newly emerged allopolyploids which does not necessarily need to exceed that of the parental species in order to outcompete them.

Speciation at a higher ploidy level

There is no reason to believe that 'normal' speciation mechanisms should work differently on tetraploid or higher ploid populations. It should, however, be considered that newly emerged allopolyploids contain a prodigious redundancy in genetic information that, at least theoretically, confers on them an adaptive power unequalled by diploid species. This plasticity may explain why all diploid species but one—*X. tropicalis*—have disappeared, parental species possibly being unfit to compete with their allopolyploid progeny. Genetic redundancy also permits a rapid differentiation between populations by randomly silencing duplicated genes and by evolving tissue or development-specific regulation between orthologous genes (Graf 1989b; Kobel 1981; Kobel & Du Pasquier 1986; Robert, Du Pasquier & Kobel 1991). *Xenopus* are indeed genetically rather diverse. Why then, despite such potential, do so few *Xenopus* species exist? Should one infer that their allopolyploid constitution provides them with an ecological versatility and a reproductive capacity which leaves no space for adaptations into microhabitats, a territory being integrally occupied by a single type? The actual distribution patterns of many of the *Xenopus* species accord with this concept, unless we have not yet fully appreciated their diversification into local entities.

Acknowledgements

I am grateful to Mr B. Barandun and S. Chraiti for technical help, to C.A. Kobel for computer assistance. Part of this work was supported by the Swiss National Science Foundation, No. 3.388–0.78.

References

Blackler, A.W. & Gecking, C.A. (1972). Transmission of sex cells of one species through the body of a second species in the genus *Xenopus*. *Devl Biol.* 27: 376–394.
Bogart, J.P. (1980). Evolutionary implications of polyploidy in amphibians and reptiles. In *Polyploidy, biological relevances*: 341–378. (Ed. Lewis, W.H.). Plenum Press, New York. (*Basic Life Sci.* 13.)
Chang, C.Y. & Witschi, E. (1956). Genic control and hormonal reversal of sex differentiation in *Xenopus*. *Proc. Soc. exp. Biol. Med.* 93: 140–144.
Colombelli, B., Thiébaud, Ch. H. & Müller, W.P. (1984). Production of WW superfemales by diploid gynogenesis in *Xenopus laevis*. *Mol. gen. Genet.* 194: 57–59.
Engel, W. & Kobel, H.R. (1983). The Z-chromosome is involved in the regulation of H-W (H-Y) antigen gene expression in *Xenopus*. *Cytogenet. Cell Genet.* 35: 28–33.
Graf, J.-D. (1989a). Sex linkage of malic enzyme in *Xenopus laevis*. *Experientia* 45: 194–196.
Graf, J.-D (1989b). Genetic mapping in *Xenopus laevis*: eight linkage groups established. *Genetics, Austin* 123: 389–398.

Kawamura, T. & Nishioka, M. (1983). Reproductive capacity of male autotetraploid *Rana nigromaculata* and male and female amphidiploids produced from them by mating with female diploid *Rana brevipoda*. *Scient. Rep. Lab. Amphibian Biol., Hiroshima Univ.* **6**: 1–45.

Kihara, H. & Ono, T. (1926). Chromosomenzahlen und systematische Gruppierung der *Rumex* Arten. *Z. Zellforsch. mikrosk. Anat.* **4**: 475–481.

Knoepffler, L.-Ph. (1967). *Xenopus fraseri* × *Xenopus tropicalis*, hybride naturel d'amphibiens anoures au Gabon. *C. r. hebd. Séanc. Acad. Sci., Paris* **265D**: 1391–1393.

Kobel, H.R. (1981). Gene mapping in *Xenopus* (Anura, Pipidae). *Monit. zool. ital. (N.S.) (Suppl.)* **15**: 119–131.

Kobel, H.R. (1985). Sex determination in polyploid *Xenopus*. *S. Afr. J. Sci.* **81**: 205–206.

Kobel, H.R. & Du Pasquier, L. (1975). Production of large clones of histocompatible, fully identical Clawed Toads (*Xenopus*). *Immunogenetics* **2**: 87–91.

Kobel, H.R. & Du Pasquier, L. (1986). Genetics of polyploid *Xenopus*. *Trends Genet.* **2**: 310–315.

Kobel, H.R., Du Pasquier, L. & Tinsley, R.C. (1981). Natural hybridization and gene introgression between *Xenopus gilli* and *Xenopus laevis laevis* (Anura: Pipidae). *J. Zool., Lond.* **194**: 317–322.

Kobel, H.R., Egens de Sasso, M. & Zlotowski, Ch. (1979). Developmental capacity of aneuploid *Xenopus* species hybrids. *Differentiation* **14**: 51–58.

Müller, W.P. (1977). Diplotene chromosomes of *Xenopus* hybrid oocytes. *Chromosoma* **59**: 273–282.

Müller, W.P. & Kobel, H.R. (1977). Spontane Polyploidisierung in der weiblichen Keimbahn bei *Xenopus* Bastarden. *Arch. Genetik* **49/50**: 188.

Nishioka, M. & Okumoto, H. (1983). Reproductive capacity and progeny of amphidiploids between *Rana nigromaculata* and *Rana plancyi chosenica*. *Scient. Rep. Lab. Amphibian Biol., Hiroshima Univ.* **6**: 141–181.

Picker, M.D. (1985). Hybridization and habitat selection in *Xenopus gilli* and *Xenopus laevis* in the south-western Cape Province. *Copeia* **1985**: 574–580.

Poynton, J.C. & Broadley, D.G. (1985). Amphibia Zambesiaca 1. Scolecomorphidae, Pipidae, Microhylidae, Hemisidae, Arthroleptidae. *Ann. Natal Mus.* **26**: 503–553.

Rau, R. (1978). The development of *Xenopus gilli* Rose & Hewitt (Anura, Pipidae). *Ann. S. Afr. Mus.* **76**: 247–263.

Reinschmidt, D.C., Friedman, J., Hauth, J., Ratner, E., Cohen, M., Miller, M., Krotoski, D. & Tomkins, R. (1985). Gene-centromere mapping in *Xenopus laevis*. *J. Hered.* **76**: 345–347.

Robert, J., Du Pasquier, L. & Kobel, H.R. (1991). Differential expression of creatine kinase isozymes during development of *Xenopus laevis*: an unusual heterodimeric isozyme appears at metamorphosis. *Differentiation* **46**: 23–34.

Schmid, M. & Steinlein, C. (1991). Chromosome banding in Amphibia, XVI. High-resolution replication banding patterns in *Xenopus laevis*. *Chromosoma* **101**: 123–132.

Schultz, R.J. (1980). Role of polyploidy in the evolution of fishes. In *Polyploidy, biological relevance*: 313–340. (Ed. Lewis, W.H.). Plenum Press, New York. (*Basic Life Sci.* **13**.)

Tymowska, J. & Kobel, H.R. (1972). Karyotype analysis of *Xenopus muelleri* (Peters) and *Xenopus laevis* (Daudin), Pipidae. *Cytogenetics* **11**: 270–278.

22 Evolutionary inferences from host and parasite co-speciation

R.C. TINSLEY

Synopsis

The distinctive and host-specific parasite fauna of *Xenopus* may be viewed as a phylogenetic unit, evolving with its host group. The parasite taxa show a high degree of parallel speciation with the recognized *Xenopus* lineages. Parasite affinities also provide an independent assessment of host relationships.

The helminth parasites which employ *Xenopus* as a final host almost all 'recognize' the separation of the 20- and 40-chromosome lineage (*Silurana*) from the *Xenopus* species with multiples of 36 chromosomes. Species infecting *Silurana* are primitive with respect to relatives amongst the *Xenopus* parasites (paralleling the status of their hosts).

Within the 36-chromosome species, a series of parasites are represented by distinct species specific to the *fraseri* subgroup, *laevis* subgroup and *muelleri* subgroup. These evolutionary correlations are robust: diverse invertebrate groups including platyhelminths and arthropods express the same taxonomic 'opinion'.

Several parasites show discontinuities in host specificity corresponding with the 36/72 chromosome boundary. *Cephalochlamys* infects *laevis* but not sympatric *vestitus* and *wittei*; *Chitwoodchabaudia* infects *vestitus* and *wittei* (with a separate species in each) but never infects *laevis*. These discontinuities may reflect combinations of genes for resistance and susceptibility brought together by allopolyploidization.

Xenopus wittei shares one species of *Protopolystoma* (Monogenea) with *X. fraseri* and *X. pygmaeus*. *Gyrdicotylus* species from *X. wittei* show a mixture of affinities with related species found in *fraseri* and *laevis*: this concurs with independent evidence that *fraseri*-like and *laevis*-like *Xenopus* were the parental forms of the hybrid *wittei*.

In some comparisons, the parasite evidence is conservative, but in other cases it implies finer levels of host evolution than are recognized in current *Xenopus* taxonomy. Two parasite groups 'recognize' the separation of western and eastern forms of *X. muelleri*. Three geographical populations of *X. wittei* are distinguished by host-specific *Gyrdicotylus* species and this might suggest that this allopolyploid host had multiple origins in the central African highlands.

Introduction

Evolutionary background

A rich and highly characteristic parasite fauna is associated with *Xenopus* (Tinsley 1981a, and this volume, pp. 233–261). A major factor determining its origins and relationships is the fundamental isolation of the host group, which is separated from other anurans both ecologically (as aquatic predators) and phylogenetically (with fossil forms comparable with present-day African *Xenopus* species already established in the late Cretaceous of South America, about 80 million years ago (Baez, this volume pp. 329–347)). It is likely that some elements of this parasite assemblage originated with the earliest pipids, since their affinities are with a common stock of anuran parasites whose descendants occur in most of the extant families. However, the influence of subsequent isolation on the evolution of the *Xenopus* parasites is reflected in a mixture of very primitive and highly specialized characters often found nowhere else in their respective groups. The host group and its parasite fauna may be considered an evolutionary unit whose distinctiveness is indicated taxonomically by the fact that almost all the parasites are represented by separate species, genera and higher taxa strictly specific to *Xenopus* (see Tinsley 1981a, and this volume pp. 233–261). Given this long and tightly-knit association, it may be predicted that parasite evolution within the evolving host genus could have led to differentiation of parasite species in parallel with the speciation of the host. In such circumstances, independent evidence on the evolutionary relationships of the parasites may provide a reciprocal guide to host phylogeny.

In the case of *Xenopus*, two aspects of host biology could influence the potential for parasite speciation. On the one hand, the extensive sympatry between host species would reduce the potential for parasite speciation because of the absence of geographical isolating mechanisms. On the other hand, the evolution of the *Xenopus* species through hybridization and genome duplication would generate instant genetic differences between sympatric host species: thus, host genetic effects could have represented a major influence in the evolution of the parasite taxa (Tinsley 1981b).

This account provides a summary of the initial findings from a project designed to determine the patterns of speciation of the metazoan parasite groups represented within *Xenopus*. The results of this project are now in press, including the designation of a relatively large number of parasite taxa new to science. The detailed taxonomic comparisons are not relevant to this volume, and since most of the species names are not yet formally published, this review is presented only in summary outline intended to provide another independent view of evolutionary relationships within *Xenopus*, complementing the more conventional comparative methods already addressed (including chapters by Kobel, Loumont & Tinsley, this volume pp. 9–33; Graf, this volume pp. 379–389). The account considers, first, a synopsis of taxonomic relationships within each of the respective parasite groups studied and, second, the implications of these relationships for interpreting the phylogenetic links amongst the *Xenopus* species.

Parasite materials

The studies of parasite speciation involved conventional morphological analysis using the characters appropriate to each respective taxonomic group (references cited below). Parasite material was derived from *Xenopus* taxa examined during fieldwork, from collections of hosts returned alive to the UK (many by very helpful colleagues), and from preserved museum collections dissected with kind permission of curators (see Acknowledgements). Although 27 parasite species have been recorded from the most intensively studied of the *Xenopus* taxa, *X. l. laevis*, records from other members of the genus are patchy and the analysis of co-speciation is correspondingly restricted. There are few parasite records from the recently described *Xenopus* species; several of these species originated from very small field samples providing no possibility of examination for parasite infection (including *X. largeni*, *X. ruwenzoriensis*, *X. longipes*); *X. gilli* is endangered and there is little likelihood of parasitological examination of field samples. Parasites attributed to *X. laevis bunyoniensis* were derived from museum material of hosts and/or parasites collected in the 1930s: this taxon may now be extinct (Tinsley 1981b). There are special problems with the *fraseri* subgroup because identification of the phenotypically-identical species requires information on chromosome number, mating call and biochemical characters. Without this information, previous records of parasite infection in the literature and all studies based on preserved museum collections cannot be allocated with certainty within the *fraseri* subgroup[1]. Despite these limitations, the data set considered is based on taxonomic analysis of 40 parasite species from 15 host species and subspecies (Fig. 22.2 and text). A guide to the geographical and host distribution of the material is illustrated for one parasite, *Protopolystoma*, in Fig. 22.3.

The host taxa

There are three extant pipid genera in Africa: *Hymenochirus* and *Pseudhymenochirus* (which are more closely related to South American *Pipa*) and *Xenopus*. Within *Xenopus*, there is a major division between one lineage (*X. tropicalis* and *X. epitropicalis*) with chromosome numbers of 2n=20 and 40 (the subgenus *Silurana*) and all other species with multiples of 18 chromosomes (the subgenus *Xenopus*). Within this latter lineage, there is evidence from morphology, cytogenetics, biochemistry and molecular biology to distinguish five subgroups of species: (1) *laevis, gilli, largeni*; (2) *muelleri, borealis, clivii*; (3) *fraseri* and related species; (4) *vestitus, wittei*; (5) *longipes* (see Kobel, Loumont & Tinsley, this volume pp. 9–33). The first two subgroups, all with 36 chromosomes, are savanna species. The others are more or less specific

---

[1] In this account, these *fraseri*-like hosts are referred to as *X. fraseri affinis*.

(a)

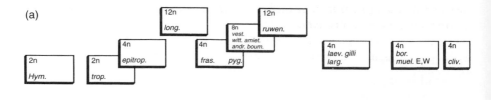

(b)

Fig. 22.1. (a) The currently recognized species of *Xenopus* together with the genus *Hymenochirus* showing the four levels of ploidy (for *Xenopus*, 2n=20, 40, 36, 72, 108) and a diagrammatic representation of relationships. The subgenus *Silurana* includes *X. tropicalis* (*trop.*) and *X. epitropicalis* (*epi.*). Other species, the subgenus *Xenopus*, comprise (1) *X. fraseri* (*fras.*) and related species *X. pygmaeus* (*pyg.*), *X. amieti* (*amiet.*), *X. andrei* (*andr.*), *X. boumbaensis* (*boum.*), *X. vestitus* (*vest.*), *X. wittei* (*witt.*), and *X. ruwenzoriensis* (*ruwen.*); (2) *X. laevis* (*laev.*) and related species *X. gilli* (*gilli*) and *X. largeni* (*larg.*); (3) *X. muelleri* (*muel.*) and related species *X. borealis* (*bor.*) and *X. clivii*. (*cliv.*); (4) the isolated species *X. longipes* (*long.*) which shares affinities with both subgenera. *X. vestitus* and *X. wittei* are shown allied with the *fraseri* subgroup but there is also evidence for recognition of a separate *vestitus/wittei* subgroup (see text). *X. clivii* is represented here in a distinct box to indicate its relative separation from the rest of the *muelleri* subgroup. *muel.* E, W relates to the proposed separation of an East African form of *X. muelleri* from the western Sahel form.

(b) Pipid hosts forming the basis of the parasite co-speciation analysis. Parasite material for *Hymenochirus* originated from *H. curtipes*; combined records are given for *X. tropicalis* and *epitropicalis* and for *X. fraseri* and *pygmaeus* respectively; *X. wittei* and *X. vestitus* are distinguished from the *fraseri* subgroup; *X. laevis* includes the subspecies *X. laevis laevis, poweri, victorianus, bunyoniensis* and *sudanensis*; records from the *X. muelleri* subgroup are separated for the three species.

to tropical forest zones: the *fraseri* group (six species with 2n=36, 72 or 108 chromosomes) is distributed from Cameroon to eastern Zaire; *vestitus* and *wittei* (both 2n=72) occur in the Central African highlands; *longipes* (2n=108) is endemic to Cameroon (Tinsley, Loumont & Kobel, this volume pp. 35–59). Evidence accumulated during the past 15–20 years demonstrates that evolution has involved unusual mechanisms of allopolyploidization and genome duplication (Kobel, this volume pp. 391–401).

Figure 22.1a depicts the 17 currently named *Xenopus* species (together with one of the other two African pipid genera, *Hymenochirus*) arranged to demonstrate the four ploidy levels and the species groupings which conform to a consensus of the comparative studies (see Kobel *et al.*, this volume pp. 9–33; Graf, this volume pp. 379–389). When species for which there are no parasite records are excluded, the spectrum of hosts contracts to that shown in Fig. 22.1b and this is used as a template in this account against which to analyse parasite speciation. There is insufficient material for comparison of parasite species represented in *X. tropicalis* and *epitropicalis*, and in *fraseri*-like species (including *X. fraseri* and *X. pygmaeus*), and records within these host groups are shown respectively as combined. There is no information for the fifth subgroup distinguished by Kobel *et al.* (this volume pp. 9–33), formed by the most recently described species, *X. longipes*, and this is omitted from all discussion.

Patterns of parasite speciation

Nematoda

Batrachocamallanus

The genus *Batrachocamallanus*, specific to the genus *Xenopus*, has four currently recognized species which exhibit various patterns of host specificity (Jackson & Tinsley 1995a). The genus contains one cosmopolitan species which occurs in more northerly *X. laevis* subspecies (*victorianus, sudanensis*), and in *X. muelleri, X. borealis, X. vestitus* and *X. wittei*. The distribution extends through a range of biotypes (savanna, wooded savanna and montane forest) from Nigeria and Cameroon in the west and Sudan in the north, south through Kenya and the central African highlands to Zimbabwe. In addition, this species has been recorded in *X. fraseri aff.* in lowland forest in eastern Zaire (Fig. 22.2a).

Other *Batrachocamallanus* species have more restricted host ranges. A second species occurs in *X. laevis*, principally in *X. l. laevis*, but also recorded from *X. l. poweri* in south-east Zaire. A third species infects *X. muelleri* (records from Ghana, Togo, eastern Zaire). None of these parasite species from the 18-chromosome lineage of *Xenopus* crosses over into the 10-chromosome *Silurana* lineage (Fig. 22.2a). Instead, *X. tropicalis* and *X. epitropicalis* are infected by a very distinctive *Batrachocamallanus* species, widely distributed from Sierra Leone along the coastal forest belt to Nigeria and Cameroon (in *tropicalis*) and from Cameroon across the Zaire Basin (in *epitropicalis*). This species is considered the most plesiomorphic, closest to procamallanine nematodes in African fishes from which the *Xenopus* parasites probably evolved (Jackson & Tinsley 1995a). However, although none of the species from the subgenus *Xenopus* has been recorded in *Silurana*, the parasite of *X. tropicalis* and *X. epitropicalis* also occurs in *X. pygmaeus* and other *X. fraseri*-like forms in Zairean lowland rainforest. *X. fraseri aff.* may therefore be infected by the cosmopolitan (largely savanna) species on the eastern edge of the Zaire forest and by the *Silurana* parasite in lowland Zaire where it co-occurs with *X. epitropicalis*. Thus the host specificity of *Batrachocamallanus* shows a strong host phylogenetic influence but, additionally, this may be moderated by ecological factors.

Camallanus

There are close parallels in the patterns of speciation of *Camallanus* and *Batrachocamallanus*. Four species of *Camallanus* are currently recognized

(Jackson & Tinsley 1995b). One has a wide host and geographical distribution within the 18-chromosome *Xenopus* lineage, infecting the subspecies of *X. laevis (laevis, poweri, victorianus, bunyoniensis, sudanensis), X. muelleri* (East and West), *X. borealis, X. wittei,* and *X. fraseri aff.* (Fig. 22.2b). Records occur more or less throughout the range of the subgenus *Xenopus*. Two species closely related to this cosmopolitan form have restricted host specificity. One infects *X. borealis* in Kenya, the other has been recorded in *X. l. laevis* in South Africa and *X. borealis* in Kenya. Curiously, therefore, *X. borealis* carries three species of *Camallanus* (two in the oesophagus and one in the intestine) although the localities of collection were geographically separate from one another. Despite their very wide geographical range, including lowland tropical forest, these *Camallanus* species do not infect the subgenus *Silurana*. *X. tropicalis* is infected by a very distinct *Camallanus* species (recorded in Ivory Coast, Togo and Nigeria) which does not cross into the 18-chromosome lineage (Fig. 22.2b). Taxonomic characters suggest that the *Camallanus* species infecting *Xenopus* are not monophyletic; instead they represent two independent colonizations from fish parasites which gave rise to one lineage in the subgenus *Silurana* and another in the subgenus *Xenopus* (see Jackson & Tinsley 1995b).

Other nematodes (not shown in Fig. 22.2)
The genus *Chitwoodchabaudia* provides a striking illustration of strict species specificity. Like many other parasites of *Xenopus*, it belongs to a monogeneric family known only from this host genus. It has been recorded only in the 2n=72 species, *X. vestitus* and *X. wittei*. In areas of sympatry with *X. l. victorianus* (Uganda and Rwanda), this 2n=36 species never becomes infected even when sharing confined habitats with heavily-infected *vestitus* and *wittei* (cf. the converse specificity of *Cephalochlamys*, below). Precise co-speciation is evident: *X. vestitus* and *X. wittei* each carry a distinct species of *Chitwoodchabaudia*, and this specificity is maintained even where these hosts are sympatric and, presumably, exposed to cross-infection (Tinsley 1981b; F.A. Puylaert & R.C. Tinsley unpubl.).

Platyhelminthes: Cestoda
Cephalochlamys
This tapeworm has a wide host and geographical distribution, with a life cycle (transmission via copepod intermediate hosts) equivalent to that of the camallanid nematodes (see Tinsley, this volume pp. 233–261). Taxonomic analysis of the genus is not yet complete; however, preliminary findings indicate a relatively high degree of correspondence between parasite and host speciation. Distinct *Cephalochlamys* occur in three subgroups of the 18-chromosome clade, in *laevis*, in *muelleri* and *borealis* and in *fraseri aff.* (Fig. 22.2c). The representatives in these *Xenopus* taxa (and in *X. clivii*, for which further study of the small samples is required) can be reliably

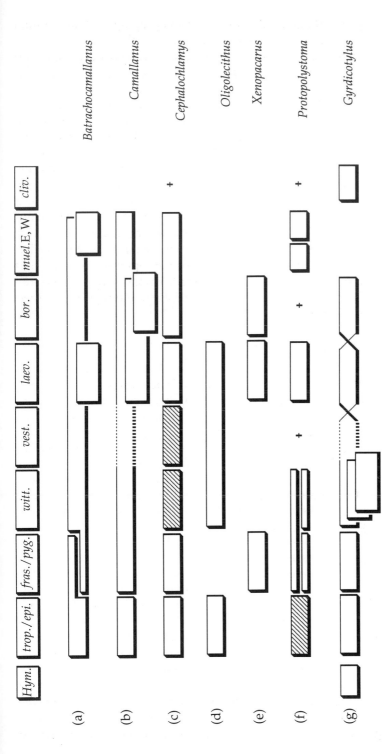

Fig. 22.2. Records of the speciation of seven parasite genera in species of *Xenopus* (and one record from a species of *Hymenochirus*, *H. curtipes*). Unshaded boxes represent a single parasite species which infects the host taxa listed above. Hatched boxes indicate conclusive evidence for the absence of infection in the corresponding host taxa. Boxes linked by X represent morphologically indistinguishable parasites which are segregated by strict host specificity. + indicates presence but small sample size precludes specific identification. Gaps (and pecked lines) reflect lack of data rather than lack of infection. Abbreviation of host names as in Fig. 22.1. Explanation in text.

distinguished from one another but are clearly interrelated. However, the representative in *X. tropicalis* shows major taxonomic differences and is set apart from the other species of *Cephalochlamys* in 2n=36 host species.

It is highly significant that *Cephalochlamys* does not infect the 2n=72 *X. vestitus* and *X. wittei*, even in regions (in Uganda and Rwanda) where both species share confined habitats (and presumably diets) with *X. l. victorianus* which are heavily infected with *C. namaquensis*. These areas of sympatry provide natural cross-infection experiments suggesting the fundamental exclusion of *Cephalochlamys* from these higher-ploidy-level *Xenopus* (see Tinsley 1981b) (Fig. 22.2c).

Platyhelminthes: Digenea

Oligolecithus

Two species are recognized. One is recorded from *X. laevis* subspecies (*laevis, poweri, victorianus, bunyoniensis, sudanensis*), extending over a geographical range of 45 degrees of latitude through savanna and wooded savanna habitats, and the same species infects *X. vestitus* and *X. wittei* in montane forest. A distinct species is represented in *X. tropicalis* in lowland tropical rainforest (Tinsley & Jackson 1995) (Fig. 22.2d).

Assessment of co-speciation in the Digenea is complicated by the influence of the molluscan intermediate host (to which digeneans generally show greater specificity than to the vertebrate final host). Discontinuities in the geographical distribution of particular snails may be a determining factor in the speciation of the parasite within its definitive host. Thus, for *Oligolecithus*, speciation corresponds with differences in geographical distribution and habitat type (whose influence is unknown) as well as the major divergence in *Xenopus* lineages between 10 and 18 chromosome clades.

Other digeneans (not shown in Fig. 22.2)

A single species of *Dollfuschella* is known, widely distributed within the 18-chromosome clade of *Xenopus* (including *X. l. laevis, X. l. poweri, X. l. victorianus, X. l. bunyoniensis, X. muelleri, X. clivii* and *X. vestitus*). Despite examination of extensive samples of *X. tropicalis* and *epitropicalis, Dolfuschella* has not been recorded in the 10-chromosome clade (although its absence must be viewed cautiously since the distribution of digenean species is generally patchy amongst *Xenopus* populations (Tinsley, this volume pp. 233–261)).

Progonimodiscus is the most widely distributed digenean infecting *Xenopus*, with records from *X. l. laevis* in southern Africa, from *X. l. victorianus, X. l. bunyoniensis, X. muelleri, X. vestitus* and *X. wittei* in Central Africa, *X. borealis* in East Africa, *X. pygmaeus* and *X. fraseri aff.* in Zaire lowland rainforest, and in *X. muelleri, X. l. sudanensis* and *X. tropicalis* in West Africa. There is continuous variation in morphometric characters through this host and geographical range but the greatest differences were found in specimens from *X. tropicalis*. Nevertheless, a conservative approach suggests that all representatives belong to a single species (J.A. Jackson & R.C. Tinsley unpubl.). The fourth digenean infecting *Xenopus* as a definitive host, *Xenopodistomum*, is so far known only from *X. l. laevis* in South Africa (Tinsley & Owen 1979).

Platyhelminthes: Monogenea

Protopolystoma

Extensive surveys indicate conclusively that *Protopolystoma* is specific to the subgenus *Xenopus* (18-chromosome lineage) and never occurs naturally in *Silurana* (10-chromosome) species (Tinsley 1981b). This exclusion has been confirmed experimentally: in *X. tropicalis* exposed to oncomiracidial infections of *P. xenopodis* from *X. l. laevis*, parasites successfully invade the kidneys but fail to survive through initial juvenile development (R.C. Tinsley unpubl.).

Within the 18-chromosome group, *Protopolystoma* has been recovered from all the host species for which adequate samples have been examined (Fig. 22.3). Within the principal subgroups of *Xenopus*, there is good evidence of parallel host and parasite speciation, although in a few cases (*X. vestitus*, *X. borealis* and *X. clivii*) the samples available are too small to assess the significance of morphometric variation (Fig. 22.2f).

Protopolystoma is represented by a single species throughout the wide geographical range of *X. laevis* (in *X. l. laevis, poweri, victorianus, sudanensis*). Experimental cross-infections demonstrate that *P. xenopodis* (from *X. l. laevis*) can produce patent infections in laboratory-raised *X. gilli*, suggesting that in areas of sympatry on the Cape this endangered species could share infections with *X. l. laevis* (cf. *Gyrdicotylus* below). In laboratory experiments, the *X. l. laevis* parasite from South Africa is also able to infect *X. l. victorianus* from Kenya and Uganda, demonstrating that the geographical (and evolutionary) separation of these host subspecies has not been accompanied by significant parasite divergence.

Protopolystoma infections in *X. muelleri* show two clearly-distinguished species, one recorded in samples from Ghana, Togo and Nigeria, and the other in hosts from Zimbabwe and Tanzania. This differentiation corresponds with the separation of two distinct forms of the host, '*muelleri*-west' and '*muelleri*-east' respectively (see Kobel *et al.*, this volume pp. 9–33) (Fig. 22.2f).

The parasite species represented in the *fraseri* subgroup are distinct from those in the *laevis* and *muelleri* subgroups and show complex interrelationships with parasites of *X. wittei* in the *vestitus–wittei* subgroup. One species of *Protopolystoma* occurs in *X. fraseri aff.* and locality records are widely distributed in lowland forest, in eastern Zaire, Gabon and Cameroon. A second species occurs in *X. wittei aff.*, with records from highland forest in eastern Zaire, Rwanda and Burundi. A third distinct *Protopolystoma* species infects *X. fraseri aff.*, *X. pygmaeus* and *X. wittei aff.* (Fig. 22.2f).

Distribution of *Protopolystoma*

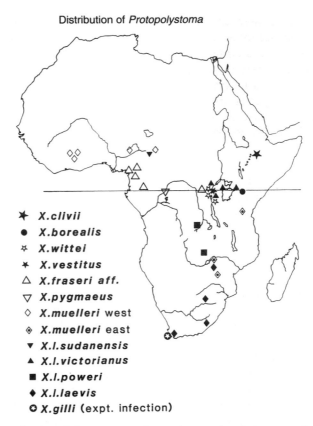

✹ *X.clivii*

● *X.borealis*

✫ *X.wittei*

✳ *X.vestitus*

△ *X.fraseri aff.*

▽ *X.pygmaeus*

◇ *X.muelleri* west

◈ *X.muelleri* east

▼ *X.l.sudanensis*

▲ *X.l.victorianus*

■ *X.l.poweri*

◆ *X.l.laevis*

◉ *X.gilli* (expt. infection)

Fig. 22.3. Geographical and host range of parasites employed in co-speciation analysis, illustrated by *Protopolystoma* (Monogenea). Some symbols represent multiple collections, others are single population samples.

This parasite has a wide geographical distribution, represented in *fraseri*-like (2n=36) hosts in the extreme east and west of the Zaire basin (in Zaire and Cameroon), in *pygmaeus* (also 2n=36) in Zaire, and in *X. wittei* (2n=72) in Rwanda. This range results in the overlap of two species of *Protopolystoma* in synhospitalic infections in both *X. fraseri*-like and *X. wittei*-like hosts, a highly unusual situation for polystomatid monogeneans. Since *fraseri* and *wittei* inhabit different biotypes, their polystomatid parasites may currently be isolated (except possibly for limited contact in overlapping margins). However, host distributions have undergone extensive change during Pleistocene climatic oscillations (Tinsley *et al.*, this volume pp. 35–59) and exchange of parasites may have occurred in the past. This complexity, including the sharing of one parasite species by both 2n=36 and 2n=72 *Xenopus* species, accords with the presumed genetic evolution of these hosts. A *fraseri*-like form is considered to be an ancestor (through hybridization with another 36-chromosome species) of *wittei* which therefore contains a set of *fraseri*-type chromosomes. This may permit the shared species of *Protopolystoma* to infect both 36- and 72-chromosome *Xenopus* alongside the separate host-specific species which also infect *fraseri* and *wittei*. In other words, *X. wittei* may have 'inherited' the shared species from its *fraseri*-like ancestor.

So far, material is lacking from the rest of the *fraseri* subgroup, particularly the 2n=72 species from west of the Zaire basin and the 2n=108 *X. ruwenzoriensis*, whose parasite interrelationships would be of considerable interest in relation to host evolutionary origins.

Gyrdicotylus

This monogenean provides a relatively comprehensive record of co-speciation supported by experimental evidence of host specificity from laboratory cross-infections (Tinsley, Harris & Jackson in prep.). The most divergent form has been recovered from *Hymenochirus curtipes*, separated by major taxonomic differences from the parasite species infecting the genus *Xenopus*. Within *Xenopus*, there is a well-marked division separating the *Gyrdicotylus* species recorded from the subgenus *Silurana* from those infecting the subgenus *Xenopus* (Fig. 22.2g). The primitive characters of the *tropicalis* parasite accord with presumed early separation of its host lineage. In contrast, parasites recorded from members of the subgenus *Xenopus* are clearly interrelated, distinguished by fine morphometric characters.

Separate *Gyrdicotylus* species have been recovered from four *Xenopus* subgroups, *laevis*, *muelleri*, *fraseri*, *vestitus–wittei*. The most distinctive of these species in the 18-chromosome lineage is that infecting *X. fraseri aff.* (Fig. 22.2g). Within the *laevis* subgroup, morphologically identical forms have been found in *X. l. laevis*, *X. l. poweri* and *X. l. victorianus*. Cross-infection experiments showed that parasites from *X. l. laevis* can successfully infect and reproduce in laboratory-raised *X. gilli* (which is sympatric with *laevis* on the Cape Peninsula) and in East African *X. l. victorianus* (demonstrating no significant divergence corresponding with the separation of these host subspecies) (cf. *Protopolystoma* above).

Two representatives of *Gyrdicotylus* have been recovered from the *muelleri* subgroup, from *X. borealis* and *X. clivii* (but not, so far, from *muelleri*). The parasite infecting *X. borealis* in Kenya is indistinguishable morphologically from the *laevis* parasite. In experimental cross-infections, worms from Cape *laevis* transferred to *X. borealis* showed altered behaviour resulting in reduced invasion success in comparison with controls using *X. l. laevis* and *X. l. victorianus* (with which *borealis* is sympatric in Kenya). Those worms which invaded *borealis* failed to reproduce and none survived more than one week (at 20 °C). The *laevis* and *borealis* parasites are therefore likely to be host-specific. However, the possibility of genetic exchange during the short period of survival in the foreign host suggests that these representatives should be distinguished as biological races rather than separate species (Tinsley, Jackson & Harris in prep.). In contrast, the *Gyrdicotylus* recorded from Ethiopian *X. clivii* is relatively distant from the *laevis* and *borealis* morphotype and this accords with the relatively isolated evolutionary position of the host within its subgroup (Fig. 22.2g).

Three forms of *Gyrdicotylus* have been recovered from three geographically separate populations of *X. wittei*-like hosts in Uganda, Rwanda and Zaire respectively. The parasites are distinguished unambiguously by morphological characters. However, these characters exhibit a series of similarities bridging the otherwise distinct features of the *fraseri* and the *laevis* parasites. The distinct status of these *wittei* forms has been confirmed by cross-infection experiments: parasites from *laevis* will not survive in *wittei*, and *wittei* parasites failed to survive in *X. l. victorianus* and *X. vestitus* (these three *Xenopus* species are sympatric in Rwanda). Moreover, even within the *X. wittei*-like hosts, the parasites from one host population (from Rwanda) failed to infect *X. wittei* from the type locality (in Uganda). It is possible that this complexity in parasite speciation reflects the complex hybrid origins of the allopolyploid host (see below).

Acari

Xenopacarus

Despite extensive searching through *Xenopus* species, I have found this ereynetid mite in only three host species: these represent three different subgroups of the 18-chromosome clade of *Xenopus*, and a distinct *Xenopacarus* species occurs in each—in *X. l. laevis, X. borealis* and *X. fraseri aff.* (Fig. 22.2e). Intriguingly, taxonomic characters suggest a closer relationship between the *borealis* and *fraseri* parasites, with the representative in *laevis* apparently the most primitive of this trio (Fain & Tinsley 1993).

Interpretation of host and parasite co-speciation

From the synthesis of the separate taxonomic analyses of *Xenopus* parasites shown in Fig. 22.2, horizontal comparisons give a summary of the speciation and host specificity for each parasite genus. The vertical patterns indicate the degree of congruence between parasite taxa represented in each of the groups and subgroups of *Xenopus* hosts.

Much of the interpretation of the parasite data set reflects evolutionary relationships which are now well established through recent genetic, biochemical and molecular studies based on *Xenopus*. Nevertheless, it is remarkable that these diverse parasite groups represent such sensitive indicators of host relationships. The most informative parasite groups in this analysis (including Monogenea and Acari) are recognized in parasitology for their relatively strict host specificity; some other groups provide more limited phylogenetic information concerning the definitive host, and it is known that the relationships of these may be influenced more profoundly by other factors (digeneans showing greater specificity for the mollusc intermediate host, for instance). Given these variations in the parasites considered, it is significant that the analysis shows a high degree of congruence. Thus, all the parasite genera except one (*Batrachocamallanus*) recognize the major evolutionary division between *Xenopus* lineages based on 10 and 18 chromosomes (the subgenera *Silurana* and *Xenopus*, respectively). In most cases, this recognition is reflected in the presence of a separate parasite species in the *tropicalis/epitropicalis* line; in other cases, a parasite group well represented in the 18-chromosome clade may be absent from *tropicalis/epitropicalis* (e.g. *Protopolystoma, Dollfuschella*). This may indicate either secondary loss or the colonization of *Xenopus* by these parasites subsequent to the split between the 10- and 18-chromosome lineages; the latter is more parsimonious. It is interesting that the phylogeny of *Camallanus* suggests that, for these camallanids, the two host lineages were colonized separately from different stocks of parasites infecting fishes. In most of the parasite groups where plesiomorphic and apomorphic character states can be established, species which occur in *Silurana* are very distinct from those in the 18-chromosome clade and clearly primitive, paralleling the status of their host group.

The analysis suggests that these parasites are likely to be relatively conservative indicators of host evolution. Thus, the forms of *Gyrdicotylus* infecting *X. laevis* and *X. borealis* are morphologically identical; however, experimental studies confirm that these parasites are actually host-specific. The relatively extensive zones of sympatry between *Xenopus* species, especially in montane regions, provide the most stringent test of parasite speciation and host specificity. In several cases, noted above, parasite species remain exclusive to particular hosts even when sympatric host species are exposed to direct and continuous cross-infection. The strict specificity of *Cephalochlamys* and *Chitwoodchabaudia* spp. *vis-à-vis* co-occurring *X. laevis, vestitus* and *wittei* is a particularly striking example.

Interpretation of parasite speciation provides relatively consistent support for the series of subgroups of the subgenus *Xenopus* identified by other techniques (summarized by Kobel *et al.*, this volume pp. 9–33, and Graf, this volume pp. 379–389). Thus, the *fraseri, laevis* and *muelleri* subgroups are clearly 'recognized' by *Cephalochlamys, Xenopacarus, Gyrdicotylus*. That the same evolutionary 'opinion' is expressed by a tapeworm, a mite and a monogenean provides confidence that the parasite evidence is robust. Both *Protopolystoma* and *Gyrdicotylus* provide strong indication of the affinity of *wittei* with the *fraseri* subgroup (Fig. 22.2f,g). Within the *muelleri* subgroup, there is evidence for the relatively distant position of *X. clivii* with respect to *muelleri* and *borealis*. Perhaps unexpectedly, several parasites strengthen the support for a relative affinity of *borealis* and *laevis* (shared *Camallanus* species, morphological similarity of *Gyrdicotylus* species). Whilst there are no natural parasite records from the endangered *X. gilli*, experimental evidence (cross-infections of *Protopolystoma* and *Gyrdicotylus* from *X. l. laevis* to laboratory raised *X. gilli*) supports the close relationship of *laevis* and *gilli* (reflecting their assignment to the same subgroup by Kobel *et al.*, this volume pp. 9–33).

From the vertical associations shown in Fig. 22.2, it is not surprising that
X. tropicalis/epitropicalis are distinguished by a column of separate species.
However, it is intriguing that *X. laevis* also carries a relatively high proportion
of parasite groups represented by *laevis*-specific species. Part of this may be
attributed to the greater detail available for *X. l. laevis* than for other taxa
(Tinsley, this volume pp. 233–261). This aside, most of the parasite genera
considered in Fig. 22.2 have a distinct species in *X. laevis*, sometimes
alongside more cosmopolitan species which also extend into other host
species. In addition, *X. laevis* supports a number of parasite genera, not
considered here, which are exclusive to *X. laevis*, without other known
representatives (e.g. *Xenopodistomum, Marsupiobdella, Pseudocapillaroides*)
(see Tinsley, this volume pp. 233–261). This evidence emphasizes the very
distinctive position of *X. laevis* within the subgenus *Xenopus* (and perhaps
the rich evolutionary biology and biogeography of the southern tip of Africa).
Regarding the relationship of *X. laevis laevis* to the other subspecies (see
Kobel *et al.*, this volume pp. 9–33), the indications are uncertain. There
is some parasitological evidence for a separation of *laevis laevis* from the
other subspecies: *Batrachocamallanus* has one cosmopolitan species infecting
a majority of the 18-chromosome *Xenopus* lineage in West, Central and East
Africa, but it does not extend into *X. l. laevis* in South Africa. Instead, *X.
l. laevis* has a distinct *Batrachocamallanus* species, but this also extends
north to infect *X. l. poweri* in south-east Zaire. The host specificity of the
camallanid nematodes may be influenced significantly by ecological factors, so
detailed extrapolation to host phylogeny may be insecure. On the other hand,
experimental evidence (from parasite invasion success in cross-infections)
provides independent assessment of subspecific relationships. This shows that
Protopolystoma and *Gyrdicotylus* from Cape *X. l. laevis* freely infect *X. l.
victorianus* from Uganda and Kenya (whereas these parasites distinguish *X.
borealis* and cross-infection is not successful). It is likely, as outlined above,
that specificity of these parasites provides a conservative assessment of host
divergence, so they do not provide sufficient resolution to assist interpretation
of the status of the *X. laevis* subspecies.

Apart from the indications of host evolutionary relationships which are now,
with the benefit of recent comparative studies, more or less predictable, the
Xenopus parasites point to some more novel conclusions. In these cases, parasite phylogeny may become a valuable predictive tool, indicating relationships
which are less well substantiated by other evidence. Thus, *Protopolystoma* is
represented by two distinct species in *X. muelleri*, one specific to the western
form inhabiting Sahel regions, the other infecting the eastern form in lowland,
largely coastal, environments. This accords with the tentative indications
from other studies that these geographically separate forms of *muelleri* are
biologically distinct.

It is intriguing that *X. borealis* in Kenya carries three different species of
Camallanus: one a cosmopolitan species, one shared with *X. laevis* and one
exclusive to *borealis*. This prompts speculation that such an unusual degree

of speciation might reflect more complex characteristics of the host. Some population samples of X. *borealis* collected in Kenya (including those which carried camallanid infections recorded in this study) have lacked the AT-rich chromosome segment normally considered characteristic of X. *borealis*. It is possible that different *borealis*-like taxa are involved and that the speciation of *Camallanus* in Kenya reflects unrecognized host differentiation. Perhaps the hybridization between X. *borealis* and X. *l. victorianus*, recorded in Kenya by Yager (this volume pp. 121–141), also has implications for the host specificity of these camallanid species.

A range of evidence points to close interrelationships within the *fraseri* subgroup (*sensu* Kobel *et al.*, this volume pp. 9–33), but the affinities of X. *wittei* and X. *vestitus* are less clearly defined. Initial genetic evidence suggested that these allopolyploid (2n=72) species may share one parental species in common (Tinsley, Kobel & Fischberg 1979), and the involvement of a *laevis*-like ancestor has been proposed (see also Tymowska 1991). The close links of *wittei* and *vestitus* are reflected in their separate subgroup status in assessments by Kobel *et al.* (this volume pp. 9–33) and Graf (this volume p. 385, scheme (b) based on globin analysis). Other evidence links X. *wittei* with the *fraseri* subgroup and separates X. *vestitus* in a relatively more isolated position (see Graf, this volume p. 385, scheme (c)). Given this uncertainty, the patterns of speciation in *Gyrdicotylus* and *Protopolystoma* provide independent evidence. Data for *Gyrdicotylus* from X. *wittei* reveal a mixture of taxonomic similarities with related parasite species in X. *fraseri*-like hosts on the one hand and X. *laevis* on the other. The speciation and host specificity of *Protopolystoma* point more or less conclusively to a strong relationship of *wittei* with the X. *fraseri* subgroup (without any resemblance to the *laevis* subgroup). However, the *Protopolystoma* of X. *vestitus* exhibits no such links with *fraseri* parasites; instead, the small samples studied indicate an affinity with representatives infecting X. *laevis*. It is interesting that the taxonomic relationships of these monogeneans point towards the same two 2n=36 *Xenopus* species as the genetic and molecular studies. More particularly, the parasite indications, that the links of *vestitus* and *wittei* may diverge in some respects, would correspond with the conclusions of Graf & Fischberg (1986) based on serum albumins (see Graf, this volume p. 385, scheme (c)).

Consideration of the age of the evolutionary events affecting the parasite groups is complicated by the fact that both ecological and phylogenetic factors may have determined parasite specificities, and 'host jumps' or secondary colonizations may disrupt patterns of co-speciation. In the case of *Camallanus*, morphological characters suggest that the subgenera *Silurana* and *Xenopus* were colonized separately by independent lineages of ancestral camallanids infecting fishes. No guide to the dating of these events can be inferred, except that the now very considerable divergence suggests relatively ancient origins. In contrast, for parasites showing a monophyletic lineage associated with their pipid hosts, there is a greater possibility of identifying pair-wise co-speciation.

The data recording the speciation of *Gyrdicotylus* are particularly comprehensive: the sequence of forms infecting *Hymenochirus, X. tropicalis, X. fraseri, X. laevis, X. borealis, X. clivii, X. wittei*, show an evolution of characters which concurs exactly with the presumed phylogenetic relationships of their hosts. The ancestors of this lineage probably lie with primitive gyrodactylids infecting swamp-dwelling fishes (including mormyrids and polypterids) which may have shared habitats with contemporary pipids (Tinsley, Harris & Jackson in prep.). The possibility that the pipid parasites result from a more recent invasion and speciation amongst already differentiated host groups seems less likely because of the clear-cut association of host and parasite phylogeny. The *Gyrdicotylus* species infecting *Hymenochirus curtipes* and *X. tropicalis* are clearly set apart from those infecting the subgenus *Xenopus*. Fossil evidence shows that *Hymenochirus*-like forms were already distinct in Africa in the Upper Cretaceous (Baez, this volume pp. 329–347).

In the case of the parasites of the polyploid host species, exactly the same problems affect attempts to estimate age of parasite speciation events as host origins (Graf, this volume pp. 379–389). Separation of the parasite species may have occurred in the ancestral host species rather than subsequent to allopolyploidization. Thus, on morphological evidence, one of the *Protopolystoma* species infecting *X. wittei* may have been inherited from one of the presumed parental species of its host, a *X. fraseri*-like form. In this case, the parasite is transferred along the evolutionary pathway followed by its host. On the other hand, allopolyploidization also creates instant changes in host constitution, and hence the environment of parasites, and these effects may be reflected in the absence of the tapeworm *Cephalochlamys* from the octoploid *X. vestitus* and *X. wittei*. Hybridization of parental hosts may bring together combinations of genes for resistance which determine parasite specificity.

The speciation of *Gyrdicotylus* in *X. wittei* prompts speculation on the mode of origin of this allopolyploid host. It is highly unusual that three distinct species of *Gyrdicotylus*, distinguished by morphometric characters and host specificity, infect the same host species. However, these parasite species occur in three geographically separate populations of *X. wittei*, in Uganda, Rwanda and Zaire (Tinsley, Harris & Jackson in prep.). This pattern would be consistent with a multiple origin of *X. wittei* which could have arisen in different areas or at different times by hybridization of the same two parental species producing lineages differing in their specific genetic combinations. The data could be interpreted as indicating a 'hybrid swarm' of *X. wittei* in the central African highlands which carries a corresponding 'swarm' of *Gyrdicotylus* with mixed affinities and host specificities.

Graf (this volume pp. 379–389) has pointed out that allopolyploid evolution of *Xenopus* species requires phylogeny to be viewed in terms not only of genetic divergence but also of the rejoining of separate lineages into new hybrid species. This review indicates that the same may also apply to the parasites of these hosts: indeed, the taxonomic affinities of the *Gyrdicotylus* from *X. wittei* point to the specific prediction (see above), that its origin may have involved rejoined lineages of parasites from *laevis*-like and *fraseri*-like hosts.

Acknowledgements

I am very grateful to Dr J.A. Jackson and Mr M.C. Tinsley for research assistance. The collection and analysis of parasites was carried out with grant support from NERC (GR3/6661 and GR9/632), the Royal Society and The Systematics Association. I am also grateful for the major contribution to the overall project made by material (of hosts and parasites) provided by Dr R.A. Avery, Dr W. Bohme, Dr V. Clarke, Dr P. Denny, Dr L. Gibbons, Dr D.I. Gibson, Mrs E. Harris, Dr P.D. Harris, Dr H. Hinkel, Professor J.L.J. Hulselmans, Mr M. Kazadi, Dr D. Meirte, Dr F.A. Puylaert, Mr M.P. Simmonds and Professor D.D. Yager.

References

Fain, A. & Tinsley, R.C. (1993). A new *Xenopacarus* (Acari, Ereynetidae) from the nasal cavities of *Xenopus* sp. (*fraseri* group), with a discussion on the evolution host–parasite. *Revue Zool. afr.* **107**: 513–517.

Graf, J.-D. & Fischberg, M. (1986). Albumin evolution in polyploid species of the genus *Xenopus*. *Biochem. Genet.* **24**: 821–837.

Jackson, J.A. & Tinsley, R.C. (1995a). Representatives of *Batrachocamallanus* n. gen. (Nematoda: Procamallaninae) from *Xenopus* species (Anura: Pipidae): geographical distribution, host range and evolutionary relationships. *Syst. Parasit.* **31**: 159–188.

Jackson, J.A. & Tinsley, R.C. (1995b). Evolutionary relationships, host range and geographical distribution of *Camallanus* Railliet & Henry, 1915 species (Nematoda: Camallaninae) from clawed toads of the genus *Xenopus* (Anura: Pipidae). *Syst. Parasit.* **32**: 1–21.

Tinsley, R.C. (1981a). The evidence from parasite relationships for the evolutionary status of *Xenopus* (Anura Pipidae). *Monit. zool. ital. (N.S.) (Suppl.)* **15**: 367–385.

Tinsley, R.C. (1981b). Interactions between *Xenopus* species (Anura Pipidae). *Monit. zool. ital. (N.S.) (Suppl.)* **15**: 133–150.

Tinsley, R.C., Harris, P.D. & Jackson, J.A. (In preparation). *Co-speciation of the genus* Gyrdicotylus *(Monogenea) with species of the anuran Pipidae in Africa.*

Tinsley, R.C. & Jackson, J.A. (1995). The genus *Oligolecithus* (Digenea: Telorchiidae) from *Xenopus* species (Anura: Pipidae), with a description of *O. siluranae* n.sp. from *X. tropicalis* (Gray) in Ghana. *Syst. Parasit.* **32**.

Tinsley, R.C., Jackson, J.A. & Harris, P.D. (In preparation). *Host specificity of* Gyrdicotylus *(Monogenea), parasitic in* Xenopus *species (Anura: Pipidae) in Africa.*

Tinsley, R.C., Kobel, H.R. & Fischberg, M. (1979). The biology and systematics of a new species of *Xenopus* (Anura: Pipidae) from the highlands of Central Africa. *J. Zool., Lond.* **188**: 69–102.

Tinsley, R.C. & Owen, R.W. (1979). The morphology and biology of *Xenopodistomum xenopodis* from the gall bladder of the African clawed toad, *Xenopus laevis*. *J. Helminth*. 53: 307–316.

Tymowska, J. (1991). Polyploidy and cytogenetic variation in frogs of the genus *Xenopus*. In *Amphibian cytogenetics and evolution*: 259–297. (Eds Green, D.M. & Sessions, S.K.). Academic Press, San Diego.

Index

Note: page numbers in *italics* refer to figures and tables